BURGESS-CARPENTER LIBRARY
ROOM 406 BUTLER LIBRARY
COLUMBIA UNIVERSITY
NEW YORK, NEW YORK 10027

URBANIZATION AND
NATIONAL DEVELOPMENT

South and Southeast Asia Urban Studies Annuals

Editors: Leo Jakobson *and* Ved Prakash

Department of Urban and Regional Planning, University of Wisconsin (Madison)
Associate Editor: Sheilah Jakobson

International Editorial Advisory Board:

JEAN CANAUX, Director, Centre de Recherche D'Urbanisme

RUTH GLASS, Director of Research, Centre for Urban Studies, University College London

JEAN GOTTMANN, School of Geography, University of Oxford

BERT F. HOSELITZ, Director Research Center in Economic Development and Cultural Change

REIKICHI KOJIMA, Director, Tokyo Institute for Municipal Research

APRODICIO A. LAQUIAN, Director of Research, International Association for Metropolitan Research and Development (INTERMET)

JOHN P. ROBIN, The Ford Foundation, Nairobi

M.S. THACKER, Bombay, India (formerly member of Planning Commission, Government of India)

ERNEST WEISSMANN, Senior Advisor on Regional Development UN Office of Technical Cooperation

EDWIN YOUNG, Chancellor, University of Wisconsin (Madison)

Volume I. SOUTH AND SOUTHEAST ASIA URBAN AFFAIRS ANNUALS

URBANIZATION AND

NATIONAL DEVELOPMENT

Edited by **LEO JAKOBSON** *and* **VED PRAKASH**

Associate Editor SHEILAH ORLOFF JAKOBSON

SAGE PUBLICATIONS, Beverly Hills

Burg.
HT
395
.A8
217

Copyright © 1971 by Sage Publications, Inc.

All rights reserved. No part of this book may be
reproduced or utilized in any form or by any means,
electronic or mechanical, including photocopying,
recording, or by any information storage and retrieval
system, without permission in writing from the publisher.

For information address:
SAGE PUBLICATIONS, INC.
275 South Beverly Drive
Beverly Hills, California 90212

Printed in the United States of America

International Standard Book Number 0-8039-0059-7

Library of Congress Catalog Card No. 74-103482

First Printing

FEB 1 8 1976 PREFACE

Despite their large and dense populations, South and Southeast Asia today
statistically belong to the least urbanized areas of the world. Nevertheless,
this region—in which we have defined South Asia as encompassing India,
Pakistan, Ceylon, and Nepal; Southeast Asia as comprising Burma, Malaysia, Singapore, Indonesia, Thailand, Cambodia, Laos, Vietnam, and the
Philippines—contains some of the great cities of the world with respect
to size and complexity of problems as well as in terms of contributions to
urban folklore and imagery. Karachi, Bombay, Delhi, Calcutta, Rangoon,
Singapore, Djakarta, Manila—all of these city names evoke colorful pic-
tures of romantic splendor or memories of human squalor and misery
beyond any terms of description. It contains also some of the most famous
new town developments of modern times: Chandigarh and Islamabad.
These two cities may represent the best developed physical executions of
the urbanistic ideas of two of the most prominent urban designers of contemporary architectural history, Le Corbusier and Doxiadis.

In historical perspective the contemporary scene reflects but a recycling
of the region's urban tradition from the ancient urban civilization of the
Indus Valley to the new towns of Mogul emperors and local rulers and the
religious centers of the many cultures this region has nurtured. The region,
then, has fascinated the imagination and elicited the interest of explorers
of the urban scene for a long time, from Vedic scribes to contemporary
analysts of social, economic and related problems.

Because of this past tradition and the growing interest in the contemporary aspects of urban affairs in South and Southeast Asia, it seemed
timely that a forum be provided to present and discuss critically the various
urban issues which this region is currently facing or may face within a
not-too-distant future. The general complexity of the urban scene and the
fragmented, widely diverse nature of urban studies suggested that the
forum be structured around a conceptual framework focusing on a set of
topically and procedurally related subject areas. It was our belief that
through this structure a continuity of coverage would be possible while still
allowing for a juxtaposition of findings, interpretations, and ideas.

The incipient state of urbanization in most of the countries of South
and Southeast Asia made the selection of the organizing framework for
this forum rather simple. It was felt that if the nations of the region were
to avoid or were to try to ameliorate some of the historical negative effects
of rapid urbanization which inevitably will result from the ongoing indus-

5

trial and technological development efforts, they would have to formulate, in advance, policies and programs for coping with the problems of urban growth. These policies and programs, however, would have to be based on careful and systematic analysis of the urbanization phenomenon, its causes, symptoms, and effects not only locally and regionally but also in an historical and world-wide perspective. Hence, planning was selected as an appropriate framework for the design of a three-volume forum (each volume appearing annually) for discussing urban affairs in South and Southeast Asia.

The word "planning" is used here in a broad, comprehensive sense—topically as well as procedurally. Our definition includes the various traditional sectors of planning—economic, social, spatial, and environmental —and the several geographic and administrative segments or levels of planning from the national level to the local levels of municipality and community. As a process, planning includes a sequential and temporal dimension, from issues which may be involved in the formulation of long-range societal goals to the immediate concerns present in remedial programs and projects. Likewise, the continua from plan conception to plan evaluation, and from the formulation of theory to its application contain temporally, procedurally, and methodologically interrelated sequences. The extent to which the various elements of this planning structure can be brought to bear on the process of public and private decision-making is the critical measure of its success and workability. It is our hope that the materials as organized and presented in this series will be helpful in stimulating constructive discussion on many of the issues around which urbanization and urban development decisions in South and Southeast Asia will have to be made.

Topically, this first volume of the projected series focuses on planning for urbanization and urban development at the national level. In the context of process the emphasis is on policy. The introductory chapter highlights the strong correlation which exists between urbanization rates and the various indicators of economic development. It suggests that urban development policies must be designed to take into account not only this general relationship but also the apparent temporal correlation between the different stages of economic and technological development and the various levels of urbanization. In the second chapter Stanislaw Wellisz discusses in more detail the relationships of economic development and urbanization. In particular, he stresses the need for more rational concepts for the use of space, for new methods of solving spatial allocation problems, and for bridging the gap between economists who neglect problems of space and physical planners who disregard economic factors.

In Chapter 3 Gerald Desmond describes the impact of national development policies on urbanization. His observations support Wellisz' general contention but go on to suggest that, until spatial considerations are introduced into the overall national economic planning process itself, "urban centers and other human settlements will continue to be derivative functions of all other decisions, unpredictable in size or shape, uncertain 'containers' of societies' highest achievements." Desmond's essay leads to an overview of the demographic dimension of urbanization in the region. The author of Chapter 4, Ashish Bose, points out many of the weaknesses of current census methods for providing data which would be useful in the formulation of urbanization policy. Specifically, the lack of adequate migration data makes meaningful forecasts about the urban population in the coming years impossible. Bose makes the important observation that one must make a distinction between the level of urbanization and its scale, and both must be viewed in the context of growth of the total population which in South and Southeast Asia is very rapid.

In each of the first four chapters, reference is made to the "state of the art" of urban study and analysis and the need for greater knowledge and improved understanding of the phenomena associated with urbanization and urban growth is stressed. Chapters 5 and 6 discuss in conceptual terms, city size and the role of cities in economic development. Brian Berry constructs his mathematical models of city size relationships within a national economic framework to suggest that there exist measurable thresholds at which changes in the distribution and growth patterns of city size occur and that these changes are closely related to economic development. An understanding of the significance of these thresholds can provide rational inputs into the formulation of policy and the making of development decisions and also reduce the reliance on such "planning fads" as exemplified by growth pole-cum-decentralization policies which seem to have persuaded both modernizers and traditionalists as well as local politicians in South and Southeast Asia. In Chapter 6, T.G. McGee looks at the role of the city as an agent of social change and in his essay develops two conceptual models of this role. One describes the role of the city in terms of traditional functions and a peasant, bazaar-based economy; the other introduces the dynamic effects of a capitalist system of production into the traditional model resulting in a constantly changing structure affecting both the role and the function of the city. This analysis leads McGee to suggest that there exists a very large gap between the preconceived roles of cities and their actual functions, and that researchers, planners, and politicians "would do well to look carefully at the empirical circumstances and structure of Asian societies rather than assume that these cities will play identical roles and functions to those of the West." At this stage of Asian history, he concludes, cities are "both catalysts and cancers, aiding and hindering the development process."

Aprodicio Laquian in Chapter 7 relates his observations of slums and squatters to McGee's models and makes some tentative optimistic assertions about the economic and social roles of these most visible "problem" areas of the larger cities of South and Southeast Asia. For one, the evidence he presents suggests that these communities contribute to economic production in many ways. Second, they do not seem to be as socially disruptive and unstable as their physical appearance may suggest. Hence, they may well fit the concept of a transitory period of urbanization as suggested in the introductory essay. Laquian observes, however, that his assertions can only be validated by more empirical studies and surveys through time. He warns against premature generalizations, particularly comparisons with slum and squatter settlements in a different era and a different place. Such comparisons are to him "at best approximate, at worst, misleading."

Chapter 8 on urban land policy, by co-editor Ved Prakash, points out some of the inadequacies of the present systems of real estate ownership, management, and taxation in coping with the pressures of rapid urban growth. In addition to new tax measures, the author asks for a comprehensive national land policy that would coordinate land use planning, public control over land, and land taxation. In Chapter 9 Tarlok Singh, a former member of the Indian Planning Commission, describes the role of government and public authorities in the shaping and execution of a national urban development policy. He concludes that "hitherto, urban development policies have been conceived in the framework of past patterns of economic and social development and of administrative behavior and organization rather than of the underlying problems and future goals." He suggests that their full integration into the process of economic and social planning has become all the more important for the future. A selective bibliography of recent materials comprises Chapter 10 and concludes this first volume. It contains a listing of the most important general contributions to the study of urbanization in the region and a country-by-country compilation of various published materials on urbanization, urban development, and urban planning.

Though by no means exhaustive in scope, the essays in this volume raise many important policy questions to which answers must be found at the national level. The contributors differ considerably in outlook and emphasis despite the fact that each chapter is part of a pre-conceived structure. It is hoped that the juxtaposition of these differences will add to the value of the discussion while at the same time exposing the difficulties inherent in the formulation of an urban policy. Concurrently, the point of concensus among all contributors is their call for more and more careful research into the problems of urbanization and urban growth so that policy can be formed on the basis of knowledge rather than fashion.

With two additional volumes to be published within the next two years, we hope to expand the forum begun with this volume. Volume II will focus on the problems created by rapid urbanization and growth at the local and metropolitan level. Some in-depth case analyses of urban planning efforts and a discussion of such issues as housing, community organization and development, and environmental improvement will be presented. The third and final volume of the series will focus on the impact of urbanization on the institutional frameworks—social, economic, legal, administrative, and political—of the societies involved.

Leo Jakobson
Ved Prakash

CONTENTS

11

URBANIZATION AND
NATIONAL DEVELOPMENT

CHAPTER 1

Leo Jakobson and Ved Prakash
URBANIZATION AND URBAN DEVELOPMENT:
Proposals for an Integrated Policy Base

Urbanization, the spatial dimension of the industrial and technological revolution of the past two centuries, has attracted the attention of scholars, social reformers and politicians since about the middle of the nineteenth century. Similarly, the study of cities—as insular objects or in systemic arrangements—as the target locus of the urbanization process extends back in time somewhat more than a century. Nonetheless, there does exist an important conceptual distinction between the study of urbanization, on the one hand, and the study of cities and city systems, on the other hand. Urbanization, by whatever definition, is a phenomenon describing a *process of change* in the situs of populations due to changing conditions in society at large. Its study entails an examination of the factors which start and sustain this process, as well as of its implications in broad general terms. In contrast, the study of cities and city life focuses on the *product of the changes* introduced by urbanization in a localized context regardless of the size of the urban system under study. Therefore, one could suggest that the study of urbanization represents the macro-level and the study of cities and urban affairs the micro-level of interrelated phenomena along a continuum, rather than a single or two separate phenomena, as is often suggested.

Until very recently, however, the study of these phenomena has taken place primarily within the disciplinary confines of the various subject areas concerned with or affected by urban growth and development, notably geography, economics, and sociology. Since the inclusion of the notion of "modernization" into the concept of development, cultural anthropologists, psychologists, and other behavioral scientists have expanded the interdis-

Author's Note: Our thanks are due to Professor Jean Fourastié for his permission to reproduce Figure 1 and Table 2, and to Dr. Bertram M. Gross and the American Academy of Arts and Sciences for their permission to reproduce Table 3.

ciplinary search for an improved understanding of urbanization and its impact on society, its members, and its ways of life. Most recently, in particular since urbanization and its consequences have become the concern of government policy in nearly all nations of the world, political scientists, public administrators, and planners of all kinds have added to the growing literature in this field. Here, as exampled in Charles Abrams studies of housing [1] and the several studies done for the Institute of Public Administration comparing metropolitan governments,[2] topical and functional questions are focused against the varied backgrounds offered by traditional disciplines.

Consequently, there does exist a diverse disciplinary literature augmented by an enormously wide range of functional and topical materials describing urbanization and the related problems of urban growth. What does not exist is a generalized, conceptual framework which would (1) adequately describe the linkages between the various components present in the processes of urbanization and urban development; (2) explain the manner in which they interact; and (3) allow for an integrated, comprehensive understanding of the not only fragmented but also frequently partial information at hand.

The need for such a general theory is, however, ipso facto a proposition of dispute. Not only do many question the need for or the feasibility of a comprehensive conceptual framework, but there are also some who a priori reject the suggestion that the processes necessarily are interlinked. For example, Simon Kuznets emphasizes that there is no inevitable technical connection between industrialization and urbanization, suggesting that it is technically possible to combine the pursuit of agriculture with urbanization, and the pursuit of modern industry with rural living, albeit at a prohibitively high cost.[3] The question he does not raise is whether there exists a rational point in the process of urbanization which could be identified by a general theory where this cost factor is no longer prohibitive. On the other side, serious efforts are currently being made to develop a general theory that would integrate our knowledge of social, economic, political, and regional (locational) phenomena.

Among these efforts one can mention the work of Walter Isard and his associates as well as that of Raymond Bauer, Bertram Gross, Albert Biderman, Robert Rosenthal and Robert Weiss.[4] Though these attempts do not specifically center on urbanization, they are mentioned in support of a growing recognition for the need of and a quest for new integrative theories in all areas which require improved understanding of the continuously increasing complexities of modern society.

The lack of an integrative theoretical framework is today most acutely felt in the developing countries, nearly all of which are attempting to guide the process of urbanization, through national development policies, into channels which would facilitate the concomitant processes of economic development and modernization, on the one hand, and reduce the social and environmental ill-effects of rapid urbanization on the other. In most cases these efforts are guided by an analysis of the experiences of the already industrialized world. Two of the key questions which one must ask are: first, does the fragmented knowledge of the unplanned processes of industrialization, urbanization, modernization, and their linkages from the western experience, suffice for the formulation of rational models for urbanization and urban development policy; and, second, do models which are derived from western experience apply to the situations and conditions encountered today by the developing nations?

In our estimation experience to date sheds little or no light on these questions. Despite some examples in the recent past which suggest that new insights and new concepts are emerging (which we will discuss later in this chapter), most planning in the developing world can still be best characterized by Kindleberger's comment in reviewing a number of World Bank reports: [5]

"Essentially, however, these are essays in comparative statistics. The missions bring to the underdeveloped country a notion of what a developed country is like. They observe the underdeveloped country. They substract the latter from the former. The difference is a program."

Bert Hoselitz uses Kindleberger's and others' reviews as a basis to suggest that a great number of scholars unfortunately are expressing views in which they envisage that "the development of underdeveloped countries depends not merely upon their adopting the economic and technological procedures of the more advanced countries, but also upon their coming to resemble them in social structure." [6] Similarly, John P. Lewis in his analysis of India's national planning effort points out that the Indian economic planners are "encumbered . . . by inappropriate economic theories learned mostly from the West." [7] In this instance, as well as in the case of missions, the suggestion is made that the experience of Western Europe and the United States has little or no transferability for aiding in the formulation of urbanization and urban development policies in the developing countries of the world today.

Fred W. Riggs also stresses the necessity of recognizing the correlation between the progressive formalization of institutional arrangements in societies on one hand, and the level of economic development on the other, in the formulation of successful policies and programs.[8] In a similar vein—though in a different context—one can speak about "past doctrine" in regard to our views about urbanization and urban development policy. This has been stated concisely in the proceedings of a recent conference on urban growth which brought together scholars, planners, and administrators from nearly all Asian countries and the United States.[9] Moreover, as we have pointed out in our comment on urbanization and regional planning in India, not only has the discussion on urbanization been dogmatic, it has also been distorted by anti-urban tradition and intellectual heritage. Thus, despite the essential and complementary nature of rural development, industrialization, and urbanization relative to economic growth, problems of economic development are typically discussed in a manner which more often than not suggests that a conflict exists between rural development and urbanization and reflects a naive desire to recapture the past.[10] As a result "the attempt to bring the whole development effort into correct spatial focus is likewise seriously obscured by a dichotomizing of development issues along rural-urban lines." [11]

It is against this background that we will attempt to respond to the questions we have raised by construing some tentative hypotheses about the nature and interaction of several factors which generate and sustain urbanization and urban growth. We will also attempt to present our hypotheses in a format which, we believe, will have direct relevance to the formulation of policy, at the national level, regarding urbanization and urban development as it reflects national goals. Although our discussion will be general, our specific focus will be South and Southeast Asia, not because of our familiarity with and interest in this region of the world, but most importantly, because we intend to suggest that within the framework of whatever general theory may emerge, there exist significant inter-regional and intra-regional differences that must be taken into account in the formulation of policy. For example, as illustrated by the data in Table 1, within India the traditional settlement pattern in the eastern parts, notably West Bengal, differs considerably from that in the more arid zones of western India, e.g., Gujarat and Rajasthan. The base for a hierarchial structure or a systemic network, to mention two possible frames of reference for the spatial component of national urbanization policy, requires, therefore, some judicious differentiation in policy.

Table 1. Size, Number and Share of Urban Population of Various Size Urban Places in West Bengal, Rajasthan, Gujarat, Madras, and India

	1,000,000+		100,000–1,000,000		50,000–100,000		20,000–50,000		10,000–20,000		5,000–10,000		UNDER 5,000	
	%		%		%		%		%		%		%	
	No.	pop.	No.	pop.	No.	pop.	No.	pop.	No.	pop.	No.	pop.	No.	pop.
West Bengal	1	67.7	3	4.9	7	5.4	31	11.1	41	6.2	50	4.2	12	0.5
Rajasthan	—	—	6	38.0	4	7.3	23	20.1	52	21.5	51	11.8	9	1.3
Gujarat	1	22.7	5	19.7	9	10.5	43	24.2	54	13.4	60	8.8	9	0.7
Madras	1	19.3	8	18.5	19	14.2	61	20.7	119	18.1	95	8.0	36	1.2
All India	7	18.8	100	25.7	141	12.2	515	19.9	817	14.3	844	8.0	266	1.1

Source: Census of India, 1961.

Before beginning our search for a general conceptual framework for urbanization policy, we would like to present a brief overview of some of the emerging concepts and attitudes on urbanization as exemplified in the proceedings of three recent international conferences on the topic of urbanization: the 1966 United Nations Seminar, "Development Policies and Planning in Relation to Urbanization"; [12] the 1967 International Union of Local Authorities symposium, "Urbanization in Developing Countries"; [13] and the "Pacific Conference on Urban Growth" held in 1967 in Honolulu.[14]

The United Nations seminar report stresses the need for a greater emphasis on the development of a comprehensive approach to urbanization and the implementation of thorough and systematic interrelation of diverse economic, social, and physical planning policies and programs. In particular, it is pointed out, urban policies and programs must be integrated in the over-all national plan for economic and social development.[15] The seminar, the report suggests, "represents a continuation of past interest and trends, but it also marked a *new stage* in United Nations action in the urbanization field." [16] The focus of the seminar papers and discussions was on the effects of population distribution and the formulation of guidelines for national urban policies. The participants recognized the crucial need to develop national urban policies conducive to economic, social, and political development. However, there was no general agreement on many basic questions regarding rapid urbanization, because the discussions of the seminar made it clear that the major issues relevant to development strategy cannot be resolved until more information is available on such factors as, for example, the roles of and the relationships among population growth, rural density, agricultural productivity, and industrial and urban development. It was recognized, however, that the physical, economic, and social problems associated with rapid urbanization were not isolated issues, but demanded a comprehensive development strategy within the context of total change in the society; they were part and parcel of the problem of under-development and their seriousness should lead to policies for guiding urbanization "in the ways most favorable to over-all development, rather than to measures which represent an *anti-urbanization attitude*." [17] The seminar proceedings conclude that the process of urbanization is largely irreversible, and that attempts to reverse or delay the process will only lead to a scattering of meagre resources. Hence, one must search for a positive urbanization policy based on improved information, in particular on the dynamic relationship between industrialization and urbanization and of specific national conditions regarding the degree and character of existing urban growth and the stage of industrial development.

Contradicting its continuous reference to the lack of adequate data to support a positive urbanization policy, the seminar took an a priori stand suggesting (1) that a policy of urbanization—covering the locational aspects of investment (both economic and social) which is most conducive to continuing over-all development—is one of "selective concentration" in a few areas with high development potential, rather than one of "dispersion," and (2) that "action is required to redirect the migratory currents, build up alternative growth-poles, implement physical, economic and social plans in the existing cities to make them a more favorable environment for development." [18]

The International Union of Local Authorities symposium focused its attention for the most part on planning, administration, and implementation of urban development at the local level. However, the urban setting—the process of urbanization—and the need for national policy and strategies for urban development were discussed and indeed provided the framework for the symposium. It was recognized that the meaning of urbanization itself is undergoing change. The concept of urbanization as a process of urban growth is being replaced by one of urbanization as a process or instrument of social change and development; that [19]

> "this reflects a change in thinking about development in general, in the sense that it is now realized that improvement of human and social conditions is a prerequisite for balanced and sufficiently rapid economic growth; [and that] as the development process advances, more and more national development policies will be influenced, if not determined, by urban development."

In reviewing the need for national policies of urban development the symposium concluded that though most developing countries have adopted national economic development plans, urban problems only too often have been relegated to the level of local issues, and that local authorities are not given the necessary political powers and financial resources to solve the problems resulting from rapid urbanization. In particular, it was noted that because of the sectoral treatment of economic development programs at the national and regional levels, the components of urban development are artificially separated from each other. The symposium further suggested that the significance of urbanization for economic and social development, demonstrated in terms of modernization and urban growth, should be taken into account in drawing up national policies and plans.[20]

Several important factors emerged from the deliberations of this symposium. First, there was a recognition that in the design of strategies for urban development and, in particular, in the reconciliation of conflicting objectives, "the decision as to which goal will be given priority depends on the historical and geographic development and on the economic and political situation of a country." [21] Second, it was observed that there may be identifiable stages in the process of urbanization which would require a differentiation of strategies at each stage. Third, it was suggested that urban development policy must be supported by judiciously selected complementary policies in the area of rural development. Finally, the symposium concluded that "any strategy that is to be applied will face at a certain time the problem of what the most strategic location would be for investment in the infrastructure for economic development. This means in fact that the element of space will have to be included into national (and higher levels of) planning." [22] At the symposium Jan Tinbergen presented a theoretical model for determining which products, including economic goods as well as social services, can be produced most competitively at the various geographic levels—nation, region, district, village—within a given framework of national targets.[23] Precedent for the Tinbergen model is found in a hierarchial model of population centers based on size and functional linkages as presented by a Ford Foundation sponsored study on small-scale industry development in India. The position advocated by the authors of this report was that "the development of well-selected towns from these various tiers, within well-selected regions, would be a feasible method of building closer rural-urban linkages and bridging the large gap between the villages and the metropolises." [24]

The Pacific Conference on Urban Growth made, in our views, the clearest conceptual distinction between what the conference report calls "prior doctrines" which "all are meant to reverse, divert, arrest, regulate urbanization" and the "unorthodox, functional doctrine" advocated by the conference which "regards rapid urbanization as of central importance in the national development process, a condition, which when properly organized, is to be encouraged rather than discouraged." [25] The interesting aspect of this conference was that its initial response followed convention: familiar anti-urban attitudes dominated the discussions which centered around decentralization of urban growth and arresting rural-to-urban migration. Only later were reservations expressed "as to the efficacy of solutions put forward during the past eleven years of international exchanges on urbanization, and with doubts as to the validity of some of their underlying premises." [26]

In presenting "the functional approach," the observation was made that [27]

> ". . . the conference was not structured as a debate, nor did it provide a forum for testing the traditional concepts against empiric evidence or produce a consensus as to the validity of the functional approach. But from the discussions facts, explanations and inferences could be extracted to support the functional position, which views the city, in particular the large city, as:
>
> a concentration of population, infrastructure, services, and markets which offers opportunities for economies:
>
> economies of scale, both in production and consumption;
>
> economies of juxtaposition, from convenient and efficient spatial and functional relationships;
>
> economies resulting from the division of labor, from specialization of function and occupation, and diversity of skills and professions, which enable cities to serve as centers of administration, transportation, com-

munication, and finance, as entrepots, factories, and markets, as sources of arts, services and recreation;

external economies (that is, economies external to the individual and common to the community, which permit all economic activities to be performed more efficiently and productively)—including economies of the infrastructure of transportation, communication, electricity, water and waste disposal, and other public facilities and services; as well as economies from the diversity and specialization of skills and services in support of production—research and development, commerce and finance, engineering, construction and maintenance, etc.—which make it efficient and profitable to do business in a modern city."

In addition to their economic role in national development the social functions of the city were stressed; specifically their role as agents for change, modernization, and innovation.

The case of the urbanization and industrialization of Japan seems to have provided the catalytic force that changed the direction of the Pacific Conference. In particular, considerable attention was given to the theory advanced by Takashi Fujii concerning the economic function of urban development in national development in explaining Japan's economic and urban growth. According to the Conference proceedings, Fujii's theory rests on the following interrelated premises: [28]

"(a) specialization of labor increases productivity; (b) the specialization of functional division of areas (such as areas devoted to industrial uses, areas devoted to cultivation, etc.) increases the efficiency of an urban economy, for it concentrates large numbers of people, skills and capital in relatively small spaces so as to facilitate communication among sectors of the economy; (c) there is a relationship between capital density and the efficiency of space; as more capital is invested in a given space, the economic efficiency of that space is increased; (d) industrialization leads to the concentrated accumulation of both capital and labor; (e) the accumulation of labor and capital leads in turn to urbanization."

It was recognized, that there may be a point beyond which increases in labor and capital in a given space begin to yield diminishing returns. It may be assumed, however, that where the optimum density of capital is exceeded, dispersal of capital to the surrounding area will occur. This leads to an expansion of areas having an optimum economic density of population. Fujii suggests: [29]

". . . that industrialization thus leads inevitably to the growth of cities, especially the large cities. And after a certain point, the large city becomes a kind of autonomous power engine in its own right; that is, as its productive capabilities are increased through the concentration of skills and capital, its consumption powers also increase, both for the products it produces and for the innumerable services required by urban producers and consumers."

Fujii's description of the large city as an "autonomous power engine" does not mean that the city becomes disconnected from its rural hinterland. On the contrary, he points out that dependency increases as cities expand, and that the growth of large cities from capital accumulation will stimulate the growth of the nation.[30]

It is of special interest that all three conferences introduced a strategic view of urbanization as a positive component in the process of national development. Furthermore, the reports stress the operational need for improved information on the nature of the linkages between urbanization, industrialization, modernization, and economic development, and for comprehensive policies and programs which, as exemplified by Tinbergen's model, cut across the various traditional economic and social policy sectors as well as relate to the hierarchial administrative structure found in all governmental systems.

In our attempts to develop a general framework for the formulation of urbanization policies in the spirit of the functional doctrines enunciated at the conferences discussed above, we have found much support in Jean Fourastié's discussions of "the [occupational] migration of the [economically] active population" and his suggestion that the current era of industrialization and urbanization is a "transitory period" in the history of mankind during which societies transform from "primary" (agriculture-based civilizations) to "tertiary" (service-occupation based civilizations).[31] Diagrammatically, Fourastié's thesis is illustrated by the percentage shifts of the economically active population according to the three-sector theory.

Though the three-sector theory has its shortcomings—which are acknowledged by Fourastié—from the point of view of measuring industrialization in relation to modernization and technological progress, on the one hand, and urbanization, on the other, it is a measurement device which we feel is still most useful in discussing urbanization in the context of national development policy.[32] However, where Fourastié defines each sector according to the degree of technological progress and output per unit of labor, we suggest that the three sectors also represent a distinct differentiation in the *location* of activities assigned to the sectors.[33] Primary activities are those where the location of an activity is determined by the location of a resource. This expanded definition would then include not only agriculture and forestry, but also mining and extraction of minerals. Secondary activities relate to the refinement and transformation of raw materials into products and units of consumption. Their location is determined by numerous factors among which the costs of transportation, labor, and marketing, and the technologies of transportation, production, and distribution play a decisive role. Notably, the secondary sector activities are less space-bound than the resource location-bound activities of the primary sector and the market-bound activities of the tertiary sector. Even those activities in the tertiary sector which satisfy the basic service needs

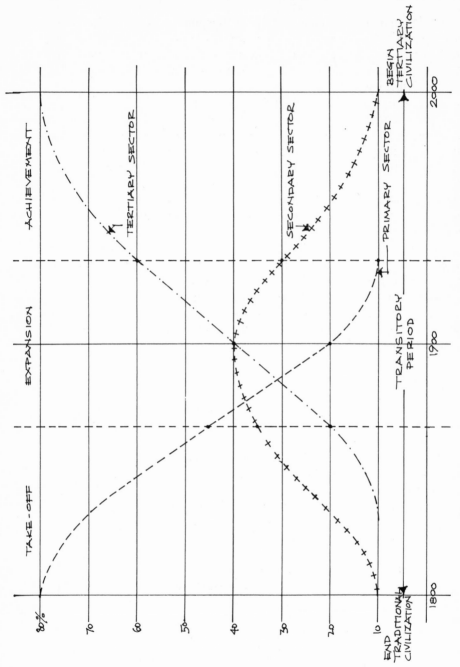

Figure 1. Percentage shifts of economically active population during Fourastié's "transitory period." *Source:* Fourastié, *op. cit.*

of the primary and secondary sector must be considered market-bound because they require minimum economies of scale for successful operation —which controls their spatial location to service markets of various minimum sizes.

Because of their independence from an a priori fixed location in national and regional space, the manipulative qualities of the variables which determine their location, and the apparent and statistical correlation between industrialization and urbanization, the secondary sector activities have been the traditional target of government intervention in the areas of urbanization policy, population distribution policy, and regional economic equalization policy. All "prior doctrine" from the utopians and social reformers of the nineteenth century to the politicians, urban and regional planners, and new towns advocates of more recent vintage, has, therefore, centered around the questions of industrial location. Economic incentives and sanctions, and improvements in transportation, utilities, and services, in combination with licensing and other regulatory controls, seem to have been the main tools for implementing these doctrines.

What appears to have escaped the attention of most planners, however, is the fact that, as Fourastié has observed, technological progress—being fastest in secondary-sector activities—affects not only the various cost variables, but importantly the job generation potential and human resource requirements within the sector itself. The human resource requirements, in turn, can change significantly the basic premises on which most traditional urbanization and urban development policy is based.[34]

It is of importance to note also that Fourastié's schema (Figure 1) divides the transitory period into three stages on the basis of the state of industrial development and its job generation capacity. The three stages are the take-off stage during which industrial job generation is great, increasing the total share of the economically active population in the secondary sector from less than 10 percent to about 35 percent. This stage is followed by one of industrial expansion characterized by fast technological progress and continued increase in the share of the secondary sector in the industrial origin of the gross domestic product. However, technological progress begins to reduce the job generating capacity of this sector, keeping its share relatively constant between 35 to 40 percent of the total economically active population. The third stage is one of industrial consolidation marked by very rapid technological progress and, consequently, a decline in the sector's relative role as a generator of added employment within the sector itself.

Fourastié's general characterization of the transitory period in contrast to the passing "traditional civilization" and the coming "tertiary civilization" is presented below, in Table 2 on page 27. A very similar tabulation can be found in Bertram Gross' discussion of social systems accounts.[35] Though developed in different contexts, the purpose of both is to improve our understanding of the various complex relationships which exist in the processes of national development and social change. A juxtaposition of these two tables is helpful not only in order to ascertain the similarities of thought and the differences of emphasis between the two authors representing two different cultures and two distinct disciplines, but most importantly for the purpose of synthesis which allows us to incorporate into the general structure proposed by the two authors the variable of urbanization and urban development—a variable given only passing attention by both.[36]

In an attempt to synthesize these two schemes one must take into account that where Fourastié discusses the period of industrialization as one of transition, connecting the preindustrial or traditional civilization with the postindustrial or tertiary civilization, Gross treats the period of industrialization in and of itself. Conceptually this means that Gross recognizes values peculiar to the industrial period which must be taken into account in the formulation of policy. He is cognizant that transitions among his three stages will occur; they are given, however, a secondary, and presumably short-lived position in his scheme. In comparison, Gross seems to view the world in a more static manner than Fourastié, whose concept is dynamic.

Our concept of the urbanization process can now be related to the two schema discussed above. It may generally be suggested that urbanization follows a rising S-curve from a level of 10 percent or less urbanization to levels of 80 percent or more urbanization, and that the urbanization curve closely follows the tertiary sector curve in Fourastié's graph. The most rapid shift then occurs in the curve's central section, which Fourastié refers to as the "expansion period." Though the correlation of these two curves is not a new observation, two important points must be stressed.

First, during and after the expansion stage of industrial development, urbanization will be sustained by continuing relative growth of the tertiary sector, albeit at a slower rate during the period of "achèvement." [37] In terms of urban planning practice, this suggests that many traditional economic-base and situs principles will have to be supplanted with new concepts centering around services and institutions as the predominant element in urban structure.

Second, we would also suggest that the spatial pattern of urbanization has distinguishing characteristics specific to each of the three periods implied by the curve. During the first period, which is coincident to Fourastié's take-off period, growth occurs in distinct nodes, at the most favorable locations.

The second period is characterized by metropolitanization, in part resulting from the emergence of internal shifts in the structure of an urban area. From a policy point of view one of the important characteristics of metropolitanization is in the fragmentation of governmental units to pro-

Table 2. General Characteristics of the Transitory Period

	TRADITIONAL CIVILIZATION	TRANSITORY PERIOD			TERTIARY CIVILIZATION
		Take-off	Expansion	Achievement	
Technological progress	Zero or very weak	Notable	Strong	Very strong	Considerable and growing steadily
Migration of the economically active population					
Primary	Stable	Decreasing	Decreasing	Decreasing	Stable
Secondary	Stable	Growing	Stable	Decreasing	Stable
Tertiary	Stable	Stable	Growing	Growing	Stable
Cost (relative movements in the long run)	Stable	Quasi-stable	Sharp shifts	Tendency towards preponderence of the stabilizing tertiary sector	Stable
Crises	Under-production in primary sector	Over-production in primary sector	Severe overproduction in primary and secondary sectors	Over-production in secondary sector	Under-production in tertiary sector
Revenues	Land tax	Income tax	Confusion in tax structure	Systematic reduction and elimination of land and income taxes	Service charges
Level of Life	Low, stable in long run but subject to short term variance; wide range of income distribution	Stagnant	Rapid Growing	Sluggishly moving towards a maximum	Very high, stable, even income distribution; persistance in scarcity of tertiary services

Source: Fourastié, op. cit., p. 212.

Table 3.

NATIONAL STRUCTURE: PREINDUSTRIAL, INDUSTRIAL, POSTINDUSTRIAL

	Preindustrial	*Industrial*	*Postindustrial*
People [a]	Low life expectancy at birth Low education	Higher life expectancy Much more education	Life expectancy above 70–75 Highly educated population
Nonhuman resources [b]	Little development of natural resources	Large-scale development (and waste) of natural resources	Large-scale conservation of natural resources
Subsystems [c]	Relatively little differentiation (fused)	Considerable differentiation (refracted)	Still more differentiation
	Agriculture more than 60 percent of labor force	Agriculture below 30 percent of labor force	Agriculture below 10 percent of labor force
		Manufacturing 15–25 percent of labor force	Manufacturing below 10–15 percent of labor force
			Services (public and private) above 60 percent of labor force
	Small government sector	Large government and mixed sector	
External relations [d]	From colonialism to independence	From empire to bloc or commonwealth	Extensive transnational, intersecting and interpenetrating relations
Internal relations [e]	Centrifugal tendencies	More integration with growth of nationalism	Less integration with growth of transnationalism
	Weak communication and transportation networks	Highly developed communication and transportation networks	Still more highly developed communication and transportation networks
Values [f]	Localism	Nationalism Cosmopolitanism Activism	Transnationalism Megalopolitanism Humanism
Guidance [g]	Restricted elites	Multiple elites National planning systems	Dispersed elites Transnational planning systems

TRANSITIONAL (between Preindustrial and Industrial)

TRANSITIONAL (between Industrial and Postindustrial)

See notes on page 29.

tect highly localized interests within the structure of an urban area. Developments of this nature can be observed in nearly all western countries, manifesting themselves with specific clarity in the United States.

The "achèvement" of urbanization during the third stage is then characterized by a slowing down of the rate of urbanization, though the metropolitan areas continue to expand into even larger conurbations on a megalopolitan scale. In a few countries the impact of this last stage of urbanization has begun and its consequences are emerging.[38]

A clear recognition of the nature of the urbanization process is important for policy purposes. First, in implying a distinct urban policy for each of the three stages in the process, and second, in recognizing that as parts of the process of urbanization these are transitory stages in the transformation of a society from a pre-urban stage to a total urban state in a postindustrial, tertiary civilization. Following are some preliminary observations offered in support of our suggestion to correlate three levels of urbanization with the processes of occupational change. We have added to the original three countries (France, Great Britain, and the United States) contained in Fourastié's analysis, Sweden, Finland, Puerto Rico, Japan, and India. This selection, though admittedly arbitrary and primarily based on easy availability of data, incorporates, nevertheless, a temporal notion. Our observations suggest that the countries selected represent randomly distributed points on a continuum beginning with the era of rapid urbanization in Great Britain in the nineteenth century to an even more rapid process of urbanization in Japan during the past few decades.

Notes to Table 3:
    a. People: a more specialized and professionalized society with "continuing education" tending to absorb at least 10 percent of the work year.
    b. Nonhuman resources: major "cybernation" systems, based on coupling of electronic computers with power-driven machinery.
    c. Subsystems: further extensions of the organizational revolution and the growth of complex clusters and constellations. More blurring of distinctions between government and nongovernment organizations, with more "mixed" enterprises. Great expansion in absolute size and relative importance of metropolitan areas and megalopoles.
    d. External relations: greater degrees of penetration of and intervention in other societies—and vice versa; with transnational legitimation of such activities and provision for cooperative transnational actions.
    e. Internal relations: high mobility and integration, together with intense conflicts and shifting conflict patterns.
    f. Values: growing secularization, with potentialities for more widespread and deeper humanism. Values reflecting growing material abundance—as contrasted with (a) "zero-sum" nonredistributive values of preindustrial societies (based on their limited productive potentials), and (b) the "scarcity economics" and "zero-sum" redistributive ideologies in industrial societies (increasingly out of keeping with growing productive potentials).
    g. Guidance: high differentiation means elite dispersion and major emphasis on polyarchic bargaining relations in social system guidance.
    Source: Gross, *op. cit.,* p. 215.

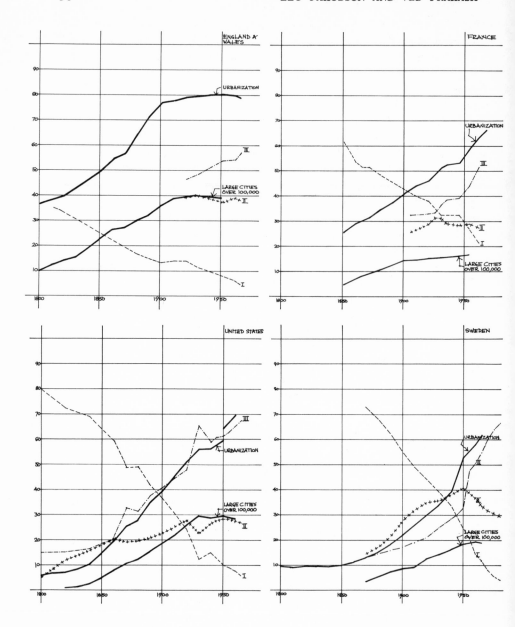

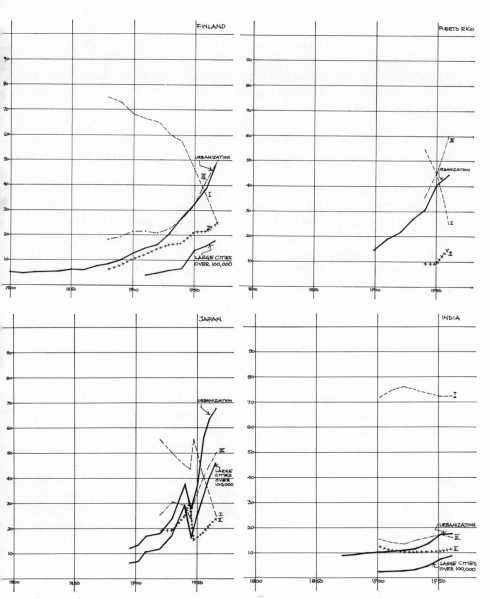

Figure 2. Urbanization rate, percent of population in cities over 100,000, and percent of economically active population in agriculture (ɪ), manufacturing (ɪɪ), and services (ɪɪɪ), for selected countries, 1800–1980.

Sources: Statistical abstracts for the various countries; Swedish projection from National Central Bureau of Statistics, "Labour Force and Population Changes by Regions in 1960–1965, with a Projection for the Period up to 1980," Stockholm, 1968.

Figure 2 shows in graphic form the occupational shift and urbanization patterns for the eight countries.[39] An examination of the trends clearly reveals the basic validity of Fourastié's conceptualization of the three stages of the transitory period although in no case has the third stage—"consolidation"—clearly begun. The before-mentioned correlation of tertiary sector growth and urbanization stands out clearly. It is of interest to note that this correlation is closest in those countries which have been industrializing and urbanizing at a later point in time, particularly since World War II, suggesting that this correlation and the more rapid rate of change as signified by the steepness of these curves is, in turn, caused by a more advanced state of technology.

Except for the unique regularity of and uniformity between the curves indicating the relative decline of persons engaged in primary-sector activities, there are considerable differences in the patterns describing the growth of the secondary and tertiary sectors. Nevertheless, in all instances the basic premise underlying Fourastié's and Gross' schema is upheld. However, these differences suggest to us that in each case the growth patterns are influenced by factors other than economic and technological.

Though at this point we do not have specific data and analyses to support our contention, our familiarity with the history and conditions of the eight countries examined allows us to hypothesize that physiographic conditions, size and density, geopolitical factors, and lastly, but not least importantly, socio-cultural forces influence the speed, and control the regularity of development in the transitory period. Many of the variables included in these groupings can and have been identified.[40] Others are yet to be recognized. However, in order to ascertain the importance of their impact on the urbanization process, much new systematic study is required.

At this point a third important measure must be introduced into any discussion of urbanization policy: the general wealth of the society or nation in question. Often per capita income or family income is suggested as an appropriate measure. We feel that this measure may be most appropriate in a discussion of, for example, national housing policy, because consumer purchasing power, or conversely, the capacity to generate savings, is of importance. Also, in local planning, where improvements are based on tax revenue, these indicators are most helpful. In a discussion of national urbanization policy, however, an individual's income contains little meaning. It is the total wealth of a nation that one will have to take into account, because urbanization is not an individual phenomenon but a collective one. This implies that it is relative to total wealth. Hence we suggest that Gross National Product (GNP) or National Income, by its industrial-origin components, may be a more appropriate indicator in the context of our discussion.

As has been suggested by many, a strong correlation exists between Gross National Product per capita and the rate of urbanization. In Figure

Figure 3. Indices of urbanization and national wealth (in constant monetary units) for Finland, France, and the United States, 1900–1970. Source: Statistical abstracts for the three countries.

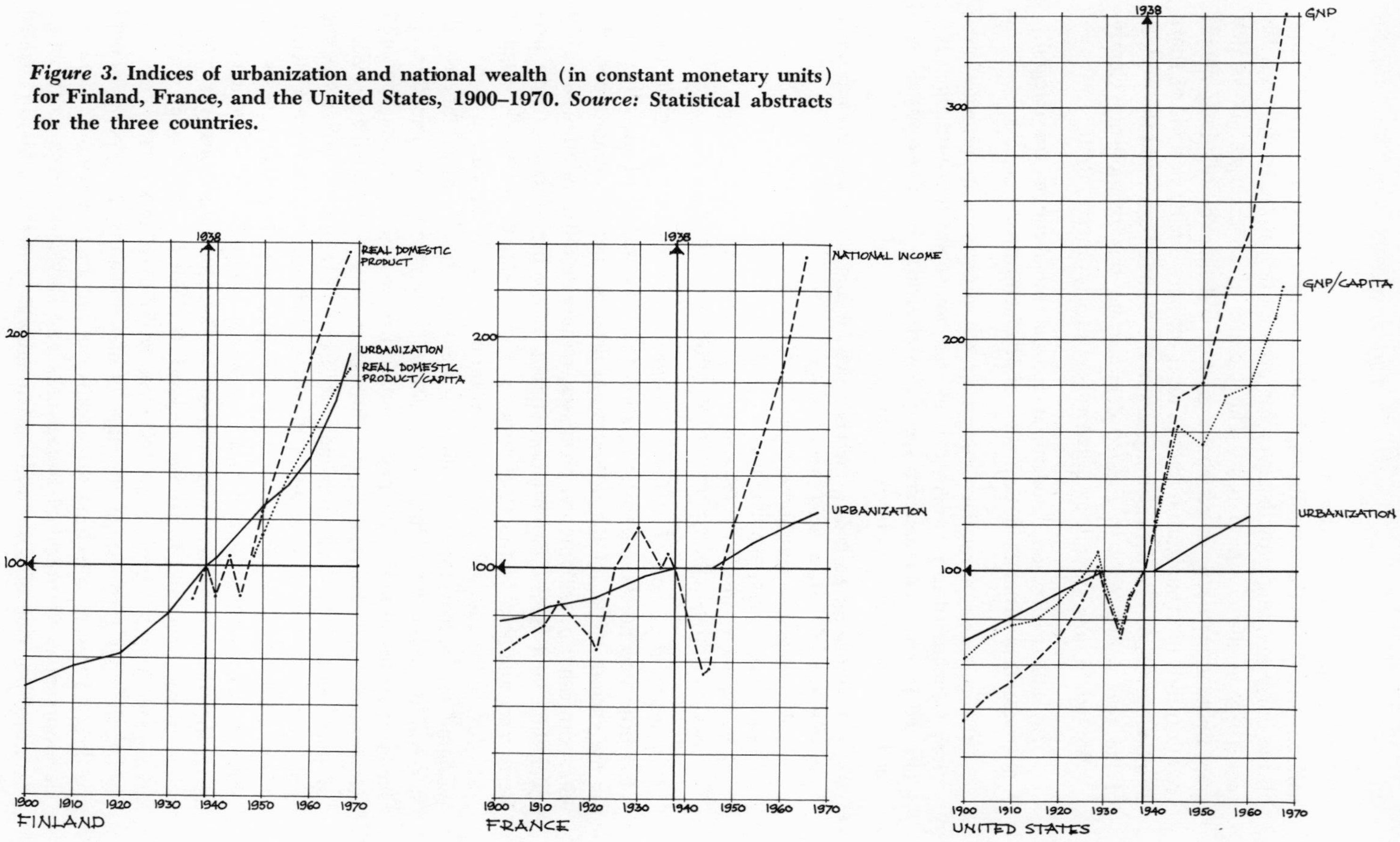

3 we have superimposed the growth index curves for total and per capita national wealth for three countries, Finland, France, and the United States, on the corresponding growth curve for urbanization using in all cases the year 1938 as the common base. The picture which emerges shows that great fluctuations occur in economic growth and that the relatively steady growth trends in urbanization tend to level off only during periods of sharp economic decline. The depression years in the United States and the periods of the two world wars in France demonstrate this relationship clearly.

We would now submit, as a general set of hypotheses, that:

(a) industrialization, modernization, and urbanization are integral, interdependent variables in the process of development;

(b) a country's relative position in the process of transformation from preindustrial to postindustrial, and the rate of this transformation, is affected by sets of factors each specifically related to the above variables; and

(c) among the basic sets, factors which facilitate or impair the process of transformation are at least the following:
physiographic—natural resource base, climate;
ecological—area, population size;
geopolitical—insularity, relation to world markets;
socio-cultural—ethnic homogeneity, norms and values.

From the point of view of national urbanization policy we can, at this time, offer little beyond the earlier suggestion that recognition of the difference in the various stages of the urbanization process, and its intimate correlation with other developmental processes, requires a differentiation in policy over time, and such differentiation should follow the general growth patterns ascertainable in the total process. For example, we would suggest that during the take-off phase urbanization policy be directed toward environmental improvement at minimal standards in the apparent growth centers, in order to reduce pre-existing environmental deficiencies, to facilitate socio-cultural adjustment, and, to follow Fujii, to encourage rapid accumulation of labor and capital through in-migration and the provision of amenities attractive to investment. A specific example of policies of the kind suggested above is found in the Basic Policies Plan for the Calcutta Metropolitan District. However, a careful reading of the plan reveals that these policies were presented in a context of "prior doctrines" which suggests that the grave urban condition in the Calcutta metropolis was recognized at the local and metropolitan level, but not in the broader context of urbanization as a process of total structural change in a society.[41]

At the aggregative stage, the stage of metropolitanization, policy should be directed toward offsetting the negative effects of fragmentation and the development of institutional frameworks and service systems to facilitate the transformation of a uni-nodal urban growth system to a multinucleated metropolitan system expressive of growing differentiation and specialization within the various subsystems operating within the metropolitan area. Again, in the Calcutta plan we can find suggestions and specific recommendations describing the nature of these kinds of policies and their validity for application in the Calcutta situation.[42] This suggests to us that urbanization policies developed at different stages of the urbanization process may well remain operational after the next stage in the process has been reached. The changing nature of the conditions at each stage may require adjustment in earlier policy. The nature of the policy formulation process may, therefore, be cumulative.

Following this model of policy formulation, policies in the diffusive phase of urbanization would be designed to augment the urban service base of metropolitan structure by the addition of the national-locational dimension. In order to overcome the negative effects or dis-economies of too large urban concentrations, and at the same time sustain the level of services achieved on the metropolitan scale, policies would seek to distribute the population and the economic activities more evenly. This distribution is the central theme of "prior doctrine," which in our opinion has been prematurely applied to solve the ill effects of rapid urbanization. We would submit that this is in part due to the fact that much of this theory was developed in England, a country which first experienced rapid, unplanned urbanization and has reached our phase of diffusion without concomitant supportive development in related phenomena, such as the modernization of social structure and development of systems serving the urban agglomeration, as well as the variables affecting these as defined by Gross and Fourastié.

The application of such concepts as new towns, growth poles, counter magnets, and dispersed centralization, as suggested by the United Nations' Urbanization Seminar, seem to us valid concepts only if the social and economic costs of their implementation are justifiable in relation to the resources at hand. Regardless of their appeal, a premature new towns policy, for example, leads only to excessive costs and a waste of scarce capital and energy without even bringing about the anticipated social returns—not to mention their total impact on the process of urbanization. This example is clearly demonstrated in the case of new towns in India.[43] Although new towns are inappropriate at the early stages of urbanization, we find at the diffusive stage such dispersal of urban activity does not contradict, but rather extends, Fujii's theory of urbanization. In Table 4 we show more clearly the nature and relation of policies formulated at the various stages of urbanization. The model (Table 4) is suggestive in nature, raising several interesting questions for research as well as conceptual debate. We have already suggested areas of critical research directed at forces affecting urbanization and urban development. The idea embodied in this model will, however, raise discussion of a nature which cannot be supported by research findings alone, because conditions of history, culture, values, and norms are involved. For example, the level of policy objectives and implementation standards is one discussion area in which our experience shows that rationality—in terms of a realistic assessment of balanced utilization of resources—is often substituted by attempts to rationalize decisions which are political or ideological. This is clearly evident in the many proposals for the building of new capital cities. The political and/or ideological motivations, however, are camouflaged by arguments demanding centrality, establishing the lack of facilities, creating room for expansion in the old capital, and similar factors.

Table 4. Urbanization Policy Model

URBANIZATION STAGE % urban	NODAL −20%	AGGREGATIVE 20–60%	DIFFUSIVE 60%+
Priority for policy focus	local	metropolitan	regional/national
Policy targets	immediate	intermediate	long-range
Policy content	remedial	normative	progressive
Level of policy objectives and implementation standards	low	moderate	high
Resource availability by sector	scarce balanced	moderate imbalanced	ample balancing
Budgetary constraints on implementation	high	moderate	minimal

We would like to see that, when decisions on urbanization and urban development policy are made, the dimension of values will become an explicit component of the decision process, so that decision-makers as well as those affected will have an understanding of the consequences: [44]

"The only way in which we can strive for 'objectivity' in theoretical analysis is to expose the valuations to full light, make them conscious, specific, and explicit, and permit them to determine the theoretical research. In the practical phases of a study, the stated value premises, together with the data (established by theoretical analysis with the use of the same value premises) should then form the premises for all policy conclusions.

I am arguing here that value premises should be made explicit so that research can aspire to be "objective"—in the only sense this term can have in the social sciences. But we also need to specify them for the broader purposes of honesty, clarity, and conclusiveness in scientific inquiry."

Myrdal's admonition would equally apply to processes of policy formulation and implementation.

NOTES

1. Abrams, Charles, *Man's Struggle for Shelter in an Urbanizing World.* Cambridge, Mass.: M.I.T. Press, 1964.

2. Among the latest additions to a growing literature in this area one can mention: Annmarie Hauck Walsh, *The Urban Challenge to Government; An International Comparison of Thirteen Cities.* New York: Frederick A. Praeger, 1969.

3. Simon Kuznets, "Consumption, Industrialization and Urbanization" in Bert F. Hoselitz and Wilbert E. Moore (eds.), *Industralization and Society.* New York: UNESCO-Mouton, 1963, p. 102.

4. Walter Isard (in association with Tony E. Smith, Peter Isard, Tze Hsiung Tung and Michael Dacey), *General Theory—Social, Political, Economic and Regional.* Cambridge, Mass.: M.I.T. Press, 1969, and Raymond A. Bauer (ed.), *Social Indicators.* Cambridge, Mass.: M.I.T. Press, 1966.

5. C.P. Kindleberger, "Review of the Economy of Turkey; The Economic Development of Guatemala; Report on Cuba." *Review of Economics and Statistics,* Vol. 34, No. 4 (November, 1952).

6. Bert F. Hoselitz, *Sociological Aspects of Economic Growth.* Chicago: University of Chicago Press, 1960, p. 55.

7. John P. Lewis, *Quiet Crisis in India.* New York: Doubleday, 1964, p. 135.

8. Fred W. Riggs, *The Ecology of Public Administration.* Asia Publishing House, 1961.

9. "The New Urban Debate." A Conference Report, Pacific Conference on Urban Growth, Honolulu, Hawaii, May 1–2, 1967. Washington, D.C.: Agency for International Development, February, 1968.

10. Leo Jakobson and Ved Prakash, "Urbanization and Regional Planning In India." *Urban Affairs Quarterly,* Vol. II, No. 1. Beverly Hills: Sage Publications, 1967, p. 53.

11. John P. Lewis, *op. cit.,* p. 184.

12. "Urbanization: Development Policies and Planning," in *International Social Development Review,* No. 1. New York: Department of Economic and Social Affairs, The United Nations, 1968.

13. "Urbanization in Developing Countries." Report of a Symposium, Noordwijk, Netherlands, December, 1967, International Union of Local Authorities. The Hague: Martinus Nijhoff, 1968.

14. "The New Urban Debate," *op. cit.*

15. "Urbanization: Development Policies and Planning," *op. cit.,* pp. 3–4.

16. *Ibid.,* p. 4 (emphasis added).

17. *Ibid.,* p. 5 (emphasis added).

18. *Ibid.*

19. "Urbanization in Developing Countries," *op. cit.,* p. 7.

20. *Ibid.,* p. 19.

21. *Ibid.,* p. 20.

22. *Ibid.,* p. 22.

23. J. Tinbergen, "Links Between National Planning and Town and Country Planning," *ibid.,* pp. 73–80.

24. The International Perspective Planning Team on Small Industries, *Report on Development of Small-Scale Industries in India—Prospects, Problems and Policies* (Submitted to the Ministry of Commerce and Industries, Government of India).

New Delhi: Ford Foundation, 1963, pp. 126–128. The study defined seven levels or tiers of urban concentrations: under 5,000—agriculture and traditional village industry; 5,000–35,000—primary market centers for villages; 20,000–150,000— primary market centers for villages and secondary centers for small towns; 150,000– 300,000—similar to above but serving a wider area; 300,000–500,000—minor regional centers; 500,000–1,000,000—major regional centers, and 1,000,000 and over—super-regional centers of national and international importance. The main distinction between the two studies is that the Ford Foundation study discusses cities exclusively, whereas Tinbergen treats regional as well as national geographic levels.

25. "The New Urban Debate," *op. cit.,* pp. 1, 2, 8.

26. *Ibid.,* p. 19.

27. *Ibid.,* p. 21.

28. *Ibid.,* p. 23.

29. *Ibid.,* p. 24.

30. *Ibid.*

31. Jean Fourastié, *Le Grand Espoir du XX Siécle,* Edition definitive. Paris: Gallimard, 1963.

32. For a recent critical discussion of the three-sector concept see Jean-Paul Courtheaux, *Classification of Economic Activities.* The Hague: IFHP-CRU, 1969.

33. Fourastié, *op cit.,* p. 42.

34. For an interesting discussion of the factors affecting locational decisions, including a regional and an intra-urban economic structure model, see Folke Kristensson, *Människor, företag och regioner.* Stockholm: Almquist & Wiksell, 1967.

35. Bertram M. Gross, "The State of the Nation: Social Systems Accounting," in Raymond A. Bauer (ed.), *Social Indicators.* Cambridge, Mass.: M.I.T. Press, 1966, pp. 154–271.

36. Fourastié mentions urbanization briefly in a short chapter discussing life style (*op. cit.,* pp. 177–187); Gross, in describing varieties in social systems recognizes "territorial entities" as a separate category in which "organized complexity . . . becomes still more formidable" (*op. cit.,* pp. 171, 173).

37. There is no precise English translation for Fourastié's term defining that phase of achievement or completion in which a "filling out" occurs.

38. For an explicit treatment of the possible consequences of megalopolitan growth see: Jean Gottman, *Megalopolis.* New York: Twentieth Century Fund, 1961.

39. We have used the data for plotting our curves as they appear in the various sources consulted, without attempting to reorganize them according to our definition of the various sectors. Similarly, we have not adjusted data on urbanization to a single definition of "urban." We feel, as does Fourastié in comparing his three-sector definition with the classic definition of Colin Clark, that the statistics will differ only marginally and that for the purposes of a general discussion of the phenomena and processes involved statistical refinement will not add to the substance of the discussion.

40. "Urbanization: Development Policies and Planning," *op. cit.*

41. Calcutta Metropolitan District, 1966–1968, *Basic Development Plan.* Calcutta: Calcutta Metropolitan Planning Organization: Government of West Bengal, 1966.

42. In addition to the Calcutta Plan, *op. cit.,* see also Kristensson, *op. cit.,* and the application of his models to the formulation of the Stockholm regional plan (Outline Regional Plan, Stockholm Regional Planning Commission, 1966).

43. Ved Prakash, *New Towns in India.* Charlotte, N.C.: Duke University Program in Comparative Studies on Southern Asia, 1969.

44. Gunnar Myrdal, *Objectivity in Social Research,* The 1967 Wimmer Lecture, St. Vincent College, Latrobe, Pennsylvania. New York: Pantheon, 1969.

CHAPTER 2

Stanislaw H. Wellisz
ECONOMIC DEVELOPMENT
AND URBANIZATION

Introduction

The current pace of urbanization of the developing countries of Asia—
comparable to that of Europe and North America in the nineteenth century [1]—is a cause for optimism but also for alarm. The positive aspect of
the development is that urbanization is usually closely associated with
increasing economic well-being. Economic development releases an ever
rising proportion of the working population from the task of providing
food to tasks satisfying other physical and intellectual wants. Twentieth-
century technology strongly favors the concentration of non-primary activities in urban areas. Such areas, in turn, are the spearhead of economic and
social progress: [2]

> "The positive association of urbanization with industrialization and eco-
> nomic growth are well known. Cities provide concentrations of popula-
> tion from which industrial labor may be drawn; they also contain a
> greater variety of skills and resources than do rural areas. Even more im-
> portant perhaps, urbanization promotes values favorable to entrepreneur-
> ship and industrial growth; in particular, cities typically tend to favor a
> propensity to analyze traditional institutions and to innovate and accept
> change since, in a relatively impersonal and fragmented setting of urban
> life, the all-embracing bonds of traditional community systems are diffi-
> cult to maintain."

The extent of urbanization is closely associated with the Gross National Product per capita—a widely used (though by no means unexcep-
tionable) index of economic development, and an association also exists
between the level of urbanization and GNP growth.[3] Thus the current
rapid pace of urbanization of the developing world, and especially of
Asia, should be taken as a welcome sign of development, and as a presage
to achieve more rapid progress in the future.

Yet the rapid pace of urbanization is regarded with alarm by many economists, sociologists, and political scientists interested in the developing world, as well as by political authorities in the countries involved. Many authorities regard cities—especially large cities—as parasitic bodies draining the countryside of people and resources into an increasingly unhealthy urban environment. The mounting costs of urbanization associated with the increasing size of urban areas compete for resources with villages and towns; they compete for the provision of productive equipment needed to give employment to the rapidly growing labor force, and that needed to increase the productivity of labor. Yet, despite the considerable resources devoted to urbanization, living conditions of major cities have deteriorated over several past decades. In Calcutta [4]

"the increase in population has not been matched by the necessary increase in developed land, with consequent land scarcities injurious both to economic growth and to individual living standards . . . obsolescence has not been matched by proper maintenance and new investment; and appropriate expansions and reorganizations of utilities and services have not been undertaken to meet population growth."

Calcutta's case is extreme, but dangerous signs of deterioration are manifest in a number of other major cities, including Bombay.[5] To stem the deterioration, greater expenditure on urban areas is necessary, but improved urban amenities only serve to attract more people from the countryside. As one critic of India's urbanization put it: [6]

"In India, towns are not likely to lead into stagnation, but they can lead into slower economic growth and political instability, because of the diversion of resources from more to less productive investments. Urban populations have more access to, and influence on, political processes. As conditions worsen, towns are likely to demand and get progressively larger proportions of the national pie at the cost of the countryside."

Public authorities in the developing countries of Asia are thus faced with a serious dilemma. Too slow a rate of urbanization, and a neglect of urban needs, might slow down economic progress and lead to a dangerous deterioration of the urban environment. Too fast a rate of urbanization might drain the countryside, resulting in the creation of unwieldy, costly— and ultimately parasitic—urban concentrations. The need for a proper balance is obvious, yet "proper balance" is difficult to define in normatively meaningful terms, and even harder to enforce.

The search for an acceptable allocation of resources between large cities and small, and between the development of urban centers and hinterland may conveniently start with a brief review of arguments and theories critical of the current state of affairs. Following this review, we shall attempt to define an efficient urbanization strategy congruent with economic and social development goals.

The Over-urbanization Arguments

Are Asia's cities too large? Proponents of the "Rank Size and Primate City" theory argue that they are. This school of thought claims that in highly developed countries the size distribution of cities follows the log-normal pattern, while in less developed countries there are one or two huge "primate" cities and too few intermediate-sized ones.[7] It is further claimed that the "primate" cities slow down the growth of smaller towns and have a parasitic effect on the economy.[8] The argument, for a period, gained considerable currency, but is not carefully grounded in statistical observation. Thus, according to Brian Berry, "there is no relationship between the type of city size distribution and the degree to which a country is urbanized," [9] and it is also incorrect to say that the "primate" size distribution is associated with early stages and the "early," and "rank size" distribution with the advanced stages of development. The rank size distribution is found in the United States and West Germany, but also in India and China; "primate" distributions are found in Denmark, Sweden, and the Netherlands, all of which are highly developed. Pakistan has an intermediate type of distribution, fitting neither description. Moreover, even if a developed nation—rank size distribution, underdeveloped nation—primate-city relationship could be established, it would in no way follow that the primate city is "bad." It could be a characteristic of a given stage of development, subject to transformation as development proceeds.

A more serious argument is based on the optimal city size concept: according to this argument the efficiency of a city, considered as a socio-economic entity, rises to a point, and thereafter it declines. The economies of city scale accruing to productive activities tend to a maximum: with progressive increases in urban size the advantages associated with an expanding labor market, expanding access to a variety of suppliers and banking facilities, and expanding number of near-by customers increase at a decreasing rate. Likewise, a city's ability to support specialized educational, health, and recreational facilities grows with size, but as the city continues to grow, the successive additions to such facilities become less and less significant. On the other hand there is considerable—though widely scattered and not properly analyzed—evidence showing that the per capita cost of providing water, sewage, urban transport, and fire and police protection rises after a critical city size (which may vary from area to area depending on topographic, socio-economic and other factors) has been reached. At some point, every city reaches an optimum size, beyond which it becomes less efficient.

It is, unfortunately, extremely difficult to determine the optimum size of any particular city: [10]

> "At some time or other most of the world's great industrial complexes have been thought to be beyond the point where economies are offset by the extra cost incurred in various ways . . . industry continues to expand in these centers, showing that many industrialists themselves still think that the major concentration retains numerous advantages."

The claim that a given city is "too large" is often based on an enumeration of the disadvantages associated with the growing size of a metropolitan area without due consideration of the countervailing advantages. Insofar as the advantages and disadvantages are a joint product (the proportions of which, to be sure, can be altered by careful planning) the positive and negative aspects should be considered together. The continued growth of a metropolis is evidence that, on balance, the positive aspects continue to outweigh the negative.

More fundamentally, objection may be raised that the concept of optimality of an urban size is irrelevant from a policy point of view, and that the relevant issue is one of alternatives. Even if it could be shown that a given city surpassed its optimal size, it would not necessarily follow that the proper policy is to limit further expansion, or to develop lesser urban centers. To use an obvious example, an overexpanded city located in an oasis may be more efficient than five optimal size cities located in a desert.

The question of an efficient pattern of urbanization cannot be settled by a "revealed preference" policy, however, for the availability of locational alternatives is much influenced by urbanization policies. A policy favoring the largest urban centers may foster their expansion beyond their economic size in absolute as well as in relative terms. It is paradoxical that while the developing countries of Asia strive to limit the size of their "primate cities" they simultaneously pursue policies favoring such cities. The case of Pakistan may be cited here, not because it is unique, but because it is typical and well documented. According to the 1951 Census the total urban population of Pakistan numbered 7.8 million (6 million in the West Wing and 1.8 million in the East Wing), while the population of Karachi, the largest city, was estimated at 1.1 million. Yet the first Pakistani Development Plan allocated to Karachi 64 percent of the total investment in water and sewerage facilities in the two Wings.[11] The Second Plan Karachi allotment came to 28 percent of the national total, with another 40 percent going to Dacca and Chittagong, the two largest East Wing cities.[12] The Karachi allotment in the Third Plan was substantially smaller, as the system was virtually completed, while Dacca and Chittagong received the lion's share.[13] The second plan also allocated to Karachi 27 percent of total road transport investment,[14] and a disproportionately large investment in other utilities and in public housing.

The availability of low-income housing and of urban social overhead facilities acts as a magnet to the population; concentrations of population meant a larger and more varied labor force, and larger markets for economic activity. Moreover, the social overhead facilities themselves (water, power, transport) accentuate the initial locational advantages of the city. No wonder, therefore, that despite the tax abatements offered to industry locating outside Karachi, the city exercised a powerful pull on industry. For instance, between April 1959 and June 1963, 55 percent of the industrial investment sanctioned by the Pakistani Government for the West Wing, and 44 percent of the national total went to enterprises in the Karachi area (with Dacca–Narayanganj and Chittagong dominating the urban economic growth of the East Wing),[15] resulting in an increasing concentration of economic activity, and fostering further concentration of population. To be sure, the successive plans paid increasing attention to the needs of the smaller municipalities, but the initial creation of conditions favorable to concentration breeds further concentration: the establishment of activities which, originally, are "foot-loose" acts as a magnet to complementary—and often also to competitive—enterprises, and once a locational pattern is established, a change may involve major economic costs.

Despite all the concern about excessive size of cities, the large urban centers receive more attention from public authorities than small ones. The human needs of large cities are more readily visible—and the pressure for action more effective—than is the case of smaller, more dispersed groups. The same holds true of economic needs. Where there is an initial concentration of industry and commerce, the felt need for social overhead facilities is greater than where such activities do not exist. Thus pressures build up for overexpansion of cities not only in the absolute (and irrelevant) sense of over-optimal size, but in the sense of economically viable alternatives.

Is Asia Over-urbanizing?

Arguments are often heard that Asia over-urbanizes not in the sense of creating unduly large cities, but in the sense of urban concentrations, large and small, outrunning the economic base of urban growth. According to Bert F. Hoselitz, principal proponent of this theory, "Urbanization in Asia has probably run ahead of industrialization, and the development of administrative and other service occupations which are characteristically concentrated in cities." This pattern of development "emphasizes the disproportion between the cost of urban growth and the maintenance of proper facilities for urban dwellers and the earning capacity of people congregated in cities." [16] Hoselitz claims also that whereas in the advanced countries cities develop because of the "pull" of urban facilities, in Asia urban growth results from a "push" of poverty-stricken rural population.

The outflow of the population from rural to urbanizing areas creates economic and social problems in the urban centers, and instead of helping the hinterland by relieving the population pressure it may actually retard its development or hasten the decline. In the words of Gunnar Myrdal,[17]

> "the localities and regions where economic activity is expanding will attract net immigration from other parts of the country. As migration is always selective, at least with respect to the migrant's age, this movement will by itself tend to favour the rapidly growing communities and disfavour the others."

Over-urbanization, in short, stands for a "perverse" stream of migration, sapping the economic strength of the hinterland, without correspondingly large benefits to urban production. Instead of being a sign of development, over-urbanization is a sign of economic illness.

In appraising the over-urbanization argument it is essential to separate fact from fancy. Asia's urbanization occurs under conditions of greater rural population pressure than was the case in nineteenth-century Europe. While the rate of growth of industrial output registered in the third quarter of the twentieth century by the developing nations of Asia is comparable to or even higher than that in nineteenth-century Western Europe, the current rate of industrial job creation barely surpasses the rate of population growth—a phenomenon explicable by the rapid rise of population as well as by the capital-intensive nature of new industrialization. As a consequence a large proportion of the in-migrants into urban areas goes into work in the "unorganized" industrial sector (dwarf enterprises, typically using primitive machinery and insignificant amounts of mechanical power per worker, largely escaping factory regulations, and subject to severe census undercounts), into trade, tertiary activities, and varieties of more or less casual work. This being true, the distinction between rural "push" and urban "pull" remains irrelevant: what causes migration is the difference between rural and urban livelihood opportunities (real or expected). The fact that migration into urban areas continues shows that there is a continuing difference between levels. If restrictive measures were taken to slow down migration, the potential migrants would be condemned to a lower living standard.

It is equally all too easy to overstate the drain of the hinterland argument. It is true, as Myrdal and others argue, that migration tends to be selective: men of prime working age form a disproportionately large faction of the in-migrants into Calcutta, Bombay, Dacca, and other major cities. Such a selective migration, to be sure, lowers the average efficiency of the workers who are left behind. It raises the wage per efficiency unit, however, over and above what it would be in the absence of the migration. Insofar as the urban workers send remittances back home (and many of the single men do help maintain their families which stayed behind), selective migration tends to raise, rather than to lower, the incomes of the farm families. It is possible, of course, that a massive exodus of the ablest, most energetic, and most productive individuals might drain the farm population of entrepreneurs and innovators willing and able to pioneer new techniques, but this is hardly the case in India or Pakistan. In both countries few of the relatively prosperous farmers migrate, whereas the bulk of the migrants consists of landless or nearly landless farm laborers who had virtually no scope for their superior talents in their native villages.

There is however an alternate interpretation of the over-urbanization argument: the policies pursued by the developing countries of Asia favor urban areas to the detriment of rural ones, causing an uneconomically fast rate of urbanization. Michael Lipton forcefully argues the case that Indian economic policy-makers have an "urban bias": [18]

> "Of *public investment,* agriculture's planned share was 20.9 percent in the Second Plan, 22.4 percent in the Third and only 18.7 percent for 1966–70 . . . desired reductions in total plan outlays have usually been achieved by cutting the share of agriculture, and . . . the share of actual outlays going to agriculture has been even less than planned."

The relatively low share of agriculture, Lipton maintains, cannot be explained in terms of higher productivity of investment in (mainly urban) industry, since "capital/output ratios are substantially lower in agriculture than elsewhere" and "agriculture uses relatively little foreign exchange and few scarce skills." [19] The investment bias is reinforced by a price policy bias, working in disfavor of agriculture, by an incentive bias reflected in higher wages to civil servants working in non-agricultural tasks, by an educational structure bias, indicated by the superiority of educational opportunities available to urban, as compared with rural, inhabitants. [20] A similar bias is shown in job-planning: "between 1951 and 1961 censuses, the proportion of workers outside of agriculture (30 percent) hardly changed, yet 6.6 million of the extra 8 million jobs created during the Second Plan were outside agriculture." A similar imbalance appears in the Third and Fourth plans. [21]

The disastrous crops of the late 1960s which seriously endangered India's continued economic progress, and the spectacular results achieved by the new wheat and (more recently) rice technology in India, Pakistan, and other developing nations of Asia gave such a forceful demonstration of the dangers of agricultural neglect—and of the potential of the agricultural sector—that the "urban bias" is very visibly smaller in planning for the 1970s than in the policies pursued in the 1950s and 1960s. Nevertheless there is little doubt that the past decades have left a legacy of excessive concentration of attention on urban areas—a legacy of a self-defeating policy. Rural neglect means a stronger "push" toward urban areas, a "push" which, in turn, generates new demands for urban investment and an increased "pull" of cities on the rural population, aggravating further the urban problems.

Is the Urban Environment Unduly Neglected?

"Over-urbanization" does not mean that the cities of the Indo-Pakistani sub-continent are unduly favored; on the contrary a progressive deterioration of conditions in a number of major urban centers is all too visible. Thus, in the Third Indian Five Year Plan we read that: [22]

> "Although efforts on an increasing scale have been made in housing during the First and Second Plans, the problem of catching up with the arrears of housing and with the growth of population will continue to present serious difficulty for many years to come. Between 1951 and 1961 there was an increase in population of nearly 40 percent in towns with a population of 20,000 or more. It was reckoned in the Second Plan that the shortage of houses in urban areas might increase by 1961 to about 5 million as compared to 2.5 million houses in 1951."

The wording is almost paralleled by the Pakistani Five Year Plan: [23]

> "In the Indo-Pakistani sub-continent urban housing first came to the forefront during the Second World War as a result of rapid urbanization. . . . By the beginning of [Pakistan's] Second Five Year Plan there was a housing shortage of 600,000 urban dwelling units. Despite the addition of 150,000 new units over the last five years, this shortage has increased to nearly a million on the eve of the Third Plan."

The feeling of urgency expressed in the plans of the two countries is not reflected in the allocation of funds for housing and for other urban needs. In both India and Pakistan housing is left almost entirely to the private sector, while the burden of providing urban amenities falls predominently on municipalities and urban Improvement Trusts. Government aid is restricted to grant-in-aid for city planning, help in financing selected major urban improvements, public building construction, and (on a very limited scale) low-income housing. Thus in the face of rising needs the Indian Second Plan allocated only 765 millions of rupees (1.67 percent of the total public planned outlay) to town planning, slum clearance, and low-income urban housing. In the Third Plan the allocation increased in amount to Rs. 1,286 million, but in relative terms it fell to 1.59 percent of the total planned public outlay.[24] The actual expenditure fell so far behind the plan as to approximately equal the Second Plan outlay.[25] The Fourth Plan once again signaled the urgency of India's urban problem and estimated "that about 23 percent of the population of India might live in urban areas by 1981 as against 18 percent in 1961" and that the addition to urban population would in all probability increase by 80 million in two decades. Once again the planned contribution of the government is insignificant: Rs.2,550 million, or 1.59 percent of the total planned outlay.[26] The Pakistani figures tell much the same story.

It could be claimed that since investment in urbanization is within the realm of the private sector and of the municipalities, the insignificance of government subsidies is no proof of underinvestment or of discrimination against cities. Yet this is to reckon without the profound impact of government policies which do much to discourage investment in housing, and to encourage other types of investment.

The urban sectors in India and Pakistan bear a disproportionately large tax burden. As noted by the Indian *Taxation Enquiry Commission,* "per capita tax is not only larger in the urban sector but increases with the degree of urbanization, for every expenditure group," and "the large contribution of the cities (particularly in the case of Calcutta, its preponderating importance), to the tax resources of their respective States . . . [contrasts with] the comparatively small share of expenditure incurred within their limits." [27] To be sure, much of the Central and State government investment goes into industry located within urban areas, but though such location brings benefits to cities by creating job opportunities and generating income, it puts an additional burden on housing and urban facilities.

The municipalities' powers of taxation in the Indian subcontinent are notoriously weak; many of them derive a major part of their income from octroi taxes and other antiquated levies. The financial weakness is compounded by administrative weakness: with some notable exceptions, the municipalities of India and Pakistan are ruled by excessively large, diffuse, and professionally unqualified bodies. The institution of municipal improvement trusts, which originated under British rule and spread since independence, saps, instead of increases, the strength of municipalities. The trusts are self-financing, but they have no power of taxation. Their activity consists of purchasing stretches of land and re-selling after development. To pay for the cost of opening a new street, for instance, a trust must make enough money through the sale of developed plots to cover the cost of acquiring the land for the street as well as the cost of street construction. As a consequence, improvement trusts tend to concentrate on projects which look like sound commercial ventures, thus putting onto municipalities the burden of less viable projects.

Government policies and institutional arrangements also tend to hamper private investment in housing. The policy of nurturing industry thus channels investible funds away from housing and toward "productive" investments, while investment in housing is made difficult by lack of mortgage money and a welter of regulations. In India commercial banks are not permitted to grant mortgages; the only institutional mortgage lender, the Life Insurance Corporation, lends to individual house-owners only, and not to speculative builders. Private mortgage money can be found for short periods only (typically 2 to 5 years) although lack of adequate legal protection for the lender makes such mortgage financing expensive and difficult to obtain. Antiquated and confusing legislation makes title search lengthy and expensive. Last but not least there is a constant threat of imposition of rent control or of expropriation of large-scale urban properties.[28] Under these circumstances private capital shuns large-scale construction and tends to flow into single-family upper-middle and upper-income housing. The needs of the majority of the urban population are neglected.

The underinvestment in urbanization is vividly reflected in the declining importance of housing in the successive Five Year plans. The Second Indian Plan estimated that private construction of all types and public investment in housing would amount to about 18 percent of total private and public planned investment. During the Third Plan period the planned proportion shrank to 13 percent, and the same ratio is foreseen for the Fourth Plan.[29] The same de-emphasis of urban investment occurred in Pakistan where the proportion of total investment going into land development, housing, and other urban projects declined from 22 percent in 1950–55 to 20 percent during the First Plan period (1955–60), and to 15 and 11 percent, respectively, in the two successive plans.[30]

The low priority assigned to urban housing is the obverse of the policy giving high priority to "productive" industrial investment. The greater the proportion of investible funds devoted to the latter, the less there is for the former. Since machines can be used to produce other machines, while housing directly produces only a consumer service, a rapid growth strategy striving at enlargement of the productive capital base seems to dictate such an order of priorities. Such reasoning is spurious, however: one could claim with equal justice that a high growth strategy should maximize the machine-to-plant ratio, on the grounds that the plant building, by itself, produces nothing. Engineers are fully aware, however, of the plant-machine complementarity; the urban housing-urban productive investment complementarity is less obvious, but no less crucial.

It is possible to strive of course to reduce the required equipment in urban infrastructure through deliberate industrial location policy, placing plants in areas where labor can be drawn from rural or semi-rural locations. In deciding on a nonurban location, however, one must weigh the savings in urban infrastructure and housing against the cost of providing the industrial infrastructure for an isolated plant (or a small plant complex) and the loss of efficiency resulting in a location remote from the urban center. Both India and Pakistan have attempted to find a middle ground by promoting industrial estates which achieve some economies of scale in industrial land preparation, utilities, and common services, such as workshops, yet which do not require a high degree of urbanization. The program's results are mixed. Estates conveniently located near major centers flourish; most of those near smaller towns attract few customers, and the basic development expenditure tends to be high relative to the investment and employment effects.[31] Very likely a more comprehensive physical planning policy of coordination of social overhead capital investment in suitable selected "development foci" could have greater success, but such a policy is yet to be tried.

India and Pakistan thus face a serious problem of underinvestment in urban housing and in urban amenities. Given the limited resources and the rapid rate of urbanization, the problem is a formidable one. It is further aggravated by the "urban bias" which makes for an uneconomically fast growth of cities: were urban job creation relatively slower, and rural job creation relatively faster, the rate of urbanization and the need for urban investment would be correspondingly reduced.

Urban Conditions: Conflicts of Standards and of Means

India and Pakistan strive to defend a decent standard of urbanization, in the face of growing needs and of limited means, by enforcing minimum standards applicable to private and public urban construction. The results of standards enforcement are inevitable: since lawful urbanization cannot proceed at a pace sufficient to satisfy demand, a large and growing proportion of the new urbanites finds shelter either by "doubling up" with the inhabitants of existing structures, or by constructing illegal shacks to house themselves and to provide space for their work. Faced with choice of evicting needy people from their shelters or of tolerating the existence of illegal structures, authorities humanely tend to ignore violations, except where they interfere too visibly with the normal life of cities and towns, or spoil the aesthetics of representational districts. Thus, in the absence of adequate means to enforce standards, the standards themselves are self-defeating, for they lead to a pattern of growth defying all regulations.

The determination of housing standards is usually approached from the point of view of individual and collective comfort and safety. Minimum floor space per capita and maximum urban density standards are primarily designed to prevent excessive personal discomfort. Other standards, including those pertaining to the durability and flammability of construction materials, sanitary arrangements, airway and street layout, etc., take into account social as well as individual factors. As a general rule higher standards are applied to larger and more densely populated areas. Thus a wooden hut might be a perfectly satisfactory village house: in case of fire a single family might, at worst, lose its abode. In an urban environment the danger of a fire engulfing a number of closely spaced dwellings prohibits the use of materials which are not reasonably fire resistant. The same holds true of water supply, sewage, and other collective facilities: a relatively primitive system, consisting of wells and night soil disposal, of surface water drains and earthen streets is acceptable for a village, but not for a town, and arrangements adequate for a town are not suited for a major city.

If standards designed in accordance with an appraisal of needs are to be enforced, means must be found to provide the requisite facilities. Neither India nor Pakistan does in fact provide the requisite funds for urbanization. To do so would require a major redefinition of priorities. Thus Britton Harris calculated that to meet India's Second Plan urbanization needs on a very modest standard would have required an investment of 20 billion rupees not provided by the Plan—an amount equal to 30 percent of the total planned private and public plan investment.[32] My own estimates show that if the entire five-year increment of the low-income urban population of East Pakistan were to be urbanized at the admittedly low standards applied to refugee housing, the Pakistani Third Plan allocation for *all* East Wing urban investments (including middle- and upper-income housing, office and commercial buildings, schools, hospital, etc.) would cover just half the cost.[33]

Are the standards too high? In humanitarian terms they clearly are not. The standard refugee house, alluded to above, consists of a single room, a verandah, a sanitary latrine, a bathing platform, but no kitchen; the total living space per family ranges from 114 square feet to 118.5 square feet depending on the type of the house. The publicly subsidized low-income houses in Calcutta and in Bombay provide somewhat better facilities but they too are far from luxurious. The street layout, the open spaces, and public facilities are also very modest. The fact remains, however, that housing meeting the official standards can be provided for only a small fraction of the people who need it.

There is another point of approach to the determination of standards, and that is willingness to pay. All people—especially the poor—face a variety of urgent needs, including the need for food, clothing, education, health, and recreation, which they must satisfy within very modest means. Given the climate of the subcontinent, housing ranks relatively low. Thus a survey of the poor families conducted in Calcutta showed that the expenditure on shelter ranged from 10 percent to 15 percent of the family budget, with the majority closer to the lower limit. The economically viable investment in urban dwelling and in the ancillary urban facilities, calculated by taking into account the income distribution of the low-income urban families, and by capitalizing the rent payments, is approximately one half of the cost of a standard low-income house in India and Pakistan.

It may be maintained, of course, that the higher-income families should subsidize the lower-income ones. To this policy there are three objections. First, the possibilities of subsidizing a majority of urban inhabitants are quite limited. Secondly, if the poor were given a free choice, they would prefer to devote a major part of the subsidy to the satisfaction of needs other than housing—hence a direct housing subsidy is socially inefficient. Thirdly, with the existing tax structure a large part of the cost of the subsidy would inevitably be borne by the low-income families themselves. Thus in India as of 1960 over 20 percent of the taxes were paid by families with monthly expenditures of Rs.100 or less, and 35 percent of taxes by families with monthly expenditures of Rs.150 or less.[34] Roughly one-third of the subsidy would be paid by the low-income families, forced thereby to cut down on their other expenditures to get better housing. They are not willing to do this voluntarily.

The alternative to heavy subsidies is the reduction of urbanization costs. If the housing standards are to be preserved, the only real possibility lies in the reduction of land costs which, next to construction, are the largest element in urban investment. Several Indian plans blame speculators for rising urban land prices (while recognizing simultaneously that urbanization and industrialization are also responsible for part of the increase) and advocate control of land prices "in the interest of low income groups." [35] Speculation, it is true, can temporarily raise land prices above the locational scarcity values, but all successful speculation merely anticipates value trends. To control land values means to transfer the land rent from the land holder to the land user. As long as user rights are transferable, the scarcity rents are capitalized, and the subsequent users pay a higher price regardless of the control. This phenomenon is widely observable in Calcutta where slum land rents are controlled under the so-called Thika Tenancy Act,[36] but where, at the same time, slum hutment rents are uncontrolled. The difference between the economic hutment rent and the legally permitted land rent accrues to the thika tenants (hutment owners) and the slum dwellers are none the better for it. In government-sponsored low-income housing where the house rents are set below the market level, sub-letting is extremely widespread: poor families alloted such housing move instead to more cramped quarters and sublet to better-to-do families willing to pay higher rents.

Even if it were administratively possible to force through and police a reduction of land rents below the scarcity costs, such a policy would lead to waste: the fact that land is put to uses in which it does not yield its scarcity rent means that a part of the potential locational rent is foregone, to the detriment of the community as a whole.

The remedy, clearly, lies not in artificial control of land rent, but in an increase of relative supply of land. Planning for rational use of existing urban space and for better access to outlying areas and the creation of alternate urban foci accomplishes this purpose. In several Indian cities, notably Delhi and Calcutta, bold development plans have been prepared, though their execution often sadly lags behind the programs.

Even with rational land use and transport planning new permanent housing remains outside the reach of a majority of urban families. The "filtering-down" process satisfies a part of the need: partially depreciated houses, originally built for upper- and middle-income brackets get "handed down" to poorer families. However, the faster the population growth, the smaller the fraction of the new low-income housing requirements which can be met through the filtering-down process. And cost calculations clearly show that new houses, built with conventional materials, are above the reach of many urbanites—no matter how modest the space and sanitation standards.

The long-run remedy must be sought in a technological breakthrough, either in the construction field for the building of cheap permanent urban structures (preferably multi-story ones, to save on land costs), or in the transportation field, to permit the suburbanization of low-income population without an excessive burden of transport costs.

An interim solution, increasingly albeit reluctantly accepted in India, consists of the construction of "controlled slums." In such settlements, the public authorities provide the layout, including streets, passages and airways, as well as surface drainage and rudimentary sanitary facilities. In some of the schemes a permanent plinth is also provided. The occupants are then permitted to build temporary hutments, much like those now "illegally" dotting the urban landscape. Such controlled slums cost a fraction of the cost of permanent structures: in Calcutta the cost of erecting a hutment is about 20 percent that of a permanent construction. The advantages of planned over uncontrolled slums are obvious. The plan can provide an acceptable level of social amenities and sanitary facilities. Moreover, while slums are receptacles for adverse social selection—all those who can afford to do so, move away from them—the layout for controlled slums can make provision for subsequent up-grading through the consolidation of adjacent plots and erection of permanent structures.[37] Far from being a segregative device, the controlled slum can be an area making provision for growth and for gradual improvement.

Is a Rational Urbanization Policy Possible?

Poverty and the mounting population pressure set limits to the improvement in urban living conditions which can be made in Asia's developing countries—above all in India and Pakistan. Attempts to improve the urban environment at general expense are bound to be largely self-defeating, for any increase in the discrepancy between rural and urban living conditions is likely to accelerate the inflow of rural migrants into urban areas. Given the relative ease of migration, the cities will be poor as long as the countryside is poor.

A rational urbanization policy must therefore start out by discarding false hopes symbolized by unrealistically high urbanization standards which cannot be maintained in practice. The gradual acceptance of the "controlled slum" idea shows that the Indo-Pakistani subcontinent is shifting away from the elitist city concept which prevailed under British rule to the concept of a city in which broad masses can afford to live.

The acceptance of modest standards does not signify the abandonment of a program for urban social overhead and public consumption facilities. It means the provision of facilities for all or at least for the vast majority of urban inhabitants. Such facilities—public supply of drinking water being the most obviously necessary, but by no means the only one—do not attract an unwarranted number of inhabitants into the urban areas, assuming the urban inhabitants themselves are to bear their cost. How to apportion the cost of a collective good to individual classes of users is, of course, a matter of social equity and of political expediency. Economics is silent at this point. It is clear, however, that an urban-rural development equilibrium can be reached only if the cities pay for their social overhead costs and the villages for theirs.

A rational urbanization policy requires an end to the urban bias in planning. The trend is clearly in the right direction since the "Green Revolution" gave such a striking demonstration of the payoff obtainable from a rural investment program. A shift in the proportion of new job creation in favor of the countryside will automatically reduce the migration to urban areas. At the same time a rational urbanization policy requires the end of an anti-"non-productive investment" bias which leads to discrimination against housing. Far from being inimical to rapid growth, investment in housing is one of the end-goals of growth. It is sometimes forgotten that machines are a means and that the end is a better fed, better housed, and better clothed population.

The realization that investment in urban-type industries is to a large extent complementary with investment in housing and urban amenities leads to a more rational allocation of investible funds to urban and non-urban production. The economic calculus should weigh the costs and benefits of urban industry-cum-urbanization against the rural industry costs and benefits. The calculus is very complex and neither economists nor physical planners are capable of giving simple and reliable guidelines. For want of such guidelines maximum reliance must be placed on the market mechanism. This means that housing standards should be geared to the ability to pay of (and should be paid for by) the individual inhabitants, while collective facilities should be geared in accordance with the ability to pay of (and should be paid for by) the communities benefiting from them. Cities which grow beyond their economic size will find it increasingly difficult to finance further expansion, while smaller towns, of a more efficient size, will be able to provide cheaper facilities. The relative costs, in turn, will provide signals for spatial planners on both the local and regional level.

Both India and Pakistan are searching for a policy of rational space use—not only within cities, but also within broader regions. In their search for a rational pattern of spatial allocation they have to overcome administrative and political difficulties, devise new methods of solving spatial allocation problems, and bridge the gap between economists, who neglect problems of space, and physical planners, who disregard the economic factors. The problems are enormous, but, despite the continuing growth of slums and the mounting shortages of urban facilities, there is hope for new and better solutions.

NOTES

1. See Kingsley Davis, "Urbanization in India: Past and Future" in Roy Turner (ed.), *India's Urban Future*. Berkeley and Los Angeles: University of California Press, 1962, pp. 8–9.

2. Irma Adelman and Cynthia Taft Morris, *Society, Politics and Economic Development*. Baltimore: Johns Hopkins Press, 1967, p. 25.

3. For a sample of 74 countries Adelman and Morris found a simple correlation of 0.66 between the extent of urbanization and per capita GNP in 1961, and a simple correlation of 0.46 between the extent of urbanization and the growth of per capita GNP 1950/51–1963/64 (*ibid.* pp. 281–283).

4. Calcutta Metropolitan Plan Organization (CMPO), *Basic Development Plan —Calcutta Metropolitan District, 1966–1986*. Calcutta: Government of West Bengal, 1966, p. 3.

5. See D.T. Lakdawala *et al.*, *Work, Wages and Well-Being in an Indian Metropolis, Economic Survey of Bombay*. Bombay: University of Bombay, 1963, pp. 33, 37–39 and *passim*.

6. Shanti Tangri, "Urbanization, Political Stability, and Economic Growth," in Turner, *op. cit.* p. 209.

7. See George Kingsley Zipf, *National Unity and Disunity*. Bloomington: 1941, and *Human Behavior and the Principle of Least Effort*, Cambridge, Mass.: 1949.

8. See UNESCO, Report of the Director General on the Joint UN/UNESCO Seminar on Urbanization in the ECAFE region, Paris, 1956.

9. Brian J.L. Berry, "City Size Distribution and Economic Development." *Economic Development and Cultural Change*, Vol. IX, 1964, p. 579.

10. R.C. Estall and R. Ogilvie Buchanan, *Industrial Activity and Economic Geography*. London: 1961, p. 107.

11. Government of Pakistan, Planning Commission, *Preliminary Evaluation Report of the First Five Year Plan for Housing and Settlements*. Karachi: December 1960, pp. 8–9, cited in John C. Eddison, "Industrial Location and Physical Planning in Pakistan" (mimeographed, n.d.) pp. 4–5.

12. Government of Pakistan, *The Second Five Year Plan*. Karachi: 1960, pp. 326–328.

13. Government of Pakistan, *Outline of the Third Five-Year Plan* [1965–1970]. Karachi: August 1964, pp. 196–197.

14. Government of Pakistan, *The Second Five Year Plan, op. cit.,* p. 306.

15. Data from Ministry of Industries and Natural Resources, Industries Division, Government of Pakistan, July 1963, Karachi.

16. Bert F. Hoselitz, "Urbanization and Economic Growth in Asia." *Economic Development and Cultural Change,* Vol. VI, 1957.

17. Gunnar Myrdal, *Economic Theory and Under-developed Regions.* London: 1957, p. 27.

18. Michael Lipton, "Strategy for Agriculture: Urban Bias and Rural Planning." In Paul Streeten and Michael Lipton (eds.), *The Crisis of Indian Planning.* London and New York: Oxford University Press, 1968, p. 85.

19. *Ibid.,* p. 144.

20. *Ibid.,* pp. 130–135.

21. *Ibid.,* p. 143.

22. Government of India, Planning Commission, *The Third Five Year Plan,* Delhi: Government of India Press, 1961, p. 680.

23. Government of Pakistan, Planning Commission, *The Third Five Year Plan, 1965–70. Karachi:* Manager of Publications, 1967, p. 363.

24. Government of India, *The Third Five Year Plan, op. cit.,* pp. 86, 681. The totals given here are net of allocations to village and plantation workers' housing.

25. Government of India, Planning Commission, *Fourth Five Year Plan, A Draft Outline.* New Delhi: 1966, p. 73.

26. Government of India, *Report of the Taxation Enquiry Commission, 1953–54.* Delhi Ministry of Finance (Dept. of Economic Affairs), 1955, Vol. I, p. 67.

27. *Ibid.,* Vol. I, p. 43.

28. For a detailed case study, see S. Wellisz, *The Private Housing Market in Calcutta.* Calcutta Metropolitan Plan Organization (mimeographed, August 1963).

29. Calculated from data given in the Indian Second Plan. New Delhi: 1956, pp. 57, 557, the Third Plan, *op. cit.,* pp. 105, 681, and the Fourth Plan, *op. cit.,* p. 41.

30. Government of Pakistan, Planning Commission, *Guidelines for the Third Five-Year Plan.* Karachi: Manager of Publications, 1963, p. 18.

31. See Government of India, *Third Five Year Plan, op. cit.,* pp. 449–50.

32. See Britton Harris, "Urban Centralization and Planned Development," in Turner, *op. cit.,* pp. 265–266.

33. For details of calculations, see S. Wellisz, *Toward an Urban Low-income Housing Policy for East Pakistan.* Cambridge, Mass.: Havard Development Advisory Service (mimeographed, February 1965).

34. The computation is based on the following sources: (1) National Sample Survey No. 69, *Consumer Expenditure by Levels of Household Expenditure,* Thirteenth Round, September 1957–May 1958, I.S.S. Calcutta, 1960; (2) Government of India, Ministry of Finance, *Incidence of Indirect Taxation, 1951–1958.* New Delhi: 1961; and (3) Government of India, *Monthly Statistical Bulletin* (various issues). The calculation assumes that all income taxes and all corporate taxes fall on families with incomes exceeding Rs.300 per month.

35. Government of India, *The Third Five Year Plan, op. cit.,* p. 690.

36. West Bengal Act II of 1949.

37. This point is made *inter alia* by Charles Abrams in his *Man's Struggle for Shelter in an Urbanizing World.* Cambridge, Mass.: M.I.T. Press: 1964, p. 97.

CHAPTER 3

Gerald M. Desmond
THE IMPACT OF NATIONAL AND REGIONAL DEVELOPMENT POLICIES ON URBANIZATION

Introduction

In dealing with a subject of such imposing dimensions it may be prudent to point out one or two hazards right at the start. A review of the limitations on unqualified conclusions might also serve to set out the objectives the writer had in mind, his approach to the problem, and the assumptions on which the work proceeds.

In addition to the normally available stock of limiting factors, e.g. the heterogeneity of the countries in the region and the lack of index comparability between countries and over time, there seems to be a remarkable divergence of understanding on the basic concepts to be dealt with. Just what constitutes a national or regional development policy, and how these differ from the operational elements of national and regional development plans was found to be a matter of such subjective interpretation that it became necessary to attempt some clarification of how these terms are understood and employed here.

Another difficulty repeatedly encountered in dealing with national plans and policies, and in the material written about them, is that of deciding what is an acceptable level of non-intention. That is, how often and how far can official goals transcend reality in order to mobilize the national spirit or, on a lesser level, to simply dress things up?

If the implications of this widespread practice were confined solely to questions of technical integrity, it need not concern us further. For the purpose of this review, however, it is important to distinguish between real and non-real intentions. This holds particularly for estimations of future trends and the influence of public policy on them.

With this in mind I have attempted to separate the first part of the subject into three categories: development planning, development policies,

57

and development decisions. In each case the term development is understood primarily in its economic sense. Sub-national or regional development
plans or policies are viewed as dis-aggregated parts of national schemes
unless specifically identified as true regional plans. Physical planning and
policies are also examined separately and, again, regional plans of an
essentially spatial or physical character are specifically noted.

The approach taken, then, is to review the development plans and
policies and, most important, the investment decisions of countries in
South and Southeast Asia in recent years to see how and to what extent
they seem to affect urbanization and urban growth in that region. Although covered in greater scope elsewhere, a brief summary of the main
demographic trends and type and location of urban growth is presented
here to provide a background for whatever conclusions may arise. The
state of spatial or physical planning is also reviewed and its current and
potential role assessed. Finally, an attempt is made to discern some patterns of future urban trends and to consider ways in which national development plans and policies can help to influence urban growth in a
positive manner.

STRUCTURE AND TREND OF URBANIZATION
AND URBAN GROWTH

Scale and Rate of Urbanization

The distinction between the terms "urbanization" and "urban growth"
is not always an obvious one. Urbanization refers to the growth of population living in urban places relative to that of the country as a whole; a
macro-consideration. The growth of individual cities and towns is a narrower and more specific observation of the specific location of growth.
Generally speaking "urbanization" includes both concepts unless indicated
otherwise in the text.

A recent United Nations Study [1] provides a thorough and up-to-date
overview of urbanization trends and projections from 1900 to the end of
the century. The present situation and especially the projected estimates
bring into immediate focus the reasons for concern and the sense of
urgency felt by planners and social scientists within the region and in the
world at large.

South and Southeast Asia, when taken together, represent the largest
and—outside of Western Europe—most densely populated region in the
world. It is also one of the world's least urbanized regions, only Africa
having a lower proportion of its population in agglomerations of 20,000
or more. Despite its low level of urbanization and the fact that the region
is urbanizing least rapidly of all, according to the United Nations projec-

tions South Asia alone by 1990 will have more people living in urban agglomerations than any of the other 21 major regions of the world.

Table 1. Projections of Urban Population 1960–2000 (millions)

Region	1960	1970	1980	1990	2000
Southeast Asia [a]	27.4	42.9	66.8	106.3	162.4
Middle-South Asia [b]	76.6	113.5	168.9	244.8	343.9
Total	104.0	156.4	235.7	351.1	506.3

a. Southeast Asia includes Indonesia, Viet-Nam, Philippines, Thailand, Burma, Malaysia, Cambodia, Laos, Singapore, Portugese Timor, Brunei.

b. Middle-South Asia: India, Pakistan, Iran, Afghanistan, Ceylon, Nepal, Bhutan, Sikkim, Maldive Islands.

Source: United Nations, *Growth of World's Population 1900–2000,* table VI-8.

By the year 2000 South and Southeast Asia will have over 500 million urban dwellers, or nearly one-quarter of the world's total urban population. It is the sheer mass of population contained in this corner of the world that produces an apparent conundrum: low level of urbanization plus slowest rate of urban growth equals greatest number of urban dwellers. Of course since the region will still have a comparatively low level of urbanization by the year 2000 (only 27 percent compared with over 50 percent in Latin America and 72 percent in North America), the majority of its enormous 1.5 billion population will still be living in villages and rural areas. However, current demographic information provides no basis to assume that a state of urban-rural equilibrium will be reached at 27 percent urban, or at 50 percent. There is some evidence that, at least in the presently industrialized countries, a degree of stabilization begins to assert itself after about two-thirds of the population are living in urban places. If this ratio were to apply also to the countries in South and Southeast Asia it is evident that the urban population of that region would surpass during the next century the total number of urban dwellers in the entire world at the present time—approximately 1,000 million.

The question of whether the countries of the region are likely to follow the pattern of the earlier industrializing countries thus looms very large in planning the physical and human environment of the future. There are really two questions involved. Are the development efforts of the countries in the region similar to those of other parts of the world? And will they have similar demographic effects? A third question that arises in considering the first two might be stated as follows: If development efforts do tend to influence the rate and direction of urbanization, can they be utilized to manipulate these trends to the benefit of the countries concerned? These and other questions are dealt with speculatively in this paper. Definitive answers are required, however, if the environment of the region is to be

prepared to accommodate human settlements on a scale beyond anything
the world has so far experienced.

Urban Growth—Impact of the Past on
Present and Future Trends

Urban centers have existed in South and Southeast Asia since ancient
times. Some of these were seats of administrative and religious leadership
—the "sacred cities" of the Angkor complex and Pagan in Burma. Later
these inland, agriculturally oriented cities were joined by coastal cities
closely identified with trade and commerce, such as Malacca and the
Malay peninsula.[2]

It was the period of colonization by European powers, however, that
established the urban network which dominates this region today. Beginning in the sixteenth century and continuing up to (and beyond) the
middle of the present century—when independence was achieved by all
but a few remnants of the former colonial empires—European interests
influenced the location, administration, and fortunes of the great urban
centers. Mainly chosen for use as ports through which indigenous raw
materials and European manufactured goods could flow, or as administrative towns in some interior locations, the urban centers established or
substantially developed by colonial powers remain the largest and most
influential cities in the region today.

This phenomenon of former colonial cities, established to serve the
trade, military, or administrative interests of a foreign power, and which
continue to occupy a primary role in the urban network of Asian countries
today, has been viewed by many writers as having a negative and distort-
ing effect on the region's further development. The overwhelming dominance of these "primate" cities, demographically, economically, and
politically is felt to reflect a "pseudo-urbanization" which differs in many
ways from the "true urbanization" of the industrialized countries of Europe
and North America.[3] One often-used index of this pseudo-urbanization is
the percentage of the labor force employed in manufacturing—a key index
of economic growth and vitality. Asian countries compare unfavorably on
this index with European and other industrial countries at comparable
levels of urbanization. Lower per capita rates of agricultural output is
seen as another example of premature and economically unhealthy urbanization. The resulting high levels of unemployment or non-productive
employment in urban centers and the attendant deterioration of the cities'
social and physical fabric may lead, it is felt, to catastrophic breakdowns
unless steps are taken to halt or divert the steady flow of migrants to the
large cities.

In this chapter some of these assumptions and conclusions are examined, within the context of the apparent development policies of countries in this region to see how such policies influence the pace and pattern of urbanization, and whether their impact will serve to improve or exacerbate urban conditions.

Before reviewing national development policies, some attention might be given to a more specific examination of the location and apparent direction of urban growth within the overall levels and projections of urbanization given above. This is useful not only because of the obviously sharper detail provided by disaggregation of gross data, but also to point out an important but often overlooked fact. Many critics of present development policies—especially of their locational aspects—feel that by re-locating economic enterprises present migration trends could be altered to bring about a more rational or balanced pattern of urban growth. It is important to note in this connection what percent of urban growth is caused by migration, as opposed to indigenous urban population growth. This ratio has a great deal to do with the efficacy of locational policies in shifting urbanization patterns.

As shown in the United Nations report, and in Table 2 here, about 13 percent of the region's population lived in urban agglomerations of 20,000 or more in 1960. The rate at which this urban percentage is growing is not particularly rapid compared to other developing regions. The ratio of urban agglomeration to total population grew less than 3 percent per year in the decade 1950–1960, compared to 7 percent for Africa and 5 percent for Latin America.[4] At the same time the annual percentage growth of these agglomerations themselves was about 5 percent.[5] These rates indicate that internal urban growth is proceeding much more rapidly than overall urbanization, from which we can conclude that natural increases in indigenous urban populations provide a large part of urban growth, while migrations from villages or rural areas, although significant, may not be as dominant a factor as has often been assumed.

Table 2. **Percentage of Total Population in Localities with 20,000 and More**

Region	1920	1930	1940	1950	1960	1970	1980	1990	2000
Southeast Asia	4	6	7	10	13	15	18	23	27
Middle-South Asia	5	6	8	11	13	15	18	21	25

Source: United Nations, *Growth of World's Population, 1900–2000,* table VI-10.

If these data accurately reflect the demographic dynamics of city growth, the possibilities for influencing or controlling the overall pace and direction of urban growth may be even slimmer than they now appear, at least in the short or intermediate term. Whether a deliberate policy of slowing or re-directing urban growth is desirable is not at issue here; whether it is feasible, and by what means, does bear directly on the central theme of this chapter.

Recent Urban Growth in South and Southeast Asia

The first part of this chapter drew some overall dimensions of urbanization and urban growth in the region. A closer examination of the structure and location of urban growth in the region as a whole and in individual countries may help to bring the discussion into more assimilable terms.

A look at Table 3 shows that the current spread of urban population in various sizes of agglomeration (urban centers) for the region (South Asia in this table) is very similiar to Europe and much less skewed toward multi-million cities and small towns than Latin America.

Table 3. Percentage of Urban Population in Agglomerations Within Selected Groups in the World and Major Areas, 1960

Major area	AGGLOMERATIONS WITHIN SIZE LIMITS (INHABITANTS):			
	20,000–99,999	100,000–499,999	500,000–2,499,999	2,500,000 and over
World Total	23	19	21	17
More developed major areas	21	19	23	16
Europe	26	18	24	9
Northern America	11	20	26	36
Soviet Union	26	23	16	9
Oceania	15	16	53	—
Less developed major areas	24	18	19	18
East Asia	22	16	24	27
South Asia	28	21	19	8
Latin America	20	13	14	20
Africa	26	26	22	7
More developed regions	21	18	22	21
Less developed regions	25	19	21	11

Source: United Nations, *Growth of World's Population, 1900–2000,* table 14.

What this means in terms of future urban growth is difficult to say. Comparisons of these rank-size distributions, however, show that the major gains have been in the large and multi-million size cities. In Europe, the total in multi-million cities declined, all of the urban growth occurring in middle- to large-size cities.

The number of large and multi-million cities in the South Asian region grew rapidly during the period 1920–1960, with most of the growth in the lesser category.

Table 4. Number of Big Cities and Multi-Million Cities in the World and Major Areas, 1920–1960

Major area	1920	1930	1940	1950	1960
Big cities (500,000 and more inhabitants)					
World Total	83	102	126	158	234
More developed major areas	62	73	85	97	126
Europe	40	47	50	52	56
Northern America	18	20	21	29	41
Soviet Union	2	4	12	14	25
Oceania	2	2	2	2	4
Less developed major areas	21	29	41	61	108
East Asia	11	13	16	22	50
South Asia	4	7	14	22	29
Latin America	5	7	8	11	19
Africa	1	2	3	6	10
Less developed regions	14	21	32	53	95
Multi-million cities (2,500,000 and over)					
World Total	7	11	15	20	26
More developed major areas	6	8	10	12	12
Europe	4	4	4	4	4
Northern America	2	3	4	6	6
Soviet Union	—	1	2	2	2
Oceania	—	—	—	—	—
Less developed major areas	1	3	5	8	14
East Asia	1	2	3	3	6
South Asia	—	—	1	2	3
Latin America	—	1	1	3	4
Africa	—	—	—	—	1
Other	3	7	9	11	11

Where and in what kind of cities has this growth occurred? One of the many interesting features of urban growth in the region under discussion is the relatively rapid growth of non-seaport cities. In 1920 this region had no large inland cities, all of the major urban centers being seaports, as might be expected in a largely colonized region. After 1940, however, the percentage of large city population in seaports declined dramatically (from 72 to 53 percent) while the number and population of large inland cities grew rapidly. A look at the actual cities where this growth has taken place may shed additional light on the reasons behind the marked decline in the primacy of seaports among the region's largest urban centers, and on other economically related trends.

Tables 5, 6, and 7 show the distribution of urban (agglomerations of 20,000 and over) populations throughout the region, with further breakdowns of distributions of increasingly larger urban size as well as specific cities in the "big city" (500,000 and over) and "multi-million city" cate-

gories. As expected there has been a steady rise in both total urban population (absolutely and as a percentage of total population) and in the
various size groups.

The growth in the number and populations of cities in the big and

Table 5. Agglomerated Population, 1920–1960 (20,000 and over)
(millions)

Country	1920	1930	1940	1950	1960
India	14.7	18.7	27.2	41.7	59.4
Indonesia	1.5	2.7	4.6	6.9	10.2
Pakistan	1.9	2.6	4.0	5.9	9.6
Philippines	0.9	1.4	2.0	3.1	4.6
Viet-Nam	0.3	0.5	0.9	2.2	4.2
Thailand	0.5	0.6	0.9	1.2	2.4
Burma	0.7	0.9	1.1	1.6	2.3
Western Malaysia	0.3	0.4	0.6	1.0	1.7
Singapore	0.4	0.4	0.6	0.9	1.2
Ceylon	0.5	0.5	0.7	0.9	1.2
Cambodia	0.1	0.1	0.1	0.3	0.5

Source: United Nations, *Growth of World's Population, 1900–2000,* table 15.

Table 6. City Population, 1920–1960 (100,000 and over) (millions)

Country	1920	1930	1940	1950	1960
India	6.4	7.9	13.4	23.1	34.0
Indonesia	0.8	1.6	3.2	5.6	9.1
Pakistan	0.9	1.5	2.3	3.7	6.6
Philippines	0.4	0.7	1.2	1.8	2.7
Viet-Nam	0.2	0.4	0.6	1.6	3.0
Thailand	0.5	0.6	0.8	1.0	1.7
Burma	0.5	0.6	0.6	0.8	1.2
Western Malaysia	0.1	0.2	0.3	0.4	0.8
Singapore	0.4	0.4	0.6	0.9	1.2
Ceylon	0.3	0.4	0.5	0.6	0.8
Cambodia	—	—	0.1	0.3	0.4
Laos	—	—	—	—	0.2
Nepal	—	—	—	0.1	0.1

Source: United Nations, *Growth of World's Population, 1900–2000,* table 14.

multi-million category is of particular interest. In India, Pakistan, and
Indonesia, the only countries in the region with more than one "big city,"
a total of 15 have been added since 1930. For the region as a whole, 22
urban centers have grown into this category. The emergence of new big
cities in the three countries, which already had a dominant city or cities,
reveals (Table 8) a dynamism and competition for growth between urban

Table 7. Big-City Population, 1920–1960 (500,000 and over) (millions)

Country	1920	1930	1940	1950	1960
India	3.6	4.0	7.8	13.7	20.4
Indonesia	—	0.5	1.5	3.0	4.8
Pakistan	—	—	0.6	1.8	3.8
Philippines	—	0.6	0.9	1.5	2.2
Viet-Nam	—	—	—	1.2	2.0
Thailand	—	0.5	0.6	0.8	1.3
Burma	—	—	—	0.7	0.9
Singapore	—	—	0.6	0.9	1.2
Ceylon	—	—	—	0.6	0.8

Source: United Nations, *Growth of World's Population 1900–2000*, table 14.

centers which is not always appreciated. Since this observation holds mainly for large countries, it is difficult to say whether the absence of national frontiers encourages urban migration or whether the sheer territorial size of these countries is bound to produce a larger number of big urban centers. In either case there does appear to be some spatial relationship as to the number of large cities within a country. To what extent country size limits the opportunities for new big cities to emerge is a question better dealt with by demographers and urban geographers. The economic implications are even less evident, but do raise further questions which are taken up in the next section.

Table 8. Population of Big Cities and Multi-Million Cities, 1920–1960 (thousands)

Major Area	1920	1930	1940	1950	1960
India					
Calcutta	1,820	2,055	3,400	4,490	5,810
Bombay	1,275	1,300	1,660	2,730	4,040
Delhi	—	—	640	1,310	2,270
Madras	525	640	765	1,355	1,700
Hyderabad	—	—	715	1,055	1,240
Bangalore	—	—	—	740	1,170
Ahmedabad	—	—	570	775	1,170
Kanpur	—	—	—	685	950
Poona	—	—	—	565	725
Nagpur	—	—	—	—	670
Lucknow	—	—	—	—	640
Pakistan					
Karachi	—	—	—	990	1,830
Lahore	—	—	—	650	1,250
Dacca	—	—	—	—	690

Table 8, Continued

Major Area	1920	1930	1940	1950	1960
Ceylon					
Colombo	—	—	—	610	810
Indonesia					
Djakarta	—	525	1,000	1,750	2,850
Surabaja	—	—	500	700	975
Bandung	—	—	—	575	925
Philippines					
Manila	—	600	900	1,475	2,150
Viet-Nam, Republic of					
Saigon-Cholon	—	—	—	1,200	1,400
North Viet-Nam					
Hanoi	—	—	—	—	650
Thailand					
Bangkok	—	500	625	750	1,325
Singapore					
Singapore	—	—	600	800	1,025
Burma					
Rangoon	—	—	—	675	900

Source: United Nations, *Growth of World's Population 1900–2000*, table V, table 14.

Outlook for the Future

On the basis of the data and analysis in the United Nations report, the outlook for South and Southeast Asia is for continued rapid growth of urban populations from 156.4 million in 1970 to 506.3 million by 2000, an average annual rate of increase (urban growth) of somewhat better than 4 percent. The increase in the percent of total population living in urban areas (urbanization) during the same period is expected to be much slower, however, only about ½ percent per year. It can be seen then that the region is only slowly changing from a predominantly rural and village settlement pattern, and will still be 73 percent rural by the end of this century. As a result of its present size and the continued (though slower) growth of its rural population, the region will have nearly 45 percent of the world's total rural population by the year 2000.

If these projections are borne out, it is clear that the region is faced with the dual problem of dealing with the physical and economic problems of a rapidly growing urban population while at the same time providing for a large and increasing rural population. When it is added that the density of rural populations in this region will be the highest in the world —three times that of Europe—the crushing responsibility facing its leaders becomes only too evident. This region cannot afford to leave events to chance, nor can it afford the luxury of learning by its mistakes. Far-reaching decisions must be made now that will vitally effect the physical and economic life of the world's largest and most densely populated region.

Some observations on the kinds of decisions being made, how they are made, and their possible effect on the distribution of populations in the region are set out in the remaining sections of this chapter.

NATIONAL AND REGIONAL DEVELOPMENT POLICIES

Economic Development and Urbanization

Increased urbanization appears to be an inevitable concomitant of economic development. The correlation is positive and universal enough to permit the assumption that high levels of economic development, measured in terms of per capita incomes and levels of output and consumption, produce high urbanization ratios. Whether the corollary is equally valid, i.e. urbanization produces economic development, is a question which generates considerable debate. A number of writers have pointed out that when compared to industrialized nations many underdeveloped countries have much higher urban ratios than their levels of economic development would seem to warrant. Hoselitz and others have shown, for example, that European countries in the early stages of their economic growth had much larger percentages of the labor force in manufacturing than the currently developing countries with lower proportions of the total population in cities.[6] Since manufacturing is a key factor in industrial growth and general development, the conclusion is drawn that there is some distortion in the present patterns of urbanization—at least in comparative terms.

At the risk of being drawn into the unprofitable "push"-"pull" argument, it does appear significant that among the developing countries of today there is a high correlation between economic growth and urbanization even though certain traditional indices might be or have been low. The experience in Latin America certainly supports this view, as does the record of the most rapidly developing countries in Asia, Korea and Taiwan. In the region under discussion, which has generally low levels in both categories, the countries achieving relative success in economic growth also have higher urban ratios, e.g. Malaysia, Philippines, and Singapore. Conversely, those with the lowest urban ratios also have the lowest levels and rates of economic development: Burma, Cambodia, India, and Pakistan. Thailand, which began with low levels in both areas, is now experiencing rapid growth in both GNP and urbanization.

At this point it is not possible to assess accurately the causal relationships between these two parallel dynamics, although some general observations do seem possible. Certainly the least understood relationship is the effect of urbanization on economic development. Much has been and can be speculated as to how development process contributes to greater urbanization, but with a few interesting exceptions little has been said about how growing urban populations can speed up economic growth.[7]

In general terms it can be stated that nearly every phase of economic development leads directly or indirectly to greater urbanization. Perhaps the most obvious and direct stimulation results from the efforts of nearly every developing country to industrialize. Increased industrial output and an increased share of industry in total output form a basic part of each country's overall development plan. To the extent that these goals are achieved, the resulting employment provided—directly and in ancillary activities—encourages further growth in urban population. Which urban areas are most affected depends on the locational choices of the decision-makers. It is observable, however, that most industrial enterprises are located in or near large urban centers. And certainly the resulting marketing, financing, and commercial (internal and export) activities are likely to take place in large cities.

The principal economic activity in the region, however, is agriculture; land ownership patterns and production methods have produced a land-bound peasantry with traditional attitudes and subsistence-level incomes. Thus, farming, together with village-oriented services, administration, and marketing operations, account for up to 85 percent of total employment in South and Southeast Asia. When one speaks of economic development, however, one envisages "change"—not maintaining traditional ways and structures. Development in agriculture implies more efficient production methods, greater capital investment, more cash crops, changing ownership patterns, new marketing and transport mechanisms.

The very process of change involved, as well as the changes themselves, opens up needs and opportunities for urban-oriented work while increasing the income of the populations engaged in primary production.

The manufacture and servicing of capital equipment, increased marketing processes—especially for export goods, and processing of primary commodities, are all basically urban rather than rural activities. And, of course, the greater the income-earning urban population, the greater demand for agricultural output and the greater the opportunity for further changes. Thus the goals of agricultural development, i.e. increased production for both domestic and export markets, if achieved, lead as surely to greater urbanization as does increased industrial production.

Once set in motion the urban trend can be further stimulated by its own momentum. As noted above, countries with the highest levels of urbanization—even where industrialization is lower than at comparable periods in European urbanization—seem to have the most dynamic economies, as well as the highest income levels. This holds true for Asian countries as well as for other developing regions. In a study of nineteen Latin

American countries for example, the rank correlation coefficient between per capita income for the period 1955–1964 and proportion of the population in urban places was 0.904.[8] The positive correlation between level of urbanization and income is not surprising. It is well known that urban occupations are more productive and yield higher incomes than rural employment. What may not be as evident is that agricultural productivity and incomes also increase with higher levels of urbanization.

Urbanization moreover, tends to be self-reinforcing because it contributes its own dynamism to the development process. To take but one example: during the period 1955–1959 household savings in India as a whole averaged 6.3 percent of disposable income while savings in urban areas were 17.3 percent. Even more striking was the difference in the form in which the savings were held. Only 2.8 percent of the savings of the country as a whole were in financial assets while urban populations held 12.5 percent of their savings in financial assets, more than half in long-term assets.[9]

The point here is that the greater propensity of urban dwellers to save (as a result of higher incomes) and their willingness to entrust these savings to financial intermediaries attracts additional investors and entrepreneurs to these areas. Such investment in turn contributes to the economic growth of the area, increasing job opportunities and new in-migration. If, as a number of writers have suggested, decisions to migrate are strongly influenced by individuals' attempts to maximize their life-time earning power, the flow will always be from low- to high-income areas, with the rate of flow a function of the income gap.[10]

This concept is fundamental to the issue at hand. If people do respond primarily to opportunities to maximize income (and the correlates of higher incomes) then it is clear that development decisions which most directly influence employment and other income opportunities will have the greatest effect on migration.[11] The corollary here is that decisions which do not have as direct or as strong an influence on incomes will have a correspondingly reduced impact on the rate and location of urbanization.

Continuing this assumption it can be postulated that a policy designed to influence or control the location and rate of urbanization must be conceived and implemented within a framework of income-affecting decisions. In the absence of such a policy the course of urbanization will be determined largely by the effects of many independent decisions on the relative fortunes of different urban areas. If correct, this is an essential principle for those who wish to direct or change current patterns of urbanization, i.e. unless plans to shape settlement patterns according to a conscious policy are able to control or influence decisions which affect incomes, they (the plans) are not likely to have any significant impact.

The key questions then are which policies and decisions have the strongest effect on investment, employment, and incomes, and what can be said about the current development policies of countries in the region with regard to the probable distribution of income opportunities. A final question is whether decision-makers are likely to be responsive to locational policies which stem from a conscious desire to control the rate and structure of urban growth if these policies are felt to be inconsistent with income-maximizing objectives.

Before examining development policies of various countries in the region which have the strongest and most pervasive impact on employment, a word on some distinctions between development plans, policies, and decisions may help to avoid mis-interpretations of ideas proffered. It is fashionable these days to discount the effectiveness, or even the reality, of many development plans. We refer to the five-year plans formally adopted as official blueprints for development strategies. In some cases these reservations are probably deserved, and in others they may be unduly cynical. It is true that there is often considerable slippage between objectives stated in the plans and policies which influence actual investments, and legal and administrative decisions. The word "influence" is used advisedly here, since there is still a further gap between accepted policies and ultimate decisions.

These response-lags naturally vary with the type of decision involved and with differences in political and economic organization among the different countries. Country size and administrative effectiveness also play a part, and there are undoubtedly many other variables. Since this elasticity cannot be measured or evaluated accurately, one can perhaps only hope to take it into account by employing a kind of flexible adjustment mechanism which helps to focus attention on the policies which seem to affect decisions most directly.

Agriculture

Easily the dominant industry in the region, agriculture, absorbs from 50 to 85 percent of the economically active populations in most developing Asian countries.[12] Compared with developed Asian countries (Japan 37.6, Australia 10.9) and others outside the region (U.S.A. 6.5, France 19.8) it is obvious that any significant change in this sector will have profound effects on population distribution and settlement patterns.

There is evidence that some dramatic changes are in fact about to occur in the region. The most important artifact of this change bears the acronym HYV ("high yield variety"; of rice, wheat, and other cereal grains). After years of laboratory development and controlled test production by governments and international agencies, new seed varieties and radically different production techniques were put into full-scale operation in selected countries in Asia. The countries where these approaches received the earliest and most intensive effort were Japan, South Korea, and Taiwan. The results so far have been very impressive, if not spectacular. Yields have been increased by an average of 50 percent in actual cultivation, and as high as 300 percent under experimental conditions.[13]

The HYV method of cultivation, which calls for far greater inputs of fertilizers, water, and intensive labor, has now been adopted by four countries in the South Asian region: India, where the new technology is part of the Fourth Five Year Plan; Pakistan; Philippines; and Ceylon. Malaysia has also a new strategy to increase rice production, but this utilizes modern techniques of multiple-cropping rather than the HYV approach. What effect is this new agricultural technology likely to have on rural employment, migration, and urban growth? A look at the experience in the two developing Asian countries using HYV provides some interesting comparisons.

Both Korea and Taiwan have achieved comparatively high levels of economic growth and per capita incomes; both have large and rapidly growing urban populations (over 30 percent in 1966, which is comparable to Japan just prior to World War II); both have large and rapidly growing metropolitan cities; both have expanding indices of industrial production; and both have very recently shown declines in overall population increase. In sum these two countries have all the signs associated with modernization, industrialization, and urbanization. Was this a cause or a result of the introduction of HYV techniques? The answer is not clear as yet. The close association of increased urbanization with increased agricultural production nevertheless points strongly to a causal relationship.

Two things can be seen, however, that must be mentioned in any discussion of the effect of development policies on urbanization. The first is that whatever the direct or indirect causal relationship between the new agricultural techniques and population movements, the decisions to go ahead were taken without much regard to this consideration. This point cannot be over-emphasized. Secondly, the type of cultivation involved, unlike the great agricultural innovations in some industrialized countries, is relatively labor intensive. Some potential implications of this factor-demand ratio have been pointed out in a recent paper prepared by the U.N. Economic Commission for Asia and the Far East which concluded that "the increases in the demand for labour that will be introduced by these technological changes are bound to alter the owner-tenant relationship in Asian agriculture and accelerate the processes of agricultural transformation." [14] Thus a transformation from subsistence farming to economically viable units of production is anticipated.

Is the long-awaited agricultural revolution at hand? Perhaps the F.A.O.'s Indicative World Plan for the region projects that by 1985 increased cereal production in India, Pakistan, and Thailand will have to be fed to livestock or exported.[15] In the meantime, it is quite possible that some of the surplus agricultural labor that has been flowing to large urban centers may be absorbed in the new technology, either in direct production or in ancillary activities. Equally possible is that the increased inputs required—fertilizer, pesticides, pumping and spraying equipment, drills, tractors, etc.—will provide a domestic market for new agro-industrial enterprises. This would lead to growth of towns and cities, favorably located and endowed, with competition for labor against the main metropolitan areas. In the long run, however, it will contribute to urban growth at large. Whether this means ultimately greater dispersion or concentration is not possible to predict.

Industry

Industrial production is still relatively insignificant compared with agriculture. Manufacturing, the most dynamic branch of industry, contributed only a meager share of GNP for most Asian countries, ranging from 5 percent in Ceylon in 1960–62 to 16 percent for the Philippines.[16] However, the region has been achieving a rapid expansion of industrial output since the mid-1950s. By 1963 developing Asian countries had increased output in mining, power, and manufacturing 340 percent over the 1953 level, compared with an 80 percent increase for Latin America and 36 percent for North America.

Since these growth rates were measured against a very small base there is some question as to the capacity for sustained growth. Indeed, in some of the larger countries (India and Pakistan) excess industrial capacity now exists despite a deliberate policy of industrialization. This has led to a growing recognition in many countries of the need to shift industrial production from the earlier policy of import substitution of consumer goods, e.g. textiles, to a renewed emphasis on import substitution of intermediate goods by increasing domestic manufactures of industrial inputs for national and regional markets. Agriculture-based industries have received particular attention along with traditional consumer manufactures. The ECAFE Economic Survey for 1968 states, "As manufacturing output continues to grow, the threshold of efficient production is crossed by increasing members of industries supplying machinery components and other material inputs for final-stage manufacturing processes." [17]

Until fairly recently the few industries which have attained appreciable size—textiles, garments, food products, and other processing operations— were labor-intensive and market-oriented. As the feasibility for engineering industries (petro-chemicals, iron and steel) develops, however, the locational factors may shift and with it the previous tendency toward ever greater concentration in metropolitan cities. Once again, it is necessary to point out that these structural changes, if they do take place, will largely be in response to market conditions and not as a result of deliberate policies to bring about changes in settlement patterns.

Trade and Export Policies

The importance of exports and their necessary counterpart—imports—in the economies of most Asian countries is well known. This is especially true of the smaller and medium size countries (Ceylon, Malaysia, Philippines, Thailand) and the city states of Singapore and Hong Kong. Although exports form a much smaller percentage of GNP in large countries such as India, Pakistan, and Indonesia, they are still important determinants of economic policy.

It is the degree of dependence on exports and on international markets that interests us here. Countries which rely on exports as a major form of income are subject not only to severe competition and cyclical economic fluctuations, but also to structural changes in market demand. Structural changes often hold the biggest threat to economic development in the exporting countries. Consider the disastrous impact on the economy of Ceylon if a cheap and acceptable substitute (or synthetic copy) were found for tea.

In the same way that international market demand influences commodity prices, it also influences domestic production techniques and limits investment choices, including locational choices. Policy makers concerned with achieving or maintaining a given rate of growth in GNP are naturally mindful of the need to adjust to international market conditions; in so doing they are less likely to endorse investment decisions based on non-economic criteria.

There are foreseeable changes in the structure of production for export, however, and these changes undoubtedly will affect investment-mix, industrial location, and, ultimately, settlement patterns. Primary exporting countries have been severely affected by declining commodity prices in recent years, and these shifts in international commodity trade have produced structural changes in the affected economies. Adjustments of economic policy to changing circumstances have taken the form of diversification of export production in some countries, and, as mentioned above, a greater domestic use of primary (and manufactured) production as inputs to further processing. This has led in turn to significant alterations in the composition of imports as well as a general decline in the import component of total expenditures. If the domestic ratio of value added to final exports continues to increase, it is reasonable to assume that this will bring about locational shifts in related economic activities—probably in favor of urban locations.

Summary of Selected Development Policies

When taken together current development policies and investment decisions in three important economic sectors, agriculture, industry, and trade, can be seen to have extremely important implications for the future location of economic activity and human settlements.

The dramatic results of the new technology in agriculture seems to have at least two probable outcomes: the further development of existing and new market-service towns; and an increase in overall urbanization as greater productivity enables fewer units of labor to produce an ever larger volume of primary goods.[18] Part of the increase in overall urbanization will undoubtedly occur in metropolitan areas.

Increasing industrial production and the development of new industrial outputs for domestic and regional markets, are also significant in determining population movements and settlement patterns. Increased output of traditional manufactured goods will continue to draw job seekers to the principal cities. But the newly developing markets for intermediate goods and primary goods intended for further domestic processing are not as metropolitan-centered and tend to locate more in response to resource availability, water, and transport systems. The marked increase in the number of non-seaport cities and the relative growth rates of medium to large towns attest to this trend toward counter-metropolitan growth centers.

Finally, changing international market conditions have had important repercussions on import and export policies, the result of which may reinforce the trends stimulated by new developments in agriculture and industry, i.e. greater urbanization, but with a large part of the growth in intermediate to large towns, in competition with the older metropolitan areas. The largest proportion of urban growth may still occur in the older centers for some time to come, however, since they already account for a majority of total urban populations, and still offer the greatest inducements to most newly establishing and expanding manufacturing and service enterprises.

Equally important to this discussion is the impression, gained in the review of economic planning literature and especially of operational policies and decisions, that, by and large, these policies and decisions are formulated without much regard to their direct and potential effects on population movements and settlement patterns. In their understandable eagerness to encourage economic development, government officials and private businessmen are keyed to economic goal-maximization which may or may not be consistent with harmonious urban or regional development, but which will certainly have a powerful impact on where people work and live.

Public and private economic policies and decisions are seen then as powerful and pervasive instruments which cause many other things to happen, including the movements of people and the growth and organization of their settlements. Is this an accurate characterization, or is it too one-dimensional? Are there not other non-economic factors which influence, control, or direct investment policies, population flows, and urban growth? Moreover should there not be a more rational basis for organizing human settlements than pursuit of economic maximization? Leaving aside the challenging but many-sided (and multi-answerable) question of what a truly rational pattern of urban settlements might be, an equally important intermediate question arises: to what extent and by what means can settlement patterns be deliberately altered, controlled, or influenced?

URBAN AND REGIONAL PLANS AND POLICIES

If future patterns of urbanization are to be influenced by conscious policies designed to achieve a more rational or at least different distribution of human settlements than present trends indicate, it follows that there must be some agreement as to: (a) what these more rational patterns might be, and (b) how the plans and policies can be implemented or enforced. Since the advocacy of planned settlement patterns, as opposed to market or political determination, has come mainly from physical planners (with strong support from sociologists, ecologists, and other social scientists) it is important to this discussion to account briefly for the state of urban and regional planning in Asia.

In most developing Asian countries, physical planning is now accepted as a tool for organizing certain aspects of urban life. It is understood that (1) layout of roads, transport lines, and community facilities for new or renewing sections of towns and cities; (2) zoning industrial and residential locations; and (3) planning and acquisition of land for future development should be done by or in consultation with plans and planners, at least for the major cities of the region. More recently, physical planners have been engaged in planning for regional development as a more comprehensive and effective framework within which to organize the physical dimensions of economic and social activities.

Many of these efforts have already produced positive results. Others have become diffused or ignored in the give and take of business and political life. In a few cases where planning received unusual emphasis and priority, dramatic physical evidence of the benefits to be gained through careful and timely planning have been produced. For the most part, however, planners have been given neither a new canvas nor a broad brush. They have more generally been assigned the task of trying to make physical sense out of often conflicting conditions imposed by decisions taken with little regard to physical or spatial concepts. The failure to introduce spatial considerations into other plans, policies, and decisions, i.e. national development plans and related governmental policies, makes the task of the physical planner difficult-to-impossible. It is argued that, to avoid the otherwise inevitably self-defeating but avoidable conflicts, spatial considerations should be an integral part of the national planning process —with the added condition that the results of such planning are best implemented in a regional context.

A review of the actual role of physical planning as a means by which to introduce spatial criteria into national development plans and policies does not yield an optimistic outlook. A United Nations Seminar on this subject held under the auspices of the Economic Commission for Asia and the Far East concluded that there was a general lack of appreciation of the role of comprehensive urban and regional planning in the overall strategy of development planning, with the result that little attention had been paid to the location and distribution of economic activities and their impact on settlement patterns. This led to plans being "spaceless" in outlook.

The Seminar members felt there was little realization that industry, transport, and power operated and interacted within the framework of a single set of spatial co-ordinates. Through overemphasis on the allocation of resources to the neglect of land use and land planning, crucial issues such as identifying nodal points for economic growth and social progress, distributing enterprises among metropolitan cities, medium-size towns, and small urban centers, and deciding the optimum scale of settlements conducive to economic growth and social advancement had all been overlooked. The Seminar expressed concern that the vital question of how best investible resources could be steered in space and time into impulse sectors and selected growth points had yet to be resolved. As physical planners were not particularly well-versed in principles of econometrics, public finance, planning law, and administration, they confined themselves to the solution of town and city problems. Economic planners, on the other hand, failed to visualize the three-dimensional implications of their economic decisions. This lack of visualization of the environmental problems often resulted in uneconomic utilization of capital resources and distorted settlement patterns. To achieve a satisfactory pattern of urban and rural settlements, the Seminar emphasized the need to harmonize social, economic, and physical planning at all levels and to close the existing gap between the formulation of development planning goals and their implementation at the regional and local level.

One is forced to conclude, however reluctantly, that in South and Southeast Asian countries (as well as the majority of others in both industrial and developing regions of the world) investment decisions taken in accord with national development policies, or in response to market stimuli or political necessities, are the overwhelming determinants of economic activities and, hence, settlement patterns. Formulations of spatially planned and—one assumes—more rational investment policies simply are not able to compete either in the national planning process, or at the point when actual investment decisions are made.

The reasons that advocates of spatial order find themselves in this relatively weak role are not difficult to discern, and probably can be summed up as follows: individuals, firms, and other decision-makers wishing to maximize income, tend to settle in physical locations which they believe to offer the greatest opportunities toward the achievement of this objective. Large cities appear to provide the optimum conditions mainly as a consequence of scale factors from both investor and job-seeker points of view. With improvements in communications and transport facilities to increase their mobility, individuals and firms both exercise income-maximization objectives by locating in large cities.

This trend is reinforced by national development policies which seek to maximize national income on a sectoral basis. The introduction and the acceptance of development planning as an official tool of national goal achievement carries with it the implicit assumption that the primary national goal is one of economic development. Thus, when faced with alternative choices, particularly when investment allocations are involved, decision-makers generally favor those choices which they believe will maximize incomes over a reasonable period of time. The locational implications of these decisions do not differ greatly from those made in the private sector, i.e. they lead toward concentration of economic activities in locations having apparent comparative advantages. More often that not this contributes directly or indirectly to further growth of existing cities, although in certain cases it does stimulate the growth of new or previously stagnant towns.[19]

Efforts to change the trend toward concentration in a relatively few urban centers in favor of a more geographically balanced pattern of urban growth can only be successful if this policy is given greater weight in development plans and policies than economic considerations, and if the implementation apparatus is sufficiently strong to overcome the thrust of private decisions.

It is necessary that regional plans and policy be given greater weight than income maximization objectives since in many cases the locational implications of decentralized growth run counter to decisions based on economic considerations in both public and private sectors. It is essential that there be strong and efficient administrative apparatus to implement national location policies if they are to overcome the natural trend toward concentration which decentralized decison-making produces.

Current practice in the region gives little evidence that either of these conditions now exists or is likely to arise in the near future in most developing Asian countries. It should be added that, as yet, there have been very few arguments brought forward to demonstrate how different settlement patterns would help achieve currently accepted national goals, or increase aggregate net benefits to individuals. This is not to say, however, that a more rational and mutually beneficial method for determining locational decisions cannot be devised—and implemented. But it is obvious that the burden rests with those who wish to introduce change no matter how enlightened their position may be.

The first and most essential step in gaining acceptance of locational criteria as legitimate and necessary components of national planning is the development of analytical tools and techniques that can compete with those which economic planners and business men have used so successfully for the last several decades. At the same time it is essential that spatial planners bear in mind that most developing countries are still overwhelmingly preoccupied with very basic economic objectives which severely limit actual and conceptional options open to them. It is therefore likely that for some time to come the main work of spatial planners will be the *ex post* organization of human settlements whose growth or demise will have been largely determined by events outside their influence.

Nevertheless, steps can be taken now, within the limitations imposed by current development priorities and the state of the planner's art, to help reduce avoidable consequences.

While spatial planners are developing the analytical techniques that will gain them rightful parity in national policy councils, more immediate results can be achieved by adapting existing administrative tools to better serve their objectives. For example, one of the most important steps in policy implementation is the budget: through the budget process in fact, policies are often changed and priorities altered in the interest of one sector or another. Although there is generally no spatial component to national development plans, the probable physical impact of various sectoral decisions can be identified once these decisions are assembled in the budget document. By calculating the probable implications on a cross-sectoral basis a "spatial budget" can be super-imposed on the actual financial document for consideration by national or regional policy officials. In this way original plans which present unforeseen spatial (or settlement) problems can be modified. The revised plans can still be carried out by the original sectoral administrators (government agencies, banks, etc.), but now further equipped with an increased consciousness and responsibility to meet not only their own sectoral objectives, but also those of the new spatial considerations.

This example describes a device to project considerable leverage— and yet is not very difficult to explain or justify. It is clear, however, that the longer-term objective should continue to be retained to introduce spatial considerations into the overall planning process itself. For until this is achieved, urban centers and other human settlements will continue to be derivative functions of all other decisions, unpredictable in size or shape and uncertain "containers" of societies' highest achievements.

NOTES

1. United' Nations, Department of Economic and Social Affairs, *Growth of the World's Urban and Rural Population 1920–2000.* 1968.
2. T.G. McGee, *The Southeast Asian City.* New York: 1967, Chapter 2.
3. *Ibid.,* p. 17.
4. United Nations, *Growth of the World's Population,* table VI-10.
5. *Ibid.,* table VI-8.
6. Bert F. Hoselitz, "The Role of Urbanization in Economic Development," in Roy Turner (ed.), *India's Urban Future.* Berkeley and Los Angeles: University of California Press, 1962. Also see Barbara Ward Jackson, "Development and the City"; Malcom Rivkin, "Urbanization and National Development," in United Nations publication, *Urbanization: Development Policies and Planning.* New York, 1968.
7. For a discussion of the economic development-urbanization argument see "Economic Aspects of Urbanization" in the United Nations report: "Urbanization: Development Policies and Planning." Important discussions of the effect of urban centers on economic growth include Jane Jacobs, *The Economy of Cities,* New York: Random House, 1969, and Lauchlin Curie, *Accelerating Development,* New York: McGraw-Hill, 1966.
8. Lowdon Wingo, "Latin American Urbanization: Plan or Process?" in Bernard J. Frieden and William Nash (ed.), *Shaping an Urban Future.* Cambridge, Mass.: M.I.T. Press, 1969, p. 124.
9. United Nations, *Finance for Housing and Community Facilities in Developing Countries.* Sales No. E.68.IV.4, p. 12.
10. Wingo, *op. cit.,* p. 125, and United Nations, ECAFE *Report of the Expert Working Group on Problems of Internal Migration and Urbanization.* Bangkok: 1967.
11. Assuming reasonably normal political conditions. Political strife, wars, and so forth, also have great influence on population movements, and during critical periods can exert a stronger force than economic considerations.
12. Exceptions are the city-states of Hong Kong and Singapore.
13. United Nations, *Economic Survey of Asia and the Far East,* 1968, p. 128.
14. United Nations, *Problems of Planning and Plan Implementation in the ECAFE Region.* E/AC.54/L.34, p. 5.
15. *Ibid.,* p. 15.
16. United Nations, *Industrial Development in Asia and the Far East.* Industrial Development News No. 1, 1965.
17. United Nations, *Economic Survey, 1968, op. cit.,* p. 111.
18. Despite the fact that the new technology is said to be capital intensive, this applies mainly in terms of labor 'input per hectare cultivated, and is not necessarily so when speaking of labor input per unit of output. The net effect on employment depends on the total area under cultivation. Labor shortages in some areas using the new methods have already led to increased mechanization for some processes.
19. For example, when the decision involves large investments in extractive industries or other heavily input-oriented industries.

CHAPTER 4

Ashish Bose

THE URBANIZATION PROCESS IN SOUTH AND SOUTHEAST ASIA

Introduction

Urbanization is an essential element in the process of economic growth and social change in South and Southeast Asia in the transformation of rural, agricultural economies into industrial, urban economies, and in the transition from tradition to modernity. An understanding of the urbanization process will not only help planners and policy-makers but will by itself accelerate the process of economic growth and social change in the Asian countries. But in order to arrive at such an understanding it is necessary for urban analysts to get out of the rut created as a result of repetitious discussions in seminars and conferences, survey reports and books—for over a decade—on the nature of the urbanization process in Asian countries. Flimsy generalizations like "over-urbanization of Asian countries," superficial analyses of "push and pull" factors affecting migration, and platitudinous prescriptions like "balanced and integrated" development of urban and rural areas, to give just a few examples, confuse the issues instead of clarifying them. In the developing countries of Asia there is need for much more data on migration and urbanization, but the greatest need is for a realistic appraisal of the Asian scene, and a clear understanding of Asian problems in an Asian setting.

Of the seven countries in the world with a population of over 100 million, five are in Asia. These "Big Five"—China, India, Indonesia, Pakistan, and Japan—have a combined population beyond 1,500 million and account for a little over 45 percent of the world's population. Even if we confine ourselves to South and Southeast Asia, we have in our region India, Pakistan, and Indonesia with a total population of over 720 million accounting for more than 21 percent of the world's population. In terms of the level of urbanization (regardless of what definition of "urban" is

81

adopted), these countries are no doubt the least urban areas of the world although in terms of absolute numbers, their urban population is sizable: India alone has over 100 million people living in her urban areas today (1969), according to the definition of "urban" adopted in the Indian census. Thus India's urban population alone roughly equals the total population of Japan, Pakistan, or Indonesia. It is worth recalling that there is no evidence in the history of economic development of a country undergoing the process of urbanization with a base population of over 500 million and a rate of natural population increase of 2.5 percent, which is

Table 1. Scale and Pace of Urbanization in South Asia (Middle-South Asia and Southeast Asia) in Relation to World Total

Indices	South Asia	World total
1. Urban population (localities of 20,000 or more inhabitants) in millions (1960)	116	753
2. Population in rural areas and small towns (smaller than 20,000) in millions (1960)	742	2,241
3. Urban population (20,000+) as a percentage of total population (1960)	14	25
4. Decennial increase (percent) in urban population (20,000+): 1920–30	26	30
1930–40	46	30
1940–50	52	25
1950–60	51	42
5. Big-city population (500,000+) in millions (1960)	42	352
6. Big-city population as a percentage of urban population (20,000+)	37	47
7. Big-city population as a percentage of total population	5	12
8. Crude tentative projections of population in millions in 1980:		
(i) Total population	1,366	4,269
(ii) Rural, small town	1,079	2,909
(iii) Urban	287	1,360
(iv) Big cities	149	725
9. Absolute increase in population in millions during 1960–80 (rough estimate):		
(i) Total population	508	1,275
(ii) Rural, small town	337	667
(iii) Urban	171	608
(iv) Big cities	107	374

Source: Computed from data given in "World Urbanization Trends, 1920–1960," prepared by Population Division, United Nations Bureau of Social Affairs, in Gerald Breese, Ed., *The City in Newly Developing Countries.* Englewood Cliffs: Prentice-Hall, 1969, pp. 21–53.

the case of India today. It is, therefore, important to consider not only the *level* of urbanization but the *scale* of urbanization. In Table 1 we present an overall picture of South and Southeast Asia in the context of the world situation.

Table 2. Population and Growth Rates in Countries of South and Southeast Asia

Country	Estimated population in 1967 (millions)	Rate of annual increase 1963–67
India	511.1	2.4
Indonesia	110.1	2.4
Pakistan	107.3	2.1 *
Philippines	34.7	3.5
Thailand	32.7	3.1
Burma	25.8	2.1
Viet Nam—North	20.1	3.1
Republic of Viet Nam	17.0	2.6
Ceylon	11.7	2.6
Nepal	10.5	1.9
Malaysia—East:		
Sabah	0.6	3.9
Sarawak	0.9	3.1
Malaysia—West:	8.6	3.0
Cambodia	6.4	2.2
Laos	2.8	2.5
Singapore	2.0	2.5

Note: The countries are ranked in order of their total population.

* According to the Population Growth Estimation Project of Pakistan, the annual population growth rate in 1963 was 2.9 percent (Population Council, *Studies in Family Planning,* No. 26, January, 1968, p. 8).

Source: United Nations, *Demographic Year Book, 1967.* New York: 1968, pp. 107–110.

There are at least three striking features of urbanization in the Asian countries: (a) most of them are undergoing urbanization under conditions of rapid and accelerating rates of population growth (Table 2); (b) the comparatively low proportion of urban population in most of these countries (Table 3) conceals the enormity of the problems of urbanization; and (c) most of these countries depend heavily on agriculture (Table 4) while their economic development plans envisage a reduction in the share of agriculture, both in respect to the working force and the gross national product and to an increase in the share of manufacturing industries. Thus implicit in their development plans is the increasing tempo of rural-to-urban migration. But most of these plans explicitly state the need for decentralization of industries and control of massive migration to urban areas, through various measures like control of location of industries. This brings about an increasing gap between the *desirable* pattern of urbanization and the *observed* pattern of urbanization.

Table 3. Proportion of Urban and City Population to Total Population in Countries of the Region

Name of country	Urban population as a percent of total population	City population as a percent of total population
India	18.0 (1961)	7.9 (1967)
Indonesia	14.9 (1961)	9.8 (1961)
Pakistan	13.6 (1961)	9.3 (1967)
Philippines	29.9 (1960) [a]	11.6 (1966)
Thailand	12.5 (1960) [a]	7.9 (1963)
Burma	10.4 (1960) [b]	—
Vietnam—North	—	3.9 (1960)
Republic of Vietnam	—	10.8 (1960)
Ceylon	18.9 (1963)	5.9 (1963)
Nepal	3.6 (1961)	1.3 (1961)
Malayasia—East:		
Sabah	14.9 (1960)	4.8 (1960)
Sarawak	15.1 (1960)	6.8 (1960)
Malaysia—West:	—	10.3 (1957)
Cambodia	10.3 (1962)	6.9 (1962)
Laos	4.0 (1959) [c]	0.6 (1962)
Singapore	100.0 (1967)	100.0 (1967)

a. These figures have been taken from T.G. McGee, *The Southeast Asian City.* New York: Frederick A. Praeger, 1967, p. 78.

b. Estimated by Homer Hoyt in *World Urbanization.* Quoted by United Nations ECAFE, *Report of the Expert Working Group on Problems of Internal Migration and Urbanization and Selected Papers.* Bangkok: 1967, p. 234.

c. Estimated by Joel M. Halpern, *Economy and Society of Laos,* Southeast Asia Studies, Yale University, 1964, table 10. Halpern's estimate is "based on maximum estimate for city of Vientiane, the only truly urban area in Laos. In reality Laos is the least urban country in all of Asia: using maximum population estimates Laos has only five cities of over 5,000 population compared, for example, to Nepal's ten. Minimum estimates would give a figure of 3 percent urbanization for Laos."

Source: U.N. Demographic Year Book, 1967, table 5, pp. 170–182, Table 6, pp. 216–223.

Given the poverty of Asian countries, the problems of urbanization become acute not only because of urban population size but also because of the inability of these economies to provide the optimum standards of economic and social overhead. It would be unrealistic, therefore, to think in terms of *optimum* standards for the urban population. The basic issue involves ensuring first the *minimum* standards for the *entire population*, for the simple reason that even if proper urban standards are achieved somehow, it would be difficult to maintain them in the face of rural poverty. The "urban preserves" of the colonial era cannot be multiplied in a welfare state.

One of the most important problems facing the developing countries is how to speed up industrialization and reduce the dependence on agriculture. The development programs of these countries have made some progress in this direction but the experience of India shows how difficult the attainment of such structural changes are, considering the limited possibilities of increase in employment opportunities in the modern manufacturing sector compared to the massive size of the labor force in the agricultural sector and the somewhat remote possibility of bringing about substantial population transfers from the agricultural to the manufacturing sector.

Table 4. Sectoral Contribution to National Product in 1960–62 and Change from 1950–52 in Selected Countries [a]

Country [b]	AGRICULTURE Share [d]	Change [e]	MANUFACTURING [c] Share [d]	Change [e]	OTHER SECTORS Share [d]	Change [e]
Philippines	32	−10	16	7	52	3
India	45	−4	16	0	39	4
Burma	40	−6	15	4	45	2
Pakistan	52	−8	14	6	34	2
Thailand	37	−1	11	1	52	0
Vietnam-South	34	5	11	1	55	−6
Cambodia	46	−10	8	2	46	8
Indonesia	61	5	8	−1	31	−4
Federation of Malaya	44	−2	6	1	50	1
Ceylon	47	−6	5	1	48	5

a. Shown as percent of aggregate product at constant prices: Percent of gross domestic product at market prices for Burma and Thailand, and of factor cost for Cambodia, Ceylon, Federation of Malaya; percent of net domestic product at factor cost for Republic of Viet-Nam; percent of net national product at factor cost for India, Indonesia, Pakistan, and the Philippines.

b. Ranked in descending order of the percentage share of manufacturing sector in the total product in 1960–62.

c. Including electricity and construction in India, and all utilities in Indonesia and Pakistan.

d. 1957–59 for Cambodia, 1958–60 for Indonesia, 1960–61 for Federation of Malaya.

e. From 1951–52 for Burma, Cambodia, Indonesia, and Thailand, and from 1955–56 for Federation of Malaya.

Source: United Nations, *Industrial Development in Asia and the Far East,* Industrial Development News No. 1. New York, 1965, p. 23.

PROBLEMS OF DEMOGRAPHIC ANALYSIS

Inadequate Data on Urban Population

Apart from the hazards of generalization in a region marked by such disparities in absolute population size, the availability of statistical material for the study of urbanization severely limits the validity of conclusions drawn for the region as a whole. As one scholar laments: [1]

> "Our knowledge of Asian urbanization is far from being proportional to its importance on the world scale. We cannot even measure with any precision what the total Asian urban population is, let alone delineate with assurance the factors responsible for its past and present growth or project its likely future shape. We are hampered both by inadequate data, reflecting imperfect census systems, and by a relative paucity of published studies."

This aspect received considerable attention from an Expert Working Group set up by the United Nations ECAFE in 1967 to consider the problems of internal migration and urbanization in the countries of the ECAFE region. The Working Group was of the opinion that "the magnitude of the problems of internal migration and urbanization in the ECAFE region was not matched either by the quantity and quality of the existing data on these phenomena or by the analyses which have so far been undertaken." [2] That group made a series of recommendations on various aspects of this problem. A more recent Working Group set up by ECAFE (May, 1969) approached the question of projections of populations of sub-national areas with special reference to the countries of the ECAFE Region. This group also pointed out the inadequacies in the existing data on internal migration and urbanization in this region.[3] A comprehensive study of the urbanization process in South and Southeast Asia must still await a more opportune time when the minimum basic data will be available on internal migration and urbanization. The discussion which follows will be heavily slanted towards the Indian experience in view of the fact, pointed out by Rhoads Murphey, that "Urbanization in India is more accurately measured and better analyzed and understood than in most other Asian countries apart from Japan." [4]

The Problem of Defining "Urban" Areas

No statistical study of urbanization is possible unless adequate note is taken of the census definition of "urban" which varies from country to country and sometimes from one census to another in the same country. In the countries of the region under study a number of definitions of "urban" prevail.[5] Commenting on the "exceptionally difficult" task of making the comparison and assessment of city growth rates and levels of urbanization in the countries of this region, McGee observes: [6]

"Some countries, such as Malaya, adopt a purely statistical definition, including all gazetted areas of over 1,000 in population as urban. Other countries, such as the Philippines, choose to define urban places on the basis of a political definition. Here the practice of establishing chartered cities allows them to be placed directly under the administrative control of the President of the Republic, thus taking power away from the provincial governor. In practice many of the cities which were to be granted charters were so small in size that it was necessary to enlarge the area, thus taking in '. . . large sectors of rural territory containing outlying villages and towns and taxable farmlands.' Thus many of the chartered cities of the Philippines classified as urban actually have substantial rural populations. Spencer, after a careful analysis of the boundaries of these chartered cities in relation to their built-up urban areas, concluded that the urban population in 1948 was over-enumerated by almost 25 percent. At the other extreme in some Southeast Asian countries the urban population has grown so rapidly in the postwar era that it has spread outwards beyond existing city boundaries. Thus the percentage of urban population in Singapore actually fell from 72.5 percent in 1947 to 63.1 percent in 1957. This was not a reflection of a decline in the urban population but of the fact that some urban population was now living in areas defined as rural by the census."

Thus the question arises: can we list a set of criteria to arrive at a common definition of urban? After discussing the limitations of adopting population size and density as possible criteria, McGee comes to the conclusion: "It would appear that the only valid definition must rest on the proportion of population engaged in non-agricultural activities, but this is difficult to apply in the Southeast Asian context because of the lack of adequate statistics." [7]

The 1961 Census of India was used for size, density, and employment, etc., in non-agricultural activities to define urban areas, and it would be worthwhile to discuss India's experience in this regard. Prior to the 1961 census, the definition of "town" was not uniformly followed in all the states of India and there was considerable scope for the use of discretionary powers on the part of the State Census Superintendents. Apart from the usual test of 5,000 population, the classification of a place as urban or rural was based on a subjective assessment by Census Superintendents of the presence of "urban characteristics." However, the common feature of the definition of towns ever since the census of 1891 has been the automatic classification of all places which are municipalities as towns.

The new definition of "town" adopted in the 1961 Census was as follows: [8]

"To qualify for an urban area, a place should first be either a municipal corporation or a municipal area or under a town committee or a notified

area committee or cantonment board. In the second place, each census has
adopted a number of census towns which do not enjoy any statutory label
of administration. This has been considered desirable in order to obtain a
truer measure of urbanization as it is usual for an administrative label to
fall somewhere behind actual achievement. These census towns were in
1961 determined on the basis of a number of empirical tests:

(a) a density of not less than 1,000 per square mile;
(b) a population of 5,000;
(c) three-fourths of the occupation of the working population should be
outside of agriculture; and
(d) the place should have, according to the Superintendent of the State, a
few pronounced urban characteristics and amenities, the definition of
which, although leaving room for vagueness and discretion, yet meant to
cover newly-founded industrial areas, large housing settlements, or places
of tourist importance which have been recently served with all civic
amenities."

It should be noted, however, that the first part of the town definition
adopted in the 1961 Census has a common feature with all earlier censuses,
namely, the automatic classification of all places which are municipalities
as towns. The three eligibility tests are, therefore, not relevant as far as the
municipalities are concerned. The detailed tables of the 1961 Census present data for area, population, and working population divided into nine
industrial categories for each town and city in India. It is thus possible to
apply the three eligibility tests laid down in the 1961 Census to each of
the 2,700 towns in India. In our analysis we have had to discard item (d)
of the 1961 Census definition of town because it cannot be statistically
studied, and in any case its impact is limited. Thus, we are left with three
attributes, namely, density (A), population (B), and economic activity
(C). We shall use the small letters a, b, and c to denote the absence of
these attributes. In other words, if "A" stands for density of 1,000 or more
persons per square mile, "a" will stand for a density of less than 1,000
persons per square mile, and so on.

On the basis of the association of these three attributes, we get the
following eight possible categories: ABC, AbC, ABc, Abc, aBC, abC,
aBc, and abc. In addition, we have a small category of unclassified towns
for which complete data are not available. A town belonging to the ABC
category satisfies all three eligibility tests. That is to say, it has a density of
more than 1,000 persons per square mile, a population of more than
5,000, and it has more than 75 percent of the working population engaged
in non-agricultural activities. Conversely, a town belonging to the abc category does not satisfy any of the three eligibility tests.

In Table 5 we give the distribution of the towns in India in 1961
according to the eight categories just described. It will be seen that out of
2,700 towns in India, 1,610 towns (i.e. 60 percent of the total number of

towns) satisfy all three eligibility tests, and their population is about 84 percent of the total urban population of India.

Table 5. Distribution of Towns in India According to Three Eligibility Tests, 1961

	Category *	No. of towns	Percent of total No. of towns	Percent of total urban population
1.	ABC	1,610	59.6	83.5
2.	AbC	130	4.8	0.5
3.	ABc	595	22.1	9.6
4.	Abc	72	2.7	0.4
5.	aBC	40	1.5	0.6
6.	abC	26	1.0	0.1
7.	aBc	155	5.7	1.9
8.	abc	28	1.0	0.1
9.	Unclassified	44	1.6	3.3

* ABC—Density over 1,000, population over 5,000 and over 75 percent of workers in non-agriculture.
 AbC—Density over 1,000, population below 5,000 and over 75 percent of workers in non-agriculture.
 ABc—Density over 1,000, population over 5,000 and less than 75 percent of workers in non-agriculture.
 Abc—Density over 1,000, population below 5,000 and less than 75 percent of workers in non-agriculture.
 aBC—Density less than 1,000, population over 5,000 and over 75 percent of workers in non-agriculture.
 abC—Density less than 1,000, population less than 5,000 and more than 75 percent of workers in non-agriculture.
 aBc—Density less than 1,000, population over 5,000 and less than 75 percent of workers in non-agriculture.
 abc—Density less than 1,000, population less than 5,000 and less than 75 percent of workers in non-agriculture.

This analysis can be further extended to distribute the ABC towns in each of the six urban classes according to population size. This is done in Table 6.

Table 6. Distribution of ABC Towns into Six Urban Classes in India, 1961

	Urban classes	No. of ABC towns	Total No. of towns	Percent of ABC towns to total
I	100,000 and over	100	107	93.5
II	50,000–99,999	133	139	95.7
III	20,000–49,999	422	518	81.5
IV	10,000–19,999	506	820	61.7
V	5,000–9,999	449	848	52.9
VI	Below 5,000	—	268	—
	Total	1,610	2,700	59.6

If we consider the two broad categories, 20,000+ towns and towns with population below 20,000, we find that in India, 14 percent of the towns in the first category fail to qualify in the A,B,C tests while in the case of the remaining towns, 51 percent fail to qualify in these tests.

We had earlier stated that the first part of the definition of "urban" treated all municipalities as towns regardless of the applicability of the three eligibility tests. However, it is possible to apply these tests to municipal areas as well as to non-municipal areas. It is thus possible to find out how many of the municipalities in India satisfy all three eligibility tests and how many do not. We may also ask how many of the non-municipal towns satisfy the three eligibility tests and how many do not. This calls for an analysis of the association of four attributes, namely, A, B, C, and M (municipal status), and for the negative attributes we have to consider a, b, c, and m. This means a total of 16 categories. Space does not permit us to give the details [9] except to say that 43 percent of the towns satisfy all four tests, and these towns comprise 73 percent of the urban population of India. A little over 16 percent of the towns with 10 percent of the total urban population are non-municipal areas which nevertheless satisfy all three tests. If we consider the effective urban (20,000+) and quasi-urban (below 20,000) towns we find that 77 percent of the towns in the former category satisfy all four tests and only 30 percent of the towns in the latter category satisfy them all.

Long-term Trends in Urban Growth— Lack of Historical Perspective

Apart from the problems of inadequate data and a suitable definition of urban areas, the study of the urbanization process in Asian countries is handicapped by a general lack of historical perspective. Most of the countries in South and Southeast Asia had a colonial past, and after the advent of independence they launched comprehensive development plans with a view to accelerating industrialization and giving better direction to the urbanization process. Most of the big cities in this region have a colonial history and their urban problems cannot be understood unless a sense of history is imparted to demographic analysis based purely on census statistics. To illustrate, ever since the first regular census was taken in 1881, almost all census reports have commented on urban growth—but these discussions are mostly descriptive and lack historical depth and statistical rigour. The Census Commissioners of India and the Census Superintendents of various Provinces and States, for understandable reasons, were more concerned with the decade immediately preceding the census for which they were responsible and their comments on urban growth were mostly confined to events in that decade alone. The presentation of statistical data was restricted to a few set tables giving the growth rates for different urban classes based on population size. Nevertheless, most of these census reports do give an idea of the urbanization process decade by decade. Some of them, however, contain considerable speculative material on the causes of the slow pace of urbanization in India in the early decades of this century. Some Census Commissioners put forward their own hypotheses on urbanization and occasionally engendered a lively controversy in successive census reports where some of the hypotheses were refuted and new ones advanced.

Among alleged causes of slow urbanization were race, rainfall, plague, and attachment to village life, while famines and the presence of pilgrims were also mentioned as factors which, by artificially inflating the urban population in the initial census year, gave the impression of slow urbanization in the following decade. World War II and the partition of India in 1947 were mainly responsible for a sudden spurt in urban growth during the decades 1931–41 and 1941–51. The 1951–61 decade was marked by rapid strides in industrialization and it was generally expected that urbanization, too, would be rapid during this decade. The 1961 census data, however, do not give any evidence of such accelerating urbanization.

In a recent paper Kingsley Davis poses the question: "Why has India's urbanization been so slow?" and proceeds to answer it as follows: [10]

"The answer, I suggest, is the relative slowness of economic development in India. Although nobody knows the past Indian rate of economic development the evidence seems to indicate that it is not likely to have been rapid, compared to that of most other countries at roughly similar stages."

Davis quotes Daniel Thorner in support of what he says. Turning to the preliminary results of the 1961 census, he is at a loss to explain "Why urbanization has not moved rapidly since 1951." Davis does give the impression that he is instinctively looking for urbanization which "will be a sign, though not an absolute proof, that economic growth has accelerated." [11]

Elsewhere [12] we have argued that urbanization in the face of rapid population growth and surplus labor—which is the case of present-day India—calls for fresh thinking on the industrialization-urbanization process. It is our contention that theoretical generalizations regarding the relationship between industrialization and urbanization are rather flimsy; empirical studies concerning the process of industrialization and urbanization lack rigorous analysis, mostly because adequate data are not available.

Urbanization Process in South and Southeast Asia

Table 7 gives the percentage variation for each of the six decades, in total, rural, and urban populations of India. In the 1901–11 decade, the rate of growth of the rural population was much higher than that of urban population, while in the next decade (1911–21) there was an absolute decrease in the rural population and a modest increase in the urban population. In the decade 1921–31 the rural population increased by 10.0 percent while the urban population increased by 19.1 percent. The next decade (1931–41) witnessed a fairly rapid growth of urban population, namely, 32.0 percent while there was only a nominal increase in the rate of growth of the rural population. The decade 1941–51 experienced the highest rate of urban growth, namely 41.4 percent, while the rate of growth of the rural population decreased in this decade compared to the previous decade. The interesting thing about the last decade (1951–61) is that while the rate of increase in rural population shot up to 20.6 percent compared to 8.8 percent for the previous decade, the rate of growth of the urban population came down to 26.4 percent compared to 41.4 percent for the previous decade.

Table 7. **Percentage (decade) Variation in Total, Rural and Urban Population of India, 1901–61**

Decade	Total	Rural	Urban
1901–11	5.8	6.4	0.4
1911–21	−0.3	−1.3	8.3
1921–31	11.0	10.0	19.1
1931–41	14.2	11.8	32.0
1941–51	13.3	8.8	41.4
1951–61	21.5	20.6 *	26.4 *

* Unadjusted. The adjusted figures after taking note of definitional changes are 19.0 for rural population, 34.0 for urban population.

Source: Ashish Bose, *Studies in India's Urbanization.* Delhi: Institute of Economic Growth, 1966, p. 149.

After making adjustments for definitional changes we find that if the same definition of urban were adopted in 1961 as was the case in 1951, the increase in the urban population during the last decade would be of the order of 34.0 percent and that of the rural population of the order of 19 percent. It may be pointed out that the abnormal influx of refugee migration was partly responsible for stepping up the rate of urban growth during the 1941–51 decade. According to our estimate such migration accounted for 6.2 percent of the urban growth, thus yielding a rate of roughly 35 percent increase in the urban population during 1941–51 due to "normal" causes. Thus, the percentage increase in the growth of urban population during the last two decades has remained very much the same even after making adjustments for the abnormal refugee migration and definitional changes in the 1961 census.

Table 8 gives the number of towns and the total urban population of India for each of the last seven census years. In 1901 there were 1,917 towns in India (as constituted today), in 1951 this figure shot up to 3,060 while in 1961, due to the application of rigorous tests, the number came down to 2,700. In terms of population, we find that during the last six decades the urban population has more than trebled: it was roughly 26 million in 1901 and 79 million in 1961. It is interesting to note that during the forty years, 1901–41, the net increase in the urban population was 18.3 million while in a single decade, 1941–51, the net increase was 18.3 million. In the last decade, the net increase was 16.5 million and after adjustments for definitional changes, 21.2 million. Expressed another way, in the *decade* 1911–21 the net increase was 2.15 million, whereas the average increase *per year* in the urban population during the 1951–61 decade was 2.12.

Table 8. Growth of Urban Population of India, 1901–61

Census year	No. of towns	Total urban population (millions)	Increase in each decade (millions)	Percent increase (decade)
1901	1,917	25.85		
1911	1,909	25.94	0.09	+0.35
1921	2,047	28.09	2.15	+8.28
1931	2,219	33.46	5.37	+19.12
1941	2,424	44.15	10.69	+31.95
1951	3,060	62.44	18.29	+41.43
1961	2,700	78.94	16.50 *	+26.43

* Unadjusted. The adjusted figure after taking note of definitional changes is 21.23 million, and the growth rate for 1951–61 is 34.01 percent.

Source: Ashish Bose, *Studies in India's Urbanization*. Delhi: Institute of Economic Growth, 1966, p. 149.

In Table 9 we present the growth rates of the "effective urban" and "quasi-urban" population of India during the last six decades. By effective urban population we mean the population of towns belonging to classes I, II, and III (i.e. population of 20,000 and over) and by quasi-urban population we mean towns belonging to classes IV, V, and VI (i.e. population below 20,000). In this table we also give the figures for 1961 adjusted for definitional changes for both these categories of urban population. This table gives evidence of a definite slowing down of the tempo of urbanization during 1951–61 compared to the earlier decade, 1941–51. The effective urban population increased by 52.6 percent during 1941–51 while it increased 42.3 percent during 1951–61. An interesting feature revealed by Table 9 is that the percentage increase of the quasi-urban population for the decades 1921–31 and 1931–41 was very much the same, namely, a little over 12 percent, but there was a substantial rise in the growth rate of the effective urban population during 1931–41 (47 percent) compared to that in the earlier decade (25 percent). The growth of the effective urban population really began after 1921. This was also true of the population of India as a whole, but the growth of the quasi-urban population showed no signs of acceleration except in the decade 1941–51 (which was considerably affected by abnormal migration of refugees from Pakistan). The slow growth of population of small towns is a phenomenon which must be noted while discussing the process of urbanization.

Table 9. Growth of Effective Urban and Quasi-urban Population, 1901–61

Year	Effective urban population	Variation (millions)	Percent variation (decade)	Quasi-urban population (millions)	Variation (millions)	Percent variation (decade)
1901	13.02	—	—	12.83	—	—
1911	13.49	0.47	3.61	12.45	−0.38	−2.96
1921	15.13	1.64	12.16	12.95	0.50	4.02
1931	18.93	3.80	25.12	14.52	1.57	12.12
1941	27.84	8.91	47.07	16.31	1.79	12.33
1951	42.47	14.63	52.55	19.97	3.66	22.44
1961	60.40	17.93	42.22	18.53	−1.44	−7.21
	60.43 *	17.96 *	42.29 *	23.24 *	3.27 *	16.37 *

* Adjusted for definitional changes in 1961 by hypothetically including, in 1961 urban population, the 1961 population of places which had enjoyed urban status in 1951 but through application of the new definition lost it in 1961.

We have given a brief statistical outline of the growth of urban population in India during the last six decades without going into the more technical aspects of demographic analysis. We have also given a few examples from old census reports of the speculations on the causes of urbanization in India. We pointed out that there has been no attempt so far to study in a comprehensive manner the role of urbanization in the process of economic growth and social change. It is unfortunate that no economic historian has ventured to undertake such a study, despite the known limitations of data; it is equally unfortunate that economists and sociologists, by and large, were diverted by the so-called socio-economic surveys of cities and towns which are mostly data-oriented, not problem-oriented. Most of the generalizations regarding urbanization are based on the experiences of Western countries in a century characterized by low rates of population growth. The political, economic, and demographic situation in the developing countries of the world to-day has very little in common with the developed countries in their pre-industrial phases. An intensive study of the Indian experience would have the additional advantage of a better understanding of the problems of countries in other parts of the world, and particularly in Asia, which have much more in common with India of the twentieth century than Europe of the nineteenth century.

Here we should like to raise two sets of questions: (1) why has the rate of urbanization slowed down in the last decade, a decade marked by rapid industrialization? Is it because industrialization has not been fast enough and has failed to keep pace with the rise in population? Is it that as a result of our planning efforts, the economic situation in the rural areas has improved, lessening the volume of rural-to-urban migration? Or is it because the large increase in the labor force in the urban areas and growing unemployment in the urban areas are warding off the potential streams of migration from the rural areas? Is it that the big cities have reached a saturation point and just cannot hold any more people? Or does the slower tempo of urbanization indicate the success of the Government's professed objective of dispersal of industries and balanced regional development? Is the lower tempo of urbanization just a statistical phenomenon which exists only in the minds of demographers and not in reality? These and numerous other related questions must be answered satisfactorily before we can say anything with finality on the process of urbanization in India during the last decade. (2) The other question is: why are the small towns (population below 20,000) growing so slowly? Is it because there is high migration from these towns to the bigger towns and cities? Or is it because of the inability of these towns to sustain themselves from the economic point of view—which again may be due to historical forces like the ruin of traditional industries, or due to the absence of adequate economic and social over-heads required by modern industries? Is the stagnation of small towns basically a statistical phenomenon arising out of definitional and other census changes, or due to the impact of reclassification of towns or the upgrading of small towns into higher urban classes with the passage of time? These and many other related questions have to be answered before we can comment with confidence on the process of urbanization in India.

Interestingly enough, our starting point, the 1901–11 decade, was characterized by slow urbanization, and our closing decade, 1951–61, is again characterized by urbanization slower than was expected, though there is a world of difference between the two time spans. In fact, each of the last six decades has been unique in its own way with a dominating theme—be it famine or plague, influenza or depression, war or partition, planned development or population explosion. Will there be a dominating theme in the decades to come? Will huge industrial conurbations spring up or will small towns blossom as centers of industry and usher in eras of "orderly urbanization"? Much depends on whether planners dream or dreamers plan.

Pakistan

As Table 10 indicates, the rate of growth of urban population in Pakistan, roughly the same as that of the rural population in the 1901–11 decade, started accelerating after 1931 and during the 1951–61 decade reached the highest figure of over 57 percent. Interestingly enough, there has also been considerable acceleration in the rate of rural population growth during 1951–61 on account of a general acceleration of population growth during this time.

Table 10. Urban and Rural Distribution of Population and Decennial Variation 1901–61, Pakistan

	URBAN		RURAL	
Year	Percent of total population	Decennial percentage variation	Percent of total population	Decennial percentage variation
1901	5.1	—	94.9	—
1911	4.9	8.5	95.1	8.4
1921	5.4	16.1	94.6	6.2
1931	6.5	32.1	93.5	7.5
1941	7.9	44.1	92.1	17.0
1951	10.4	41.9	89.6	5.1
1961	13.1	57.4	86.9	20.6

Source: Mohammed Afzal, "Migration to urban areas in Pakistan." International Union for the Scientific Study of Population, *Contributed papers, Sydney Conference,* Australia, 21 to 25 August 1967, p. 693.

Indonesia

In Indonesia, the census data are available for 1930 and 1961. During this period the urban population increased by 129 percent whereas the total population increased by 60 percent. The big cities of Djakarta and Bandung were six times larger in 1961 compared to 1930. According to Tangoantiang "an important reason why people move to Djakarta is the prevalence of lawless disturbances in the interior." [13] According to another study, by Senaprawira and Prodjorahardy on the basis of place of birth data in the 1961 census of Indonesia, it can be said that 49 percent of the total population of Djakarta Raya was comprised of migrants.[14]

Ceylon

In Table 11 we present the growth rates of urban population for Ceylon.

Table 11. Growth of Urban Population in Ceylon

Year	Percent of population	Percent increase in the preceding period
1901	11.6	—
1911	13.2	31.1
1921	14.2	17.5
1931	13.9	15.6
1946	15.3	38.8
1953	18.2	44.1
1963	18.9	35.5

Source: The figures for 1901 to 1953 are taken from
B.L. Panditharatna, "Trends of Urbanization in Ceylon
1901–1953." *The Ceylon Journal of Historical and Social
Studies,* Vol. 7, No. 2, July–December, 1964, p. 205. The
figures for 1963 are taken from United Nations *Demographic Year Book, 1967.* The figures given by Panditharatna are inclusive of the population of town councils.
The figures for urban population for 1963 given in the
country statement of Ceylon presented at the Asian Population Conference was 1,586,053 (see Indian Population
Bulletin, No. IV, January, 1967, New Delhi: 1968, p. 29)
whereas the U.N. *Demographic Year Book* gives the figure as 1,997,930.

Thailand

Table 12 gives an idea of urban growth in Thailand.

Table 12. Urban Growth in Thailand

Year	Percent urban	AVERAGE ANNUAL INCREASE (PERCENT)	Urban population	Rural population
1947	9.9			
1960	13.1	1947–60	5.0	3.0
1967	14.4	1960–67	4.8	3.1

Source: Sidney Goldstein, "Regional Differences in Urbanization
in Thailand, 1947–1967" (mineographed paper, 1969).

In Thailand's urbanization the role of Bangkok-Thonburi as a primate
city is at once noticeable. According to Goldstein, greater Bangkok's population in 1947 (781,700) was 21 times greater than that of Chiengmai, the
next largest place. By 1960, the capital's population had increased threefold (1,800,700), equal to 27 times the population of Chiengmai. In
1967 the 2,614,400 persons living in greater Bangkok outnumbered the

residents of Chiengmai 32:1. As Goldstein puts it: "Bangkok's urban primacy is among the most striking in the world." [15]

UNEXPECTED TRENDS IN URBANIZATION

Soon after the first results of the 1961 Census of India were out, we drew attention [16] to the unexpected phenomenon of a slower tempo of urbanization in a decade (1951–61) of rapid industrialization. We examined several possible reasons for the phenomenon, like the absence of large-scale refugee migration, the reaching of a saturation point in several cities, progress in rural areas, decentralization of industries, better transport facilities and increased commutation, growth of conurbations, and the presence of surplus labor in urban areas. We concluded that the main reasons for the slower tempo of rural-urban migration were: (1) a large increase in the urban labor force on account of an accelerating rate of natural increase in population for the last four decades; (2) the existence of a large backlog of unemployed and underemployed persons in the urban areas tending to block the streams of rural-to-urban migration; and (3) the increasing trend toward migration from small towns to big cities.

In a recent paper Zachariah estimates that the net rural-urban migration in India during 1941–51 was 8.2 million, while it was 5.2 million during the decade 1951–61. As he puts it: "It is surprising that rural-urban migration decreased by about 37 percent at a time when the country had successfully completed two five year plans. . . ." [17] Among the causes, he lists the possibility of structural changes in the character of migration in the sense that urban-to-urban and rural-to-rural migration might have become more important than before.

In Pakistan, some surprise has been expressed at the small volume of migration from rural areas to the city of Karachi which witnessed a growth rate of as high as 79.6 percent during 1951–61. According to a recent survey of Karachi, only 9 percent of the total workers in that city came from the rural areas of Pakistan.[18] To quote Khan,[19]

". . . various explanations may be given for this low participation of labor from rural Pakistan in the industrial employment of Karachi. First, a previously urbanized population having some industrial experience has been the major source of labor supply to the newly established factories in Karachi. Second, the journey from the rural areas to Karachi is long and costly. Third, the population of the Punjab in West Pakistan has a fascination and respect for armed services which channels much labor into that activity. Finally, climatic conditions and surrounding social environments (particularly language) encountered in Karachi also discourage movement to the locality to some extent."

Factors Affecting Migration

It has become almost a ritual to analyze causes of migration in terms of "push and pull factors." A corollary of the push and pull thesis was the "over-urbanization" thesis propounded by Kingsley Davis, Philip M. Hauser, and others. Sovani has effectively demolished the "over-urbanization" thesis in one of his recent papers. We have also argued against migration analysis in terms of push and pull factors. One of the most recent critics of push and pull factor analysis is Gino Germani. We shall briefly present the viewpoint of Sovani and Germani, in addition to our own.

According to Sovani: [20]

> "A common and widely accepted view about rural-urban migrants in Asian countries is that they are 'pushed' rather than 'pulled' into the urban areas. . . . This is such a widely accepted view of rural-urban migration that it takes quite an effort even to notice the evidence that does not fit into this picture."

Sovani argues on the basis of data collected in Orissa State in India and also takes into account the findings of several socio-economic surveys of Indian cities. He compares the income distribution of households in rural areas which send out migrants and those which do not send out migrants. In one of the three districts he studied, the migration from the highest income brackets is considerable. He concludes: [21]

> "On the whole there is an indication that household income is an important factor which influences out-migration but it is doubtful whether it is the or the most decisive one."

Turning to the occupation of migrants before migration, Sovani finds that the proportion of households sending out migrants from among the agricultural laborers (who are more or less landless) was small compared to households with other occupations. "As landless labor belongs to the poorest section of the rural areas, it seems strange that the 'push' to the urban areas does not operate more strongly on it." [22]

Sovani then proceeds to compare the unemployment rates among migrants to the cities and the non-migrant population of cities. Several city surveys have revealed that the incidence of unemployment was higher among residents. One explanation of this has been offered by the author of the Calcutta Survey who says: [23]

> "Unemployment has always been lowest among the migrants. The most probable explanation is the fact that a large number of such migrants who are unable to secure any job in the city after staying here for a certain period go back to their native places or some other places. Otherwise it is difficult to explain the persistently low unemployment rates among this group of employment seekers."

Commenting on this, Sovani says: [24]

> "If this is true generally, it throws a new light on rural-urban migration in India because it would show that it is not such a blind phenomenon as results in over-migration to urban areas because of the rural 'push'. The movement seems to be much more cautious and discerning and reversible. *It seems a carefully calculated move and retreat is an integral part of it.*"

Elsewhere [25] we have drawn attention to the phenomenon of "turnover migration" in India. We have argued that push and pull factors must be interpreted in the over-all demographic context. Under conditions of a high rate of *natural increase in population* not only in the rural areas but in the urban areas as well (as a result of high urban birth rates and rapidly declining urban death rates), the push factor operates in the urban areas also. In India and Pakistan the urban labor force is quite sizable, the urban unemployment rates are high, and there already exist pools of under-employed and unemployed population in the labor force—pools which act as deterrents to fresh flows of migration from rural areas to urban areas. We have called this the "push-back" factor. If new employment opportunities are created in the urban areas, the first persons to offer themselves for employment are the marginally employed persons already residing there unless, of course, special skills are required.

In support of our argument we may point out that, in urban India, the unemployment rate was 8.2 percent among the resident (non-migrant) population while it was 6.4 percent among the migrant population.[26] In urban Delhi it was found that while 4.2 percent of the working force in resident households was unemployed, only 2.7 percent of the working force in migrant (excluding refugee) households was unemployed.[27] In Bombay City it was found that 7.1 percent of the resident males were unemployed while only 4.5 percent of the immigrant males (non-refugees) were unemployed.[28] At the same time, it is true that more people come to cities in search of employment rather than *better* employment. On the basis of NSS data we estimated that for every 100 persons migrating to urban areas in India because they have found better employment, 254 persons come in search of employment.[29] These two factors taken together, namely, the larger number of persons coming to cities in search of work compared to those who come having procured better employment, versus the lower unemployment rate among the migrants, indicate the operation of the push factor both at the rural and urban ends. In other words, there are not always clear streams of migration from rural to urban areas but there is probably a high rate of turnover migration indicating push to and fro.

There is also another type of urban "push." In the absence of social security measures, except in a limited sense and for a small section of the

population, there is a constant push factor in urban areas operating when-
ever a migrant in a city falls sick, is unemployed, or retires from service.
This is especially true of the migrant who comes to the city in expectation
of a job but fails to get it. He is inevitably pushed back to his village or to
some other city in search of employment.

In regard to the push factor in rural areas, we may also interpret
"push" in a literal sense to indicate the dash and dynamism on the part
of some sections of the population. In a study of migration in Punjab
(India), Gosal lists among the causes of migration, "the adventurous spirit
of the Punjabis whom distance does not deter in the pursuit of economic
opportunities." [30] But he also points out that the areas from which outmigration has been the highest are where density of population is the
highest, per capita cultivated land the smallest, and where large parts
have been affected by water-logging. This indicates the combination of
positive and negative aspects of the push-factor.

In a recent paper on migration and acculturation, Germani says: [31]

> "It is usual to analyse rural-urban migration in terms of push-pull factors.
> . . . While this approach may be quite useful in certain respects, it must
> be recognized that it has the risk of over-simplifying the process, reducing
> it to a kind of mechanical balance of external impersonal forces."

He lays the greatest emphasis on the psycho-social level of analysis of
causes of migration.[32]

> "Not only criteria of what must be considered bad or good conditions,
> attractions or repulsions are to be found in the norms, beliefs and values
> of the society of origin but also attitudes and behaviour patterns which
> in this society regulate migration."

Germani, however, emphasizes that "we are not reducing the causes of
migration solely to a psychological process; what we are trying to point
out is the need to use a psychological and a normative context in order
to understand the working of the objective factors." [33]

In this connection, it would be more fruitful to look for the *decisive
factors* in rural-urban migration which really trigger off the process of
migration. Under conditions of mass illiteracy and unorganized spread of
employment market information, mobility of labor is severely restricted.
There may be rural unemployment or under-employment and at the same
time there may be cities with unfulfilled demands for labor. It does not
necessarily follow that this will result in rural-urban migration. It is our
contention that, except in the case of contract labor and similar forms of
labor recruitment, individual migration of labor from rural to urban areas
(especially long distance migration) depends largely on the presence of
friends or relations of fellow-villagers in the city of migration. Unfor-

tunately, the socio-economic surveys of cities in India did not collect data to substantiate or nullify this viewpoint. However, there is one recent study based on the survey of rural labor in Bombay City by Kunj Patel which supports our contention (Table 13).

Table 13. **Distribution of Workers of Rural Origin in Bombay (sample survey) According to Relations Working in Mills**

Particulars	No. of workers	Percent
Close relations	333	66.6
Relations	75	15.0
Villagers	86	17.2
No relations/no villagers	6	1.2
Total	500	100.0

Source: Kunj Patel, *Rural Labour in Industrial Bombay.* Bombay: 1963, p. 29.

Patel comments that "Just when a lead is given, the villager does not leave his home on his own. Only when there is a close relation or friend to take him to Bombay, does he leave the village with him, and it is through his influence that the newcomer gets a job, in whichever industry the former happens to be employed." [34] It is also interesting to note that when workers were asked if they had any of their fellow-villagers working in Bombay, 99 percent replied affirmatively—and many of them in mills.

Thus in order to get a better insight into the causes of migration it is necessary to know more about the method of labor recruitment. Here again there is need for careful case studies with a historical perspective. Unfortunately such studies are rare in developing countries. Reference must be made here to recent work: *The Emergence of an Industrial Labour Force in India—A Study of the Bombay Cotton Mills, 1854–1947,* by Morris David Morris. Commenting on the various hypotheses about the commitment of rural labor and related questions, Morris says: [35]

"Much of the literature tends to base interpretation on hypothetical psychological and sociological propositions which themselves are highly suspect. Moreover, the methodology is questionable. The historical argument typically rests on scattered fragments of evidence taken indiscriminately from all areas of the country and from all sorts of industry, seasonal and perennial, large-scale and small-scale; it relates to all kinds of labour, casual and permanent, unskilled and skilled. . . . I have suggested that satisfactory studies of the creation of an industrial labour force cannot be made on an all-India basis."

The so-called "pull" factor to cities in terms of the glamour of city lights, also cannot stand close scrutiny. In a study of factory workers of

Poona City, Richard D. Lambert presents interesting data on commitment of labor (Table 14).

Table 14. Percent of Workers Committed to Factory, Occupation, and City (Poona)

Type of attachment	Percent
Factory	76.2
Occupation	46.0
City (Poona)	4.3

Source: Richard D. Lambert, *Workers, Factories and Social Change in India*. Bombay: 1963, p. 84.

As Lambert points out: "The comparatively low proportion of workers who would insist on working in Poona shows how much stronger the employment commitment is than the residential commitment." [36] The glamour of city lights is basically the "glamour" of better employment in cities.

To conclude, macro-analysis of the causes of migration have doubtful value in arriving at meaningful generalizations on the process of rural-urban migration.

In view of the non-availability of data on rural-urban migration in the censuses of this region except for India, one has to look into the indirect evidence of characteristics of migrants to urban areas. In Table 15 we compare the age distribution of the total population and the urban population in a number of Asian countries. In almost all these countries the proportion of males in the productive age group (15–59) is higher in the urban areas compared to that of the total population, suggesting the predominance of adult males in the migration streams to urban areas. Another evidence of this is the sex ratio of urban areas in these countries where the number of males is higher than the number of females, as will be evident from Table 16. There are, however, exceptions to this general observation on the sex ratio in Asian cities: in the million-plus city of Surabaja in Indonesia there were more males than females in 1930 but in 1960 the position was reversed—the female population exceeded the male population by 5 percent in 1961. Commenting on this Tangoantiang observes: "At present we are unable to explain this phenomenon. With more census statistical data available to us in the future, we may find the key to the mystery of this 'change of sex' in Surabaja." [37] In India a detailed examination of the sex ratio in 113 cities with a population of over 100,000 in 1961 was undertaken by Desai, and he too was handicapped by the lack of data on the birth and death rates in cities. He concludes: "It is not possible, therefore, to explain fully the variations in sex ratio that have taken place in India during the last six census decades." [38]

Table 15. Age Distribution of Total and Urban Population in Selected Countries

Country	Age group	MALE Total	MALE Urban	FEMALE Total	FEMALE Urban
India (1961)					
	15	40.9	37.3	41.4	41.0
	15–59	53.6	58.3	53.0	53.8
	60+	5.5	4.4	5.9	5.1
Indonesia (1961)					
	15	43.1	40.5	41.1	39.7
	15–59	54.3	57.4	56.2	57.7
	60+	2.6	2.1	2.7	2.6
Pakistan (1961)					
	15	44.0	38.5	45.1	44.9
	15–59	49.6	56.2	49.4	50.1
	60+	6.4	5.3	5.5	5.0
Ceylon (1963)					
	15	40.6	35.1	42.4	39.8
	15–59	52.4	58.4	51.5	53.6
	60+	7.0	6.5	6.1	6.6
Nepal (1961)					
	15	41.2	34.2	42.4	39.8
	15–59	53.8	58.4	51.5	53.6
	60+	5.0	6.5	6.1	6.6
Malaysia—East Sabah (1960)					
	15	42.9	42.8	44.2	45.6
	15–59	53.2	53.6	51.9	50.0
	60+	3.9	3.7	3.9	4.5
Sarawak (1960)					
	15	45.1	45.1	43.8	45.7
	15–59	49.5	48.9	51.1	45.0
	60+	5.4	6.1	5.1	9.3
Cambodia (1962)					
	15	44.3	44.1	43.3	44.7
	15–59	50.9	53.0	51.8	51.4
	60+	4.8	2.9	4.9	3.9

Source: United Nations, *Demographic Year Book, 1967,* Table 4, pp. 170–182.

Table 16. Sex Ratio in Urban Areas in Selected Countries

Name of country	Year	Sex ratio (males per 1000 females)
India	1961	1184
Indonesia	1961	1001
Pakistan	1961	1289
Ceylon	1963	1185
Nepal	1961	1124
Malaysia—East		
Sabah	1960	1133
Sarawak	1960	1038
Cambodia	1962	1024

Source: Computed from Table V in United Nations, *Demographic Year Book, 1967,* pp. 170–182.

Table 17. Million-Plus Cities in South and Southeast Asia

Rank	City	Country	Year	Estimated population (thousands)
1.	Bombay	India	1967	4,903
2.	Calcutta a (U.A.) *	India	1967	4,765
3.	Djakarta b	Indonesia	1961	2,973
4.	Delhi (U.A.)	India	1967	2,874
5.	Karachi (U.A.)	Pakistan	1967	2,721
6.	Bangkok-Thonburi c (U.A.)	Thailand	1967	2,614
7.	Singapore	Singapore	1967	1,956
8.	Madras	India	1967	1,927
9.	Lahore (U.A.)	Pakistan	1967	1,674
10.	Saigon	Vietnam (Rep. of)	1965	1,485
11.	Bangalore (U.A.)	India	1967	1,473
12.	Ahmedabad (U.A.)	India	1967	1,414
13.	Manila	Philippines	1966	1,402
14.	Hyderabad (U.A.)	India	1967	1,328
15.	Kanpur (U.A.)	India	1967	1,139
16.	Surabaja	Indonesia	1961	1,008

* U.A. stands for urban agglomeration.

a. The delimitation of the Calcutta agglomeration leaves scope for different estimates. The estimated population is much higher according to other definitions of the Calcutta Industrial Region.

b. The figure for Djakarta is taken from Tangoantiang, "Growth of Cities in Indonesia, 1930–1961" (mimeographed paper, 1967).

c. The figure for Bangkok-Thonburi is taken from Sydney Goldstein, "Urban Growth in Thailand, 1947–67." *Journal of Social Sciences,* Vol. 6, April 1969, pp. 100–118.

Source: Compiled from United Nations, *Demographic Year Book, 1967.*

The Role of Big Cities

In Table 17 we list the million-plus cities of South and Southeast Asia ranked according to their latest estimates. It will be seen that India has eight such cities, Pakistan two, and Indonesia two. Some countries do not have a single million-plus city, as is indicated in Table 18.

Table 18. Capital City with Population below One Million

City	Country	Year	Population
Rangoon	Burma	1957	821,800
Phnom-Penh	Cambodia	1962	393,995
Colombo	Ceylon	1963	510,947
Kuala Lumpur	Malaysia	1957	316,230
Katmandu	Nepal	1961	121,019
Vietiane	Laos	1962	162,297
Hanoi (U.A.)	Vietnam—North	1960	643,576

Source: United Nations, *Demographic Year Book, 1967.*

The Role of Small Towns

A neglected field in demographic analysis concerns the role of small towns in the urbanization process. In a recent paper [39] on this subject we have analyzed the data for India and Pakistan and have drawn attention to the stagnation and decay in a number of small towns in these countries. Our study leads to the conclusion that all is not well with small towns. The slow growth, stagnation, and decay of a large number of small towns is a phenomenon which has to be studied in the context of economic history. Such stagnation in an era of planned industrialization deserves serious attention from planners and policy-makers in India and Pakistan.

CONCLUSION

In conclusion we may point out that the study of the urbanization process in South and Southeast Asia is greatly handicapped by lack of even the basic cross tabulations in censuses with regard to migration of urban population in most of the countries of this region, and also on account of the paucity of analytical studies even on the basis of whatever limited data are available. There are, of course, obvious limitations of generalizing for a region marked by such wide disparities in respect of total population, proportion of urban population, number of million-plus cities, the rate of growth of urban population, and so on. Notwithstanding these disparities, one can view the urbanization process in these countries in the wider context of economic growth and social change, and analyze the role of cities in accelerating economic growth and social change. Most of the countries in this region had a colonial past, and after the advent of independence these countries have launched comprehensive development programs. All these plans aim at a higher tempo of industrialization which will result in an increase in the pace of urbanization. Further, it is also necessary to consider the modernizing role of cities in the economic, political, and social life of these countries. In the transformation of traditional rural, agricultural societies into modern, urban, industrial societies, the cities indeed have a vital role to play.

We have also distinguished between the *level* of urbanization (as measured by the proportion of urban to total population) and the *scale* of urbanization in terms of absolute size of the population concerned. India, for example, has an *urban* population of roughly 100 million which is the *total* population of some of the largest countries in the world. The biggest city in this region—Calcutta conurbation—has a population of over 6 million compared to Kathmandu, the capital of Nepal, with a population less than 150,000. This region has the whole gamut of urban proportions, ranging from 4 percent in Nepal and Laos to 100 percent in Singapore. There is, however, one common feature in this region. All these countries are undergoing a process of urbanization in the face of rapid population growth ranging from 2 to 3 percent per year, with the rate of urban growth between 3 to 6 percent per year. In view of the lack of data, it is not possible to make any meaningful forecasts about the urban population in the coming years. Besides, such an exercise is not necessary at this stage in view of the censuses taken around 1970 in most of these countries. We must make an urgent plea here for the inclusion of questions on place of birth and migration in forthcoming census questionnaires, and also for the tabulation of data separately for rural and urban areas with a cross classification of as many demographic, economic, and social characteristics as possible. Only then will it be possible for urban analysts to conduct studies on the urbanization process in South and Southeast Asia with greater confidence.

NOTES

1. Rhoads Murphey, "Urbanization in Asia." in Gerald Breese (ed.), *The City in New Developing Countries—Readings on Urbanism and Urbanization.* Englewood Cliffs: Prentice-Hall, 1969, p. 58.

2. United Nations, *Report of the ECAFE Expert Working Group on Problems of Internal Migration and Urbanization,* held at Bangkok, Thailand, 24 May–5 June 1967. Bangkok: 1968, p. 71.

3. United Nations, *Report of the ECAFE Working Group on Projections of*

Population of Sub-national Areas, held at Bangkok, Thailand, 14–23 May 1969. Bangkok: 1969.

4. Rhoads Murphey, "Urbanization in Asia," *op. cit.,* p. 59.

5. The following definitions of "urban" for different countries, except India (discussed in detail in this paper), are given in *Demographic Year Book, 1967,* p. 3:

Cambodia. Towns.

Ceylon. Municipalities, urban councils, local board areas and towns proclaimed under the Birth and Deaths Registration Ordinance.

Indonesia. Municipalities, regency capitals, and other places with urban characteristics.

Malaysia

Sabah: Towns of 3,000 or more inhabitants, *i.e.* Sandakan, Jesselton, Tawaua, Kudat, and Victoria (Labuan).

Sarawak: Kuching municipality and towns of 3,000 or more inhabitants, i.e. Miri, Simanggang, Bintulu, Sarikei, and Lutong.

Nepal. Cities of 10,000 or more inhabitants in identifiable agglomerations with essentially urban characteristics, *i.e.* Kathmandu, Lalitpur, Bhaktapur, Biratnagar, Nepalguni, and Birguni.

Pakistan. Municipalities, civil lines, cantonments not included within municipal limits, any other continuous collection of houses inhabited by not less than 5,000 persons and having urban characteristics and also a few areas having urban characteristics but fewer than 5,000 inhabitants.

Philippines. Bague, Cebu, and Queson cities; all cities and municipalities with a density of at least 1,000 persons per square kilometer; administrative centers, barrios of at least 2,000 inhabitants, and those barrios of at least 1,000 inhabitants which are contiguous to the administrative center, in all cities and municipalities with a density of at least 500 persons per square kilometer; administrative centers and those barrios of at least 2,500 inhabitants which are contiguous to the administrative center, in all cities and municipalities with at least 20,000 inhabitants; all other administrative centers with at least 2,500 inhabitants.

Singapore. City of Singapore.

Thailand. Municipal areas.

Viet-Nam, North. Cities.

6. T.G. McGee, *The Southeast Asian City.* New York: 1967, pp. 76–77.

7. *Ibid.,* pp. 77–78.

8. *Census of India 1961, Vol. I, Part II-A(i),* p. 51.

9. For details see, Ashish Bose, "Urban Characteristics of Towns in India—A statistical study." *Indian Journal of Public Administration,* Vol. XIV, No. 3, 1968, pp. 449–465.

10. Kingsley Davis, "Urbanization in India: Past and Future," in Roy Turner (ed.), *India's Urban Future.* Berkeley and Los Angeles: University of California Press, 1962, p. 8.

11. *Ibid.,* p. 9.

12. Ashish Bose, "Urbanization in the Face of Rapid Population Growth and Surplus Labour—the case of India." *Indian Population Bulletin,* No. 3, 1968.

13. Tangoantiang, "Growth of Cities in Indonesia 1930–1961," (mimeographed paper, 1967), p. 2.

14. Sotijan Senaprawira and Prodjorahardy, *"Migrants in Djakarta-Raya: Analysis of Census Place of Birth Data"* (mimeographed).

15. Sidney Goldstein, "Regional Differences in Urbanization in Thailand, 1947–67" (mimeographed paper, 1969).

16. Ashish Bose, "Population Growth and the Industrialization Urbanization Process in India," *Man in India,* Ranchi, Vol. 41, No. 4, October–December 1961.

17. K.C. Zachariah and J.P. Ambannavar, "Population Redistribution in India: Inter State and Rural-Urban," in Ashish Bose (ed.), *Patterns of Population Change in India, 1951–61.* New Delhi: 1967, pp. 93–106.

18. Mohammad Irshad Khan, "Industrial Labour in Karachi." *The Pakistan Development Review,* Vol. III, Part 2, pp. 598–599.

19. *Ibid.*

20. N.V. Sovani, *Urbanization and Urban India.* Bombay: 1966. See Chapter I, "The analysis of over urbanization," pp. 1–13. This was first published in *Economic Development and Cultural Change,* Vol. XII, No. 2, January 1964.

21. *Ibid.*

22. *Ibid.*

23. *Ibid.*

24. *Ibid.*

25. Ashish Bose, "Urbanization in the Face of Rapid Population Growth and Surplus Labour—The Case of India," *op. cit.*

26. Computed from data given in Government of India, National Sample Survey, No. 53, *Tables with Notes on Internal Migration.* Delhi: 1962.

27. V.K.R.V. Rao and P.B. Desai, *Greater Delhi—A Study of Urbanization, 1940–1957.* Delhi: 1965.

28. D.T. Lakdawala *et al., Work, Wages and Well-Being in an Indian Metropolis: Economic Survey of Bombay City.* Bombay: 1963.

29. Computed from data of the National Sample Survey, *op. cit.*

30. Gurdev Singh Gosal, "Redistribution of Population in Punjab During 1951–61," in Ashish Bose (ed.), *Patterns of Population Change in India, 1951–61, op. cit.,* pp. 107–129.

31. Gino Germani, "Migration and Acculturation," in UNESCO, *Hand Book for Social Research in Urban Areas.* Paris: 1964.

32. *Ibid.*

33. *Ibid.*

34. Kunj Patel, *Rural Labour in Industrial Bombay,* Bombay: 1963.

35. Morris David Morris, *The Emergence of an Industrial Labour Force in India—A Study of the Bombay Cotton Mills, 1854–1947.* Berkeley: 1965, p. 4.

36. Richard D. Lambert, *Workers, Factories and Social Change in India.* Bombay: 1963.

37. Tangoantiang, *op. cit.*

38. P.B. Desai, *Size and Sex Composition of Population in India, 1901–1961.* Delhi: 1969.

39. Ashish Bose, "The Role of Small Towns in the Urbanization Process of India and Pakistan." Paper Presented at the International Union for the Scientific Study of Population Conference, London, September 1969.

CHAPTER 5

Brian J.L. Berry

CITY SIZE AND ECONOMIC DEVELOPMENT: Conceptual Synthesis and Policy Problems, with Special Reference to South and Southeast Asia

In the developing countries, and particularly in South and Southeast Asia, a conceptual rift relating to the role of cities in economic development separates two groups of urbanists and planners, the "modernizers" and the "traditionalists."

The pace of urbanization in these countries in the last quarter-century has made quite painfully obvious their manifold societal inadequacies in diet, housing, health, and sanitation. Unrest among urban populations with these inadequacies has been increased by improvements in communications media that have served to highlight discrepancies between present conditions and what might be achieved. It is a resulting increase in concern with the welfare functions of society that has led to the conflict between the two policy-making groups. The modernizers argue that continued concentration of economic growth in large cities is necessary to capture economies of scale and to accumulate externalities in the form of social and economic overhead and infrastructure because these, in turn, are the prerequisites for the further growth needed to provide resources required to overcome the societal inadequacies. The traditionalists who contest them argue that the inadequacies are a product of severe dis-

Author's Note: A variety of people have contributed to this paper, and I wish to thank each of them: Gerald Pyle for drawing the maps and contributing the materials on the Philippines and the Portuguese urban system, Yeu-man Yeung for the graphs on Malaya, Roger LeCompte for the graph on Denmark, Howard Spodek for the historical series on India, Kenneth Rosing for an initial input to the References, Cherukupalle Nirmaladevi for her insights into the policy debate, and Fred Hall for his careful reading and critique of the entire manuscript. All errors of course remain my responsibility.

economies of scale—of the concentration of growth and development in a few large cities at high population densities, that the "primacy" of large cities reflects "over-urbanization," that it leads to "parasitic" draining of the vitality of society at large, and breeds a continuous state of "hyper-urbanization" that can only be combatted by strategies of deliberate de-centralization.

The debate centers on the shape of the urban hierarchy and the re-lationships of different hierarchic forms to economic growth. "Primacy" is equated with conditions under which the largest cities are "too big," developing in consequence malignant hydrocephalic "gigantism." But such conclusions exceed the limits of current knowledge about the relationships between city size and economic development.

The following pages are devoted, therefore, to a consideration of the issues needing resolution. Are there, for example, differences in the form of urban hierarchy from one country to another? Are differences, if any, related to level of economic development or systematically to the rates and regional patterns of growth? These questions will be asked generally and exemplified by materials relating to the countries of South and South-east Asia. We look first at the general issue of whether a typology of city-size distributions is indicative of a typology of urban hierarchies and consider the generative processes giving rise to different distributions. Dis-tributional change is then related to growth. Finally, we explore the rela-tions of cities and the diffusion and spatial patterning of development, together with recent trends in the policy debate.

URBANIZATION AND ECONOMIC GROWTH
IN THE DEVELOPED COUNTRIES

That there is currently a correlation between the level of economic development of countries—however indexed, the degree to which they are urbanized, and the extent to which their populations are concentrated in large cities is unquestioned. Table 1 provides clear evidence.

Common Growth Characteristics
of the Developed Countries

These systematic differences have emerged because a series of common growth characteristics have served to differentiate the "developed" countries from more "traditional" polities and from their own prior traditional states. All the developed countries have experienced high rates of increase in per capita product (15–30 percent per decade) accompanied by substantial rates of population growth (over 10 percent per decade). Much of the economic growth has been sustained as a result of improved production

Table 1. Level of Development Related to Urbanization, Late 1950s

	GROUPS OF COUNTRIES BY PER CAPITA INCOME					
	$1,000 over	*$575 to 1,000*	*$350 to 575*	*$200 to 350*	*$100 to 200*	*Under $100*
	(1)	*(2)*	*(3)*	*(4)*	*(5)*	*(6)*
Percent of total population in urban areas (recent census)	68.2	65.8	49.9	36.0	32.0	22.9
Percent of population in communities of more than 100,000 about 1955	43.0	39.0	35.0	26.0	14.0	9.0
Expectation of life at birth, 1955–58 (years)	70.6	67.7	65.4	57.4	50.0	41.7
Percent of private consumption expenditures spent on food, 1960 or late 1950s (36 countries)	26.2	30.5	36.1	37.6	45.8	55.0
Percent of population, 15 years and over, illiterate, 1950	2.0	6.0	19.0	30.0	49.0	71.0

techniques, largely improvements in the quality of inputs and greater efficiency traceable to increases in useful knowledge and improved institutional arrangements. All sectors of their economies, manufacturing in particular, have participated in the increasing efficiency, while at the same time sustained growth has involved changes in the relative sector importance. A declining share of total product is attributable to agriculture, rising shares to manufacturing and public utilities, and rapidly increasing shares to personal, professional, and governmental services. Significant changes have occurred in the structure of final demand. These changes have been both the effect and the cause of changes in the productive process, and have included important shifts in the regional allocation of resources and increasing size of production units as both product and labor have shifted from smaller to larger firms and organizations. These, of course, have led to increased concentration of people in cities. Shifts in capital allocation, in product and in labor, have in turn depended upon rapid institutional adjustments and mobility in factor inputs, and it is here that urbanization has played a critical role in facilitating shifts in population and labor force, both between and within regions, and by type.

In short, as the "developed" polities of the world have modernized, the size of their production units has increased along with the number and complexity of production decisions. Market mechanisms have expanded in scope and extent to bring together production and consumption through use of transportation, communication, financing, policing, and related services, providing order for increasingly impersonal processes. Progressive division of labor, increasing scale, regional specialization and the complexities of articulating intensifying marketing networks have all bred greater population concentration in larger cities and higher degrees of urbanization than is characteristic of more traditional societies.

Regional Implications of the Growth Process

Rapid increases in total product have also implied greater pressure on natural resources, while increasing scale and concentration imply wide differentials in types and rates of growth among different social and economic groups in different regions, the interregional mobility of labor and capital, and the emergence of a geographic pattern of *core* (or *heartland*) and *periphery* (or *hinterlands*).

Large-scale industry has tended to concentrate in a limited number of cities in a limited region that serves as a polity's industrial heartland and, because of the large numbers of industrial workers employed, the center of national demand. Such a concentration develops a self-generating momentum as complementary services and activities are established, each helping the other to pyramid the productive process; increasing numbers of workers further concentrate the scale of the local market and even more strongly pull to themselves activities seeking optimal national market access.

This cumulative causation extends outwards to the hinterlands, for once the core-periphery pattern is set the core region becomes the lever for development of peripheral regions, reaching out to them for their resources as its input requirements increase, stimulating their growth differentially in accordance with its resource demands and the resource endowment of the regions.

The result of this core-centered pattern of growth and expansion is regional differentiation—the *specialization* of regional roles in the national economy. Specialization, in turn, determines the entire content and direction of regional growth. Since regional economic growth is externally determined by national demands for regional specialties, and organized geographically in a pattern of industrial heartland—together with hinterlands specialized in resource subsectors—the nature of these specialties, alternative sources of them, and changes in the structure of demand therefore determine in large measure the nature and extent of regional growth. This extends to the secondary support needed by export industries—housing, public facilities, retail establishments, service facilities and the like. The size of the multiplier effect of exports on "local" or "residentiary" growth, however, depends upon *local* expenditure patterns and income distributions, patterns of ownership, and political organization. Among the relevant issues raised are whether earnings are retained locally or transferred outside, or whether the basic industry generates a middle class. Local decisions can of course also help shape the future of a regional economy to the extent that the local economy is "closed" from external influence or when local decision-making becomes an element in national decisions. But basically, it is external factors that create growth opportunities or lead to decline. Opportunities have to be perceived and seized by imaginative leaders; otherwise they are lost, for ultimately growth can be traced back to individual location decisions about particular business establishments, and the size and nature of the entrepreneurial class thus assumes a critical role.

Cities Articulate Relations Among Regions

Cities are the instruments whereby specialized sub-regions are articulated in a national space-economy. They are the centers of activity and of innovation, focal points of the transport network, locations of superior accessibility at which firms can most easily reap scale economies and at which industrial complexes can obtain the economies of localization and urbanization. Agricultural enterprise is more efficient in the vicinity of cities. The more prosperous commercialized agricultures encircle the major cities, whereas the inaccessible peripheries of the great urban regions are characterized by backward, subsistence economic systems.

Two major elements characterize this spatial organization of the developed countries:

(a) *A system of cities,* arranged in a hierarchy according to the functions performed by each.

(b) Corresponding areas of urban influence or *urban fields* surrounding each of the cities in the system.

Generally, the size and functions of a city and the extent of its urban field are proportional. Each region within the national economy focusses upon a center of metropolitan rank, and it is the network of intermetropolitan connections that articulates the whole. The spatial incidence of economic growth is a function of distance from the metropolis. Troughs of economic backwardness lie in the most inaccessible areas along the intermetropolitan peripheries. Further sub-regional articulation is provided by successively smaller centers at progressively lower levels of the hierarchy —smaller cities, towns, villages, etc.

Impulses of economic change are transmitted in such a system simultaneously along three planes:

(a) Outward from heartland metropoli to those of the regional hinter-lands.

(b) From centers of higher to centers of lower level in the hierarchy, in a pattern of "hierarchical diffusion."

(c) Outward from urban centers into their surrounding urban fields.

Part of the diffusion mechanism is to be found in the operation of urban labor markets. When growth is sustained over long periods, regional income inequality, for example, should be reduced because the higher the capital-labor ratio in a region, and the higher the employment level of the unskilled at any wage rate and at any given social minimum, therefore, the smaller the number of involuntary unemployed. Any general expansion in a high-income area, such as a heartland metropolis, will reach a rising floor to the wage-rate first. Some industries will be priced out of the high-income labor market and there will be a shift of that industry to low-income regions, i.e. to smaller urban or more peripheral areas. The significance of this "filtering" or "trickle-down" process lies not only in its direct but also in its indirect effects. If the boom originated in the high-income region, as is very likely, the multiplier effects will be larger in the initiating region, although the relative rise in income may be greater in the underdeveloped region. But the induced effects on real income and employment may be considerably greater in the low-income region if prices there are likely to rise less and/or if the increase in output per worker could be greater. Both are likely, because of decreasing cost due to external economies stemming from urbanization of the labor force. If the boom can be maintained, industries of higher labor productivity will shift units into lower-income areas, and the low-wage industries will be forced to move into even smaller towns and more isolated areas.

A Rank-Size Relationship Will Characterize the Urban System of an Integrated, Developed Polity

The significance of these diffusion mechanisms is that if economic growth is sustained over long periods it results in progressive integration of the space economy. Regional differences in levels of welfare are progressively eliminated, since demands for and supplies of labor are adjusted by outward flows of growth impulses through the urban hierarchy and the inward migration of labor to central cities. Troughs of economic backwardness are reduced in intensity, and each area finds itself within the fields of influence of a variety of urban centers of a variety of sizes. Sufficient growth impulses will move through the system so that each region and each city can expect to grow at about the same rate as the nation, although local factors may cause, when viewed from the national perspective, seemingly random variations above and below this national expectation.

These conditions add up to the satisfaction of a "law of proportionate effect" in which growth of cities is proportional to their size, that is, that the average growth-rate is the same for cities of each level in the hierarchy, although, again, there may be random variations from the expectation.

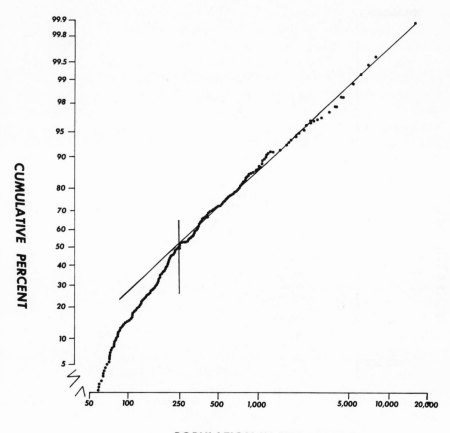

POPULATION IN THOUSANDS

Figure 1. Lognormal distribution of U.S. labor markets in 1960. The observations are "consolidated urban regions" and "functional economic areas," determined by commuting criteria.

Source: Developed and reported in Brian J.L. Berry, Peter G. Goheen, and Harold Goldstein, *Metropolitan Area Definition: A Re-Evaluation of Concept and Statistical Practice.* Washington, D.C.: U.S. Government Printing Office, Bureau of the Census Working Paper No. 28, 1968.

Appendix B provides the mathematical proof, evidently needing restate-
ment in spite of two decades of relevant literature, that when such growth
conditions obtain for a system of cities, their frequency distribution by
size is "lognormal," and that if the distribution by size is lognormal, the
system of cities will form a "rank-size distribution." (The purpose of
Appendix A is to refute a related myth that appears in the literature.)

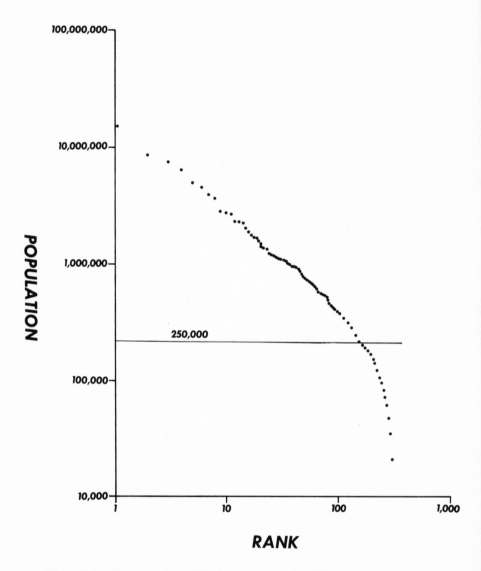

Figure 2. Rank-size graph of U.S. labor markets in 1960.
(For source see Figure 1.)

Appendix C shows that this condition is perfectly consistent with a regular urban hierarchy in which there is a given ratio of numbers centers of any particular level to those of the next higher level. Appendix D relates the slope of the line in the rank-size graph to whether the number of cities is expanding at a significant rate over some minimum "threshold" size for viability, and Appendix E relates the shape of the lognormal distribution to the Lorenz coefficient of concentration and the Lorenz coefficient to the degree of hierarchical organization of the system as measured by the span of control of lower by higher order centers and the number of levels in the hierarchy. Generally, the more highly skewed the lognormal distribution, the greater the concentration of population in larger cities, and the lower the ratio of lower level to higher level centers.

The two methods of plotting the relevant graphs are shown in Figures 1 and 2, using data for the United States in 1960, because the United States is the classic case and base of comparison. In Figure 1 the cumulative distribution of urban centers by size is plotted on graph paper that has the cumulative percentages arranged according to a normal probability scale on the ordinate, and population of the urban centers arranged logarithmically on the abscissa. To the extent that the points form a straight line, the distribution is lognormal, and this is true down to a population of a quarter of a million in the American case. A variety of authors have suggested that 250,000 is the minimum threshold size for self-sustaining growth at about the national U.S. rate today (i.e. for the law of proportionate effect to hold); a high proportion of the smaller urban centers display low incomes, slow growth and substantial cyclical sensitivity, and sizable outmigration of the better educated and more active segments of their population.

Figure 2 is the companion rank-size graph in which logarithm of population is plotted against logarithm of rank, after the urban centers have been arrayed in decreasing order of size. Again, the linearity of the larger centers and the change in slope at around 250,000 population are evident.

CITY SIZE AND ECONOMIC GROWTH IN SOUTH AND SOUTHEAST ASIA

Other Bases of Hierarchical Organization

In every country, of course, urban centers are organized functionally into a hierarchy; the hierarchy is the instrument whereby society and polity, as well as the economy, are integrated over space. In traditional societies, the organization is based more or as much on purposive sacerdotal, juridical, military, or administrative principles as on economic grounds. And it is increasingly asserted that in the most developed coun-

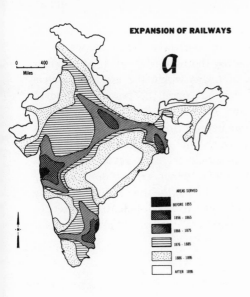

EXPANSION OF RAILWAYS

a

AREAS SERVED

- BEFORE 1855
- 1856 - 1865
- 1866 - 1875
- 1876 - 1885
- 1886 - 1895
- AFTER 1895

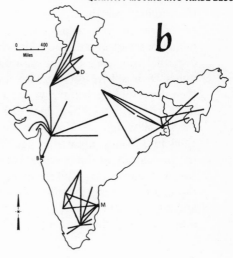

FUNCTIONAL REGIONS BASED UPON TOTAL QUANTITY MOVING INTO TRADE BLOCKS

b

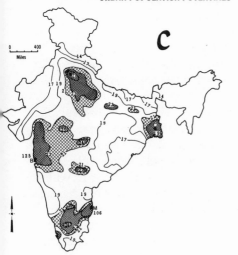

URBAN POPULATION POTENTIALS

c

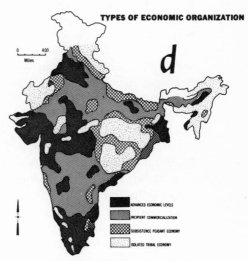

TYPES OF ECONOMIC ORGANIZATION

d

- ADVANCED ECONOMIC LEVELS
- INCIPIENT COMMERCIALIZATION
- SUBSISTENCE PEASANT ECONOMY
- ISOLATED TRIBAL ECONOMY

Figure 3. Colonial imprints on India. (a) Date of railroad construction. (b) Functional regions based on commodity flows. (c) Relative access to national markets: urban population potentials in 10,000's per mile, 1960. (d) Types of economic organization in India, 1960.

Source: (a) R.P. Misra, *Diffusion of Agricultural Innovations.* Mysore: Prasaranga, University of Mysore, 1968, p. 15; (b) and (c) Brian J.L. Berry, *Essays on Commodity Flows and the Spatial Structure of the Indian Economy.* Chicago: University of Chicago, Department of Geography Research Paper No. 111, 1966, pp. 161, 157, respectively; (d) Adapted from Joseph E. Schwartzberg, "Three Approaches to Mapping Economic Development in India." *Annals of the Association of American Geographers,* Vol. 52, 1962, pp. 455–468.

tries, under post-industrial conditions, new purposive political-economic considerations are tending to replace the narrower "spontaneous" economic bases of organization introduced by the industrial revolution. Whatever the principles, of course, status of towns within the hierarchy determines their sphere of influence and depends upon the power residing in their level of the hierarchy; status and sphere will, however, vary from level to level depending upon the organizing principle. City size remains the simplest and best index to this power. Differences in the numbers of cities in different population-size classes should therefore reveal differences in the nature of hierarchical organization (either *cross-nationally*—between countries at a given time, or *longitudinally*—for a given country through time), and these in turn should relate to developmental differences and variations in the degree of urbanization and the proportion of population concentrated in the largest cities.

Effects of Colonialism

The situation is further complicated in South and Southeast Asia, however. Each polity has had a long, rich urban history, with many cities serving variously as religious, ceremonial, military, administrative, and marketing centers until well into the nineteenth century. Extension of Western colonial control took place at this time by establishment of ports that served as administrative centers, foci for colonial exploitation of raw materials and distribution of imports, and generally as "head links" with the mother country and the world community. Thus, the great cities that dominate the region's urban hierarchies today are creatures of colonial intervention.

Indian data reveal with clarity the structuring effects of British colonial rule on South Asian regions. The pace and paths of development are shown in Figure 3a, which maps the expansion of the railroads inland from Bombay, Calcutta, and Madras. These three metropoli, along with Delhi-New Delhi, have come to dominate four great sub-regions (3b). Within each region, relative access to growth impulses drops off with distance (3c), and as a consequence the more commercialized village and town economies are found in the areas of good access, while isolated tribal economies prevail in the inaccessible peripheries (3d). Nonetheless, the Indian subcontinent is large and varied enough so that its cities *in the aggregate* conform to a rank-size distribution (Figure 4). However, its subregions do not. The entire Bengal region that formerly focussed on Calcutta definitely does not display the near-straight line of the rank-size rule (Figure 5), and neither do its two post-independence parts of West Bengal (Figure 6) and East Pakistan (Figure 7). On the other hand, West Pakistan shows much greater linearity, as do the two halves of Pakistan added together (Figure 8).

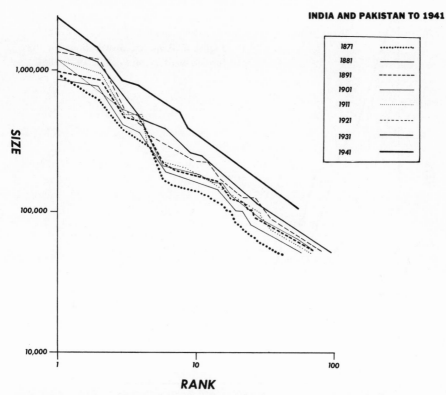

Figure 4. Rank-size distributions for the Indian subcontinent, 1871–1941.
(Figures 4–8 prepared by Howard Spodek.)

Primacy and Oligarchies: Systematic Deviations of the Largest Center(s) from the Rank-Size Distribution

It is the deviations from the rank-size distribution that have excited the greatest interest in the developing nations. A rank-size distribution should be characteristic of any well-defined system of cities in which growth has, as a minimum, obeyed the law of proportionate effect for some period of time. This has been satisfied, for example, in Malaya where recent growth has been proportional to size of city (Figure 9) and where there has been progressive convergence on the rank-size pattern over the years, with the exception of the second world war (Figures 10 and 11). Hamzah Sendut has an interesting discussion of this case in his 1966 paper "City Size Distribution in Southeast Asia" (see References).

A "well-defined system of cities" has been interpreted in many ways. To G.K. Zipf it implied that group of cities unifying a given polity, and others have added the notion of interdependence and a high degree of

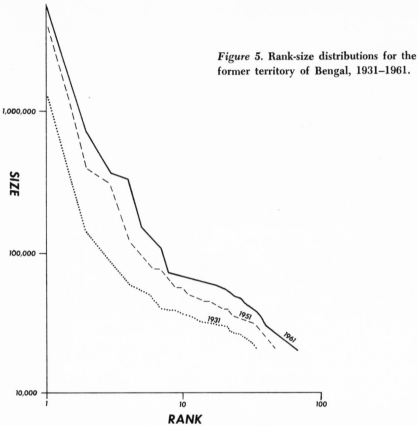

Figure 5. Rank-size distributions for the former territory of Bengal, 1931–1961.

closure of the polity—a maximum degree of internal connections and a minimal amount of interaction across the boundaries of the political system. But systematic deviations of the largest city or cities from expected populations based upon the regression line $\log P_r = k - q \log r$ have been noted in many cases.

Figures 12 and 13 illustrate the diversity of relationships in South and Southeast Asia. The mainland states of Burma, Indochina, and Thailand, which developed as relatively "inward-looking" rice-bowl food suppliers of the tropical colonial countries and which have low levels of economic development and urbanization, show the greatest deviations from rank-size. On the other hand, the island nations (Indonesia, Philippines, together with Malaya), each more "outward-looking," developed by emphasizing mineral exploitation and the production of crops for the metropolitan economies. They have many towns, higher levels of development and urbanization, and increasing conformity to the rank-size rule.

Where the actual population of the largest city exceeds the expected, a condition of "primacy" is said to exist. Colin Clark uses the additional term "oligarchy" to describe situations such as Japan, India, Australia, or Brazil, where the towns over 100,000 population have a bigger share of the total urban population than would be expected from the straight-line relationship, but where at the same time the primacy of the leading city is kept in check. A particularly striking example is provided by the Portuguese colonial system in which the head-links form one oligarchic rank-size regime and centers functioning at the local levels of the hierarchy form another (Figure 14). Clark also defines as "counter-primacy" either declining primacy of the largest city through time, or increasing negative deviation of the population of the largest city from predictions based on the rank-size rule under conditions of national planning directed at achieving such outcomes as goals. Good examples of counter-primacy in which there has been increasing convergence of the city-size distribution on the rank-size pattern are provided by the Philippines (Figures 15, 16, 19, and 20) and, for comparison, Denmark (Figure 17). On the other hand, Buenos Aires has maintained its primacy in Argentina for the last century, to cite a Latin American case (Figure 18).

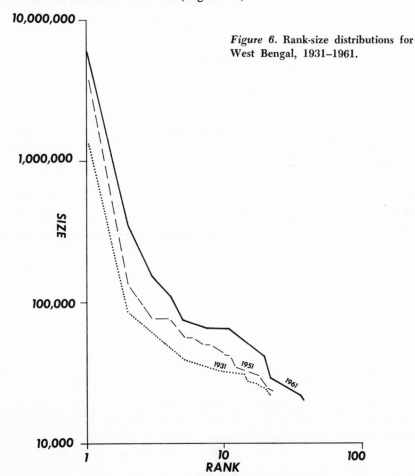

Figure 6. Rank-size distributions for West Bengal, 1931–1961.

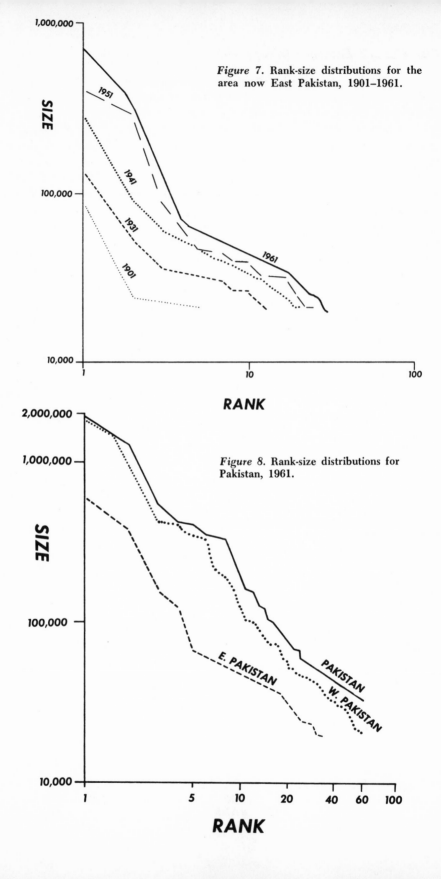

Figure 7. Rank-size distributions for the area now East Pakistan, 1901–1961.

Figure 8. Rank-size distributions for Pakistan, 1961.

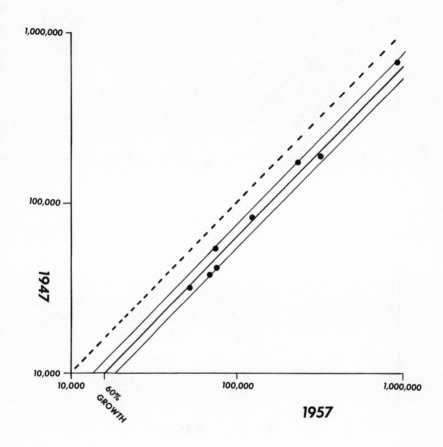

Figure 9. Sizes of the eight largest cities in Malaya, 1947 and 1957, reveal conformity to Gibrat's law of proportionate effect.

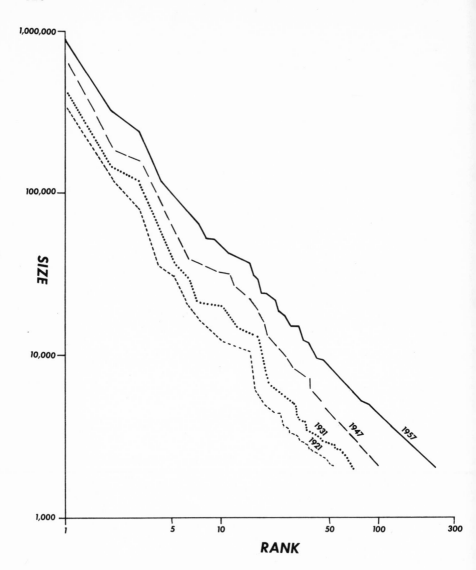

Figure 10. Rank-size distributions for Malayan cities, 1921–1957.
(Figures 10–11 prepared by Yeu-man Yeung.)

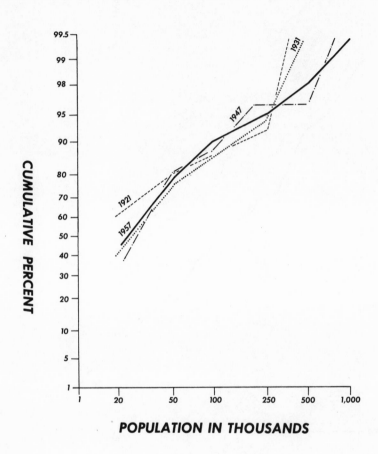

Figure 11. Lognormal plots of the cumulative distributions of Malayan cities, 1921–1957.

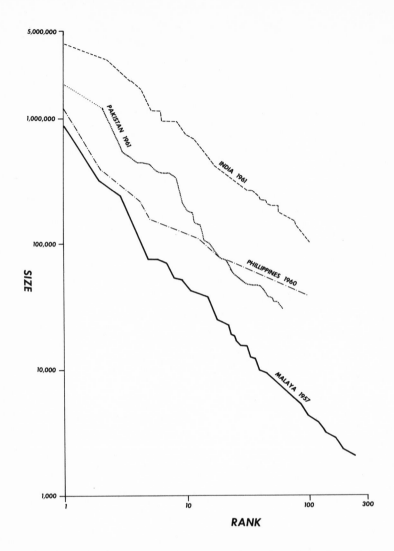

Figure 12. Rank-size distributions of several South and Southeast Asian countries, latest census date.

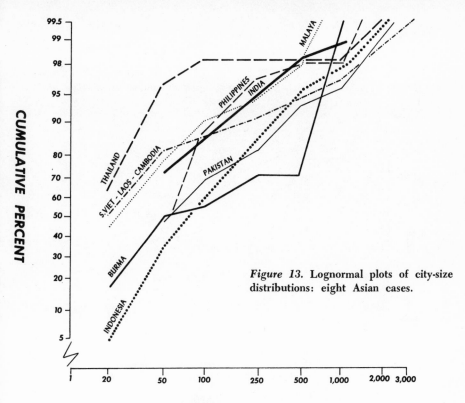

Figure 13. Lognormal plots of city-size distributions: eight Asian cases.

POPULATION IN THOUSANDS

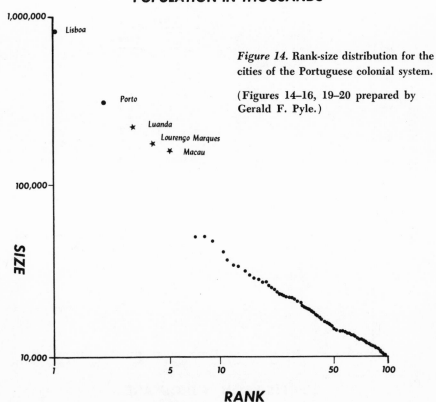

Figure 14. Rank-size distribution for the cities of the Portuguese colonial system.

(Figures 14–16, 19–20 prepared by Gerald F. Pyle.)

SIZE

RANK

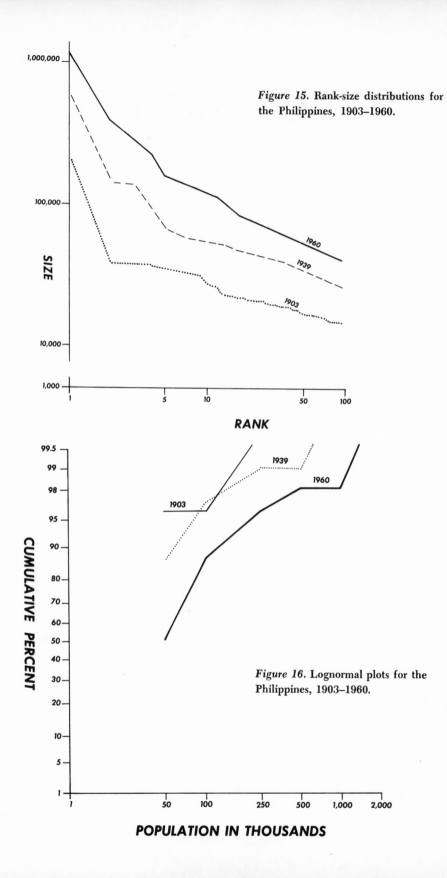

Figure 15. Rank-size distributions for the Philippines, 1903–1960.

Figure 16. Lognormal plots for the Philippines, 1903–1960.

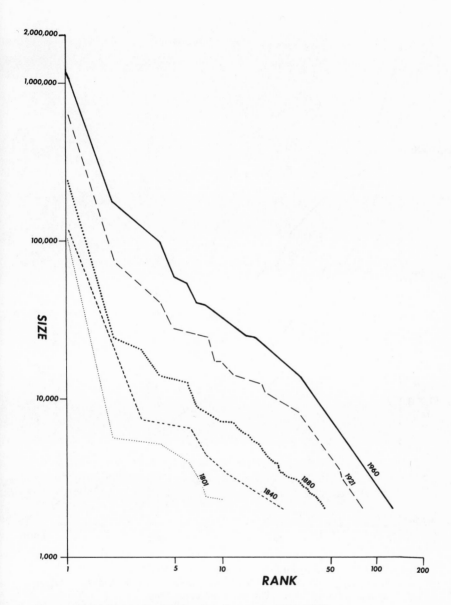

Figure 17. Rank-size distributions for Denmark, 1801–1960.
(Prepared by Roger LeCompte.)

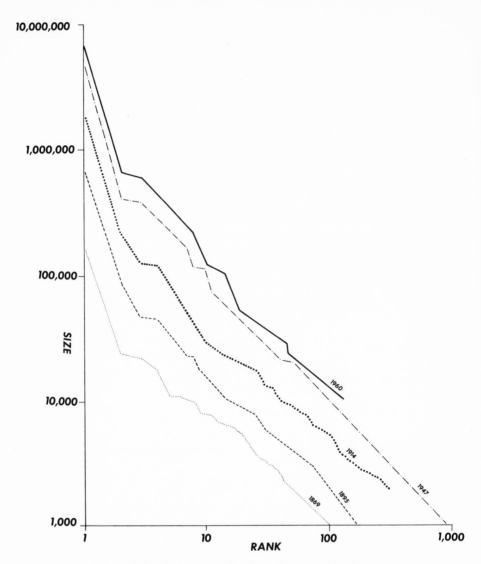

Figure 18. Rank-size distributions for Argentina, 1869–1960.
   Source: Cesar A. Vapnarsky, "Rank Size Distribution of Cities in Argentina." Unpublished M.A. thesis, Cornell University, 1966.

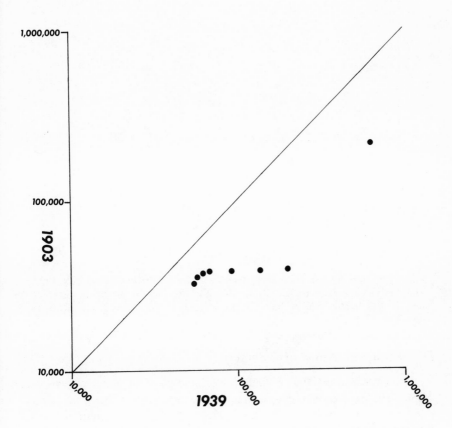

Figure 19. Sizes of the eight largest cities in the Philippines, 1903 and 1939. The growth rate of the top four far exceeds that of the next four in rank.

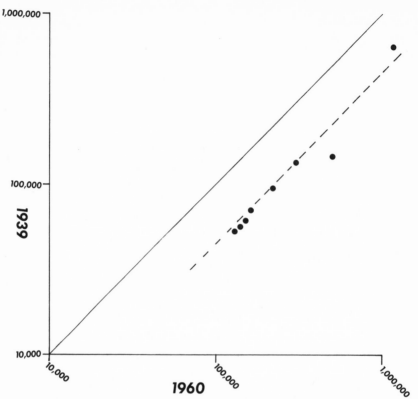

Figure 20. Sizes of the eight largest cities in the Philippines, 1939 and 1960.
There is much greater conformity to Gibrat's law, and the accelerated growth
rate of the smaller places has led the city-size distribution much closer to
rank-size.

The Interpretations of Primacy

The idea of primacy as initially formulated by Mark Jefferson was very
simple. He argued that everywhere "nationalism crystallizes in primate
cities . . . supereminent . . . not merely in size, but in national influence." He assessed the degree of eminence of cities within countries by
computing the ratios of size of the second- and third-ranking cities to that
of the largest place. But immediately after Jefferson's papers had appeared,
Zipf directed attention to the entire system of cities, focusing in particular
on the case of the rank-size rule in which the exponent $q = 1.0$ and
$K = P_1$ (see Appendix D), so that the expected populations of the largest
cities were the harmonic series P_1, $P_1/2$, $P_1/3$. . . . Such, he argued, was
the situation to be expected in any "homogeneous socio-economic system"
that had reached a harmonious equilibrium state.

It remained for discussants at a series of post-war UNESCO conferences on urbanization in Asia and the Far East, and in Latin America,

to put the two together. Cases deviating from the rule were to arise from "over-urbanization" of the economies of lesser-developed countries because of "excessive" in-migration and superimposition of limited economic development of a colonial type, creating "dual economies" characterized by "primate cities" that tend to have "paralytic" effects upon the development of smaller urban places and to be "parasitic" in relation to the remainder of the national economy. Southeast Asia has since been cited as the example *par excellence* of primacy.

Obviously, each of the words in quotation marks involves a value judgment, but nonetheless the idea of the primate city from this point on was firmly established as a dictatorial deviation from expectations about hierarchical organization derived from the rank-size rule, with strong pejorative connotations. The value-laden attitude is illustrated in the following quotation from Lampard's work:

> "Its growth and maintenance have been somewhat parasitical in the sense that profits of trade, capital accumulated in agriculture and other primary pursuits, have been dissipated in grandiose urban construction, servicing, and consumption. . . . The labor and enterprise which might otherwise have been invested in some form of light manufacturing or material processing . . . are drawn off to the great city by the attractive dazzle of a million lights."

Measurement of primacy has subsequently become a textbook exercise in urban studies (see e.g. the volume by Jack P. Gibbs). To planners such as John Friedmann and Tomas Lackington, primacy cannot be separated from the idea of "hyperurbanization":

> ". . . a society exhibiting a high measure of disequilibrium between its levels of urbanization and per capita income is likely to experience serious internal tensions. . . . [This] hyperurbanization . . . takes the form of increasing concentration of urban activities. This may be established in several ways . . . the degree of urban primacy [may be measured] according to . . . the relation of the largest to the second urban complex [details of Chilean data omitted]. . . . Another possible measure of primacy is in relation to the rank-size rule which describes a standardized distribution of a country's population among its cities."

To many planners and policy-makers in the developing countries "gigantism" of the largest cities is a characteristic to be feared, even when the principal city is small relative to cities in other areas.

The Occurrence of Primacy

Linsky, improving on a previous contribution of mine, suggests on the basis of the ratio P_1/P_2, that there are five typical situations in which primacy is found: in small, densely populated countries; in countries with low per capita incomes; when economies are based on exports or agriculture; in countries with high population growth rates; and in countries with a history of colonialism. Our illustrations already suggest that primacy may be greater in the inward-looking mainland countries of Southeast Asia, as opposed to the island countries that participated more fully in the international metropolitan economy. The nature of the colonial experience and differences in colonial policy may nevertheless be of some significance. The more recent work of El Shaks, in which a general coefficient of concentration is used to measure primacy, suggests that there is a near "normal" curve of primacy on economic development. That is to say, primacy is rare in very underdeveloped countries, rises during the take-off stage, and decreases thereafter. Interestingly, J.G. Williamson has suggested that if regional income inequality is plotted against economic development the result is also a bell-shaped curve. By extension from the findings of Linsky and El Shaks, the greater the size of country, the longer its history of urbanism, the greater the complexity and interdependence of its economic, administrative, and political structures; and the greater its degree of modernization and urbanization, the greater the probability that it will have a rank-size distribution of city populations.

These world-scale findings are cross-sectional; they deal with many countries at one point in time. The evidence is stronger for changes through time. That increasing urbanization and/or economic growth are correlated with a progression toward rank-size has been shown for Spain (Lasuen), Portugal (Alves Martins), Poland (Dziewonski), and for Israel (Bell). Figures 4–7, 10, 15, 17, and 18 provide additional data, much of it for South and Southeast Asia. In the Philippines, the growth rate of Manila far exceeded that of smaller places between 1903 and 1939 (Figure 19) but between 1939 and 1960 there has been convergence on rank-size and on the law of proportionate effect. Such convergence on rank-size may be interpreted as the progressive achievement of a steady-state—of an urban hierarchy in which there is a regular increase in numbers of centers with each succeeding lower level, together with overall conformity to a national growth norm which, once achieved, is maintained in succeeding periods of growth.

How, then, is the normal distribution of primacy on economic development to be interpreted? One insight comes from the entropy measure discussed in Appendix E. From a situation of low degrees of urbanization and modernization with a high degree of entropy (many small, equally sized urban places, or a lower level steady-state), primate cities provide the largest discrete reductions in entropy possible (i.e. the greatest gains in systematic order through organization) as the system is moved from one level of organization to another. Jefferson's sense of achieving supereminence in national importance thus seems central. Once the quantum leap has been made, there can be a gradual convergence on a new steady-state (lognormal if the same proportionate growth processes are extended over the entire system of cities).

Reasons for Primacy

While interesting, the entropy interpretation does not provide any insight into the mechanisms by which the patterning of growth by urban hierarchies results in primacy in early stages of development and leads to convergence on the rank-size rule later on. The reason is to be found in the filtering mechanism that produces hierachical diffusion.

This mechanism works poorly, if at all, in many parts of the world. Instead of development "trickling down" the urban-size ratchet and spreading its effects outwards within urban fields, growth is concentrated in a few metropolitan centers, and a wide gulf between metropolis and smaller city is apparent. Rather than articulation there is polarization. Why?

There are two classes of reasons, one institutional, the other functional. Under colonial rule, most empires were controlled by holding key cities and strategic points that linked the colonial net. For colonial powers extending and consolidating their authority in alien social and geographical territories, cities were their necessary base of action. British rule in India, for example, centered on capital and provincial cities both for maintaining an integrated and authoritarian administrative structure and for securing the economic bases of its power—the collection of taxes and control over the export of raw materials and the import of British manufactured goods. Elsewhere in Southeast Asia McGee writes that the structure of the colonial economy did not permit the cities to be generative of economic growth. The colonial cities were subordinate to the metropolis and world trade, acting as foci for the alien middleman and effectively inhibiting economic growth.

However, the colonial cities did help create modern indigenous elites and give birth to nationalist movements, and they restructured traditional economic patterns to focus on themselves, initiating massive inward flows of rural migrants. The resulting high rates of city population outstripped rates of economic growth after independence, and confounded the trickle-down process. The result was to worsen the way of life outside the metropolis and simultaneously increase the urban labor supply at all levels, but particularly among the unskilled at a pace that has never permitted economic expansion to meet rising floors to metropolitan wage-rates; thus growth cannot decentralize naturally. This systematic confounding of trickle-down mechanisms is particularly acute where, either because of the very small size of a political unit or because of a critical dependency of the polity upon external trade relationships, there is a single national focus and sizable alternative growth centers are absent. Only after extended periods of continued economic growth at rates greater than population growth have the smaller developed or developing countries been able to reestablish hierarchical diffusion of growth impulses, and only under these conditions does primacy then abate.

A series of simulation experiments by Poul Pedersen confirms these results. Starting with a simple central-place hierarchy he postulates that innovation diffusion proceeds from the metropolis to other centers according to a "threshold" gravitational concept. The potential for diffusion of an innovation from one urban center to another is given by

$$D(P) = P_iP_j/d_{ij} \propto$$

If, for a given $\propto$, this quantity exceeds a given threshold value, both the "giving" and "receiving" center will grow at a certain rate for a period of time. After this time, the gravity terms are re-evaluated and a further diffusion of growth impulses takes place, and so on for a series of "generations" of the model. He concludes (1) that an urban hierarchy will converge on rank-size via intermediate conditions of primacy as diffusion of innovations proceeds down the urban hierarchy and outwards spatially from the metropolis, with the law of proportionate effect only operating after an innovation has reached a given urban place, and (2) that spread effects dominate diffusion sequences when distance-decay gradients ($\propto$'s) are steep and interaction is locally circumscribed, but as distance frictions are reduced, diffusion becomes increasingly hierarchical and relative diffusion times are shortened. Therefore, not only do new growth impulses filter down the urban hierarchy, but the speed of diffusion is hastened, so that the entire urban system above threshold benefits from the same growth impulses reasonably rapidly.

THE POLICY DEBATE

Ironically, the causes of primacy suggest that in the debate on urban policy the modernizers are the conservatives and the traditionalists are the radicals.

The Views of the Modernizers

The finding that primacy is normally distributed with respect to level of development and that, given time, there appears to be a point when filtering mechanisms reassert themselves, leads the modernizers to conclude that policies directed at efficient national growth will suffice, for urban and regional structure will follow along and ultimately reach a harmonious steady-state. Conservatively they say "Do nothing, for time will take care of the situation." To be sure, they admit, in the early stages of development the advantage lies with the developed centers, the large cities of the national heartland, or the centers of colonial penetration which enjoy the existence of overhead facilities, external economies, political power, spatial preferences of the decision-makers, immigration of the more vigorous and educated elements from the underdeveloped regions, flows of funds from the land-wealthy in the hinterlands to the financial markets in the cities, and a variety of other factors. These factors lead to polarization, to concentration in the large cities, and to increases in the differences of regional incomes. Trickle-down mechanisms will reassert themselves, they go on, when the spread of literacy and bureaucratic practices improve knowledge in and about the backward areas, when the opening of transportation routes to reach these areas as markets for the developed centers also opens them as possible locations for productive activities, and when universal education and standardization of all aspects of life permit an integration of the space-economy—and by making externalities more nearly comparable everywhere, the more distant opportunities become more accessible and more interesting for development. Thus, in this view, in the early stages of development there will be increasing disparity between developed and underdeveloped regions, but there will be a tendency toward equalization as the economy reaches maturity.

Further, the modernizers argue, it is by no means clear that primacy or, indeed, size itself have the evil implications normally assumed by the planners of South and Southeast Asia. They cite increasingly strong evidence that some large threshold size must be reached for an urban center to free itself from the uncertainties of a narrow export basis for growth—in effect by achieving the necessary internal scale for self-sustaining growth. In North America a variety of studies points to a size of not less than 250,000 (see the threshold in Figures 1 and 2). In India, a study of the costs of urban infrastructure for industry completed by Stanford Research Institute reveals a very strong localization of industry within different size-ranges consistent with the labor-market formulation of trickle-down processes, and cost curves generally reversed-J in shape that declined steeply to the 130,000-200,000 size class and remained level thereafter. Another Indian study by Cherukupalle Nirmaladevi reveals systematic changes in demographic characteristics up to a size of 250,000, but no significant differences explainable other than by variations in manufacturing specialization thereafter.

The Views of the Traditionalists

Such materials fail to impress the traditionalists, however. They argue that continued change within the framework of present economic systems is insufficient and that radical changes in the nature of urban systems must be introduced by fiat to overcome continuing constraints of size, low rates of growth, export orientation and colonial inheritance that conspire to confine modernizing influences to primate cities with little foreseeable hope for filtering to take place.

They attack the modernizers using Kuznets' arguments that historical experiences of developed nations are irrelevant in Asia today. The developed countries of today were economically ahead of the "rest of the world" when they commenced modern economic growth and the underdeveloped countries are the worst off economically at the present time. The world of the eighteenth and nineteenth centuries had freer trade policies, freer opportunties for international movements of population and less political and economic barriers than the world of today. Modern economic growth was associated with skewed income distributions in its initial phases both with respect to wage earners and profit earners on the one hand and with respect to agriculturists as against manufacturers on the other. Economic growth thus involves a transfer of resources from agriculture to industry and proceeded at the expense of agriculture. It involved a transfer of resources to the upper middle classes which have a greater propensity to save and hence to invest. The transfer was also from urban wage earners. The extent to which resources transfer to a richer class (with a greater propensity to save) is politically feasible today is debatable.

Further, in the underdeveloped countries today institutional settings which are the products of modern economic growth of the developed countries have preceded the process of growth. The political institution of democracy, for instance, often precludes exploitation of the proletariat for the process of growth, a feature which was not uncommon in the initial stages of growth of the west. Welfare measures like minimum wages, regulation of hours of work, prohibition of child labor, are all institutions which in certain ways do not make for efficient economic development of the sort that took place under private entrepreneurship. This is compounded by the coexistence of highly developed and underdeveloped countries with rapid means of communication and transportation that enable peoples of the underdeveloped countries to copy consumption patterns of the developed countries. This "demonstration effect" is asymmetrical, i.e. it applies to consumption but not to investment and saving patterns, thus diverting resources from investment to conspicuous and other types of consumption. Parallel "derived development" projects divert investments to conspicuous and monumental structures rather than to productive ones.

The Planners' Compromise: Growth Centers

For each of these reasons, an increasing number of urban and regional planners have begun to argue for programs of decentralization into new growth centers. They coopt the traditionalists by pointing to the advantages of being latecomers in the development process. "Mistakes" made in the western nations can be avoided and limited resources diverted to more productive uses. For example (feeding on the traditionalists' fears) they say that decentralized patterns of urbanization, with emphasis on medium sized cities, probably will cost less in infrastructural investment while avoiding the social perils of gigantism. They go on to point out to the traditionalists the institutional advantages that did not exist heretofore. There is greater acceptance of the role of the public sector and of organized bureaucracies in economic development. International agencies provide assistance to the underdeveloped areas in planning their economic growth. There can be greater rationality in both the process and the outcome of development. New regional structures can be created that overcome the constraints of the past and the limitations of the present.

To the modernizers, the planners sing another tune, however. Strategies of "downward decentralization" are argued in this arena less to contribute to equity goals in a restructured society than to further rapid growth serving efficiency goals by creating a system of alternative growth centers. These, it is said, bring new or underutilized resources into the development process. Further, allocation of public investments to simulate the filtering process will bypass current constraints by creating alternative magnets for migrants, reducing pressures on the large-city labor market, hence enabling the "natural" filtering process to reassert itself as long as overall national growth is maintained at high levels and attempts to reduce the natural rate of population increase are successful.

Both modernizers and traditionalists appear to have been persuaded by these arguments, and the viewpoint also is attractive to local politicians concerned with obtaining the greatest possible share of development funds for their constituencies. As a result, growth pole *cum* decentralization policies appear to have achieved the status of the latest planning fad. As with all such fads, evaluation must now await the test of time and practice.

MATHEMATICAL APPENDICES

Appendix A. Proof that S People Randomly Distributed in N Cities Will Not Result in the Rank-size Distribution

The situation to be considered is one in which a total urban population of size S is distributed randomly among N cities. Average city size will be $\bar{P} = S/N$. If a pure random process operates in which the probability that an urban resident will locate in a given city is $1/N$, the probability that a city will have a population P will be given by the binomial probability law:

$$\text{pr}(P) = \binom{S}{P} \left(\frac{1}{N}\right)^{P} \left(\frac{N-1}{N}\right)^{S-P} \tag{A1}$$

For large values of N, this expression has the Poisson distribution as its limiting form:

$$\mathrm{pr}(P) = \frac{e^{-\bar{P}}\bar{P}^{P}}{P!} \tag{A2}$$

This skewed distribution has led some authors to conclude that typically skewed arrangements of city sizes are purely random in origin. However, when S is large relative to N, the Poisson converges on the normal distribution. Expected city populations are then a uniform $\bar{P}$, deflected from this value only by a random normal deviate. *When a large number of urban dwellers is distributed at random among a large number of urban centers, the outcome is therefore a set of city sizes varying randomly from a common norm.* "Randomness," of course can be interpreted as a situation in which there are many factors present working in many ways to affect the sizes of cities, with none of them individually strong enough to pattern the outcome.

Appendix B. Proof that the Rank-size Distribution is the Product of a Growth Process Embodying the Law of Proportionate Effect

For convenience assume an initial situation in which the urban population is distributed randomly among a set of cities (this is for convenience only; it is unnecessary), and the average population of the cities is relatively small. Further, assume that the populations of the cities are growing as a result of the operation of a large number of independent growth influences, that in very small time intervals city growth is small relative to city size, and that absolute growth is proportional to city size (i.e. the growth rate is the same for each size class of cities; this is called *Gibrat's law of proportionate effect*). Let P_t be the population of a city at time t. Then we can write that relative growth G_1 is:

$$G_1 = (P_1 - P_0)/P_0 \tag{B1}$$

i.e.
$$P_1 - P_0 = G_1 P_0 \tag{B2}$$

$$P_1 = G_1 P_0 + P_0 = (G_1 + 1)\, P_0 \tag{B3}$$

Following $t = n$ time intervals, because G_1 is independent of P_0:

$$\sum_{t=1}^{n} G = \sum_{t=1}^{n} [(P_t - P_{t-1})/P_{t-1}] \tag{B4}$$

but because we are dealing with small increments, this can be approximated in continuous form as:

$$\int_{P_0}^{P_n} dp/P = \ln P_n - \ln P_0 \tag{B5}$$

which yields:

$$\ln P_n - \ln P_0 = \sum_{t=1}^{\overset{n}{}} G_t \qquad (B6)$$

or:

$$\ln P_n = \ln P_0 + G_1 + G_2 + \ldots + G_n \qquad (B7)$$

so that of course:

$$P_n = P_0 \, (1 + G_1) \, (1 + G_2) \ldots (1 + G_n) \quad (B8)$$

By the additive form of the central-limit theorem, log P_n is asymptotically normally distributed and hence P_n is asymptotically lognormally distributed. *Cities growing according to the law of proportionate effect ultimately develop a lognormal size distribution.*

The probability density function (p.d.f.) of cities according to their populations is therefore

$$f(x) = (1/\sqrt{2\pi}) \exp. \,[\tfrac{1}{2} \, x^2] \qquad (B9)$$

where x is the standard normal deviate

$$x_i = (\log P_i - \mu)/\sigma \qquad (B10)$$

and μ and σ are the mean and standard deviation of the log P's, respectively. For negative values of x one uses the fact that $f(-x) = f(x)$.

In equations B9 and B10 the linear relationship between the log P's and their standard normal deviates, the x's, is evident. The quantile relationship is log $P_i = \mu + x_i \sigma$. The cumulative distribution function (c.d.f.) is

$$F(x) = \int_{-\infty}^{\times} \frac{1}{\sqrt{2\pi}} \exp. \,[-\tfrac{1}{2} t^2] \, dt. \qquad (B11)$$

For negative values $F(-x) = 1 - F(x)$. Equation B11 shows that a linear relationship exists between the logarithm of the cumulative proportion, $F(x)$, of cities and the standard normal deviates. From equations B9 and B11, therefore, there is a linear relationship between the logarithm of the cumulative proportion of cities $F(x)$ and the logarithm of their population log P:

$$P = a \, [F(x)]^b \qquad (B12)$$

which is a Pareto distribution with intercept a and exponent b.

Now define rank r to be the ordinal position of the city of population P_r in a ranking of n cities in decreasing order of size $P_1, P_2, \ldots, P_r, \ldots P_n$. This index of rank is formally equivalent to $N(P_i)$, the number of cities of population P_i or greater, and, of course $1 - N(P_i)/n$ is the cumulative proportion of cities of population P_i or less, or $F(x)$. Hence:

$$P = a \, [1 - \frac{N(P_i)}{n}]^b = a \, [1 - r/n]^b \qquad (B13)$$

and

$$n - r = (a/n)P^b \qquad (B14)$$

But since $1 - r/n$ is the complement of $F(r)$, rank itself will bear a linear inverse relationship to the log P_i's. Hence,

$$Pn = Kr^{-q} \qquad (B15)$$

This is *Zipf's rank-size distribution*. In the classical (U.S.) case $K \approx P$, and $q \approx 1.0$. *A rank-size distribution of cities thus arises when cities grow over a span of time according to Gibrat's law of proportionate effect; an equivalent statement is that the steady-state is a lognormal distribution of city sizes.*

Appendix C. Demonstration that a Regular Central-place Hierarchy Generates a Rank-size Distribution of City Sizes when City Populations are Affected by a Random Normal Deviate

A rank-size distribution can also arise from a randomly disturbed central-place hierarchy. Let P_1 be the population of the lowest level of centers in a regular central-place hierarchy, m the rural population served by a center of such level, P_n the population of a center of level n, M_n the total market area population of such a center, and k the ratio of centers of level $n - 1$ to those of level n. Assume

$$P_1 = u(P_1 + m) \qquad (C1)$$

so that $u/1 - u$ is the urban multiplier, i.e.:

$$P_1 = \frac{u}{1 - u} m \qquad (C2)$$

and:

$$P_n = uM_n \qquad (C3)$$

then, following from the nesting relationship of higher and lower level centers:

$$M_n = kM_{n-1} + M_{n-1} - P_{n-1} + P_n \qquad (C4)$$

Verbalizing, equation C4 says that the total population served by a center of nth level comprises the populations served by k centers of level $n - 1$ (i.e. kM_{n-1}), plus the total population which the nth level center serves in its role as the $k + 1$st $n - 1$st level center (i.e. M_{n-1}), less the population it would have if it were only a center of level $n - 1$ (i.e. P_{n-1}), plus its population because it is a level n center (P_n).

From equation C4, it follows that

$$M_n = (\mathrm{k} + 1)M_{n-1} - uM_{n-1} + uM_n \qquad \text{(C5)}$$

$$M_n = \frac{1 + k - u}{1 - u} M_{n-1} \qquad \text{(C6)}$$

$$= \frac{(1 + k - u)^{n-1}}{1 - u} M_1 \qquad \text{(C7)}$$

Thus

$$P_n = \frac{(1 + k - u)^{n-1}}{1 - u} P_1 \qquad \text{(C8)}$$

Both center size (P_n) and total market area population (M_n) therefore are exponential functions of rank in the hierarchy; the change in size from one rank to another is the constant growth factor $(1 + k - u)/(1 - u)$. *The pure hierarchy model results in city populations of each hierarchical level arranged in steps along the rank-size graph.* For example, if $u = 0.5$ and $k = 4$, the growth factor $(1 + k - u)/(1 - u)$ is 9 and from level to level center populations increase by approximately an order of magnitude. If the largest center has a population of ten millions, there is room for seven levels in the hierarchy.

Introduction of a disturbance term randomly deflecting the populations of centers of each level from the expected value for that level leads to convergence of the hierarchical steps and the rank-size distribution. As either k (the number of centers of lower level dependent on a higher-level center, i.e. the degree of centralization of control) or u (essentially the degree of urbanization) increases, so does the growth factor, and therefore also the degree of skewness of the city-size distribution. Hierarchical organization, urbanization, and the city-size distribution are therefore completely interrelated.

Appendix D. An Extension of the Lognormal Distribution to an Expanding System of Cities; The Yule Distribution

Urban growth obeys Gibrat's law if year-to-year changes in city sizes are governed by simple Markov process in which probabilities of the size change in any period are independent of the city's present size—generally, if the expected percentage change in the population of cities in each size stratum is independent of stratum and therefore the same for all strata, under a further assumption that the number of cities, n, is fixed. A simple way to test whether this condition holds is to plot the scatter diagram of cities by size for time t and time $t + 1$ on doubly logarithmic paper. If the regression line slopes up to the right at an angle of 45 degrees, the law of proportionate effect holds. Displacement of the intercept indicates the mean growth rate of the system of cities. Strictly, the plot should be homoscedastic, but the result will be the same if relative variances (and therefore the transition matrices) differ by size strata, and even if expected rates of change of individual cities are serially correlated over certain time periods. Changes in variance by size stratum are consistent with the observation that the relative variance of urban growth rates characteristically decreases with size, with the distribution rapidly converging on the national urban mean growth rate for centers of metropolitan status.

Often, it will be useful to add a second assumption to that of the law of proportionate effect, following from the idea that there is some *minimum threshold size* for a city, and that new cities are "born" as smaller places grow to exceed the threshold. A simple check for such a threshold is to look for a "break" to a steeper slope in a rank-size graph at a population beneath which the number of smaller places rapidly "tails off." Such a "break" is suggested at approximately 250,000 population for the functional economic areas of the United States in 1960 in Figure 1. This second assumption is that of a constant "birth rate" of new cities moving above the threshold level, and it differentiates the *Yule distribution* (named after the British statistician G. Udny Yule, who first proposed it) from the lognormal distribution. Translated into rank-size form, the Yule distribution can be written

$$N(P) = KP^{-q}b^P$$

This, of course, is the rank-size distribution, with the difference that as a result of the second assumption, $q = 1/(1 - S_c/S_T)$. S_T is the total (absolute) population growth of all cities in the system during some specified time period, and S_c is that part of growth attributable to the entry of new cities reaching threshold size in the period. Thus S_c/S_T is the fraction of total growth attributable to new cities. When S_c is small relative to S_T, q should converge on a value of 1.0, and unity is therefore the value of the exponent to be expected if one is dealing with a fixed number of cities (lognormal case) as opposed to the case in which the number of cities is growing. If S_c is substantial, as in the case of a system of cities rapidly expanding in numbers above threshold, q will exceed unity. If, on the other hand, the number of centers in the system is shrinking, either because of absolute losses or because the threshold size for admission to the system is being systematically raised, S_c will be negative, and q will be less than 1.0. In general, a value of q exceeding unity implies proportionately more small places, whereas q's of less than one arise with increasing concentration of the urban population in larger cities.

Appendix E. Relationships between the Lognormal Distribution, the Lorenz Coefficient of Concentration, the Gini Coefficient, and an Entropy Concept

The mean and variance of the log P's are $\mu + \sigma^2$, respectively, and the relative positions of the arithmetic mean, median, and mode of the P's are

$$\bar{P} = \exp. [\mu + \tfrac{1}{2}\sigma^2] \tag{E1}$$

$$P_{\text{med}} = e^\mu \tag{E2}$$

$$P_{\text{mode}} = \exp. [\mu - \sigma^2] \tag{E3}$$

The upper and lower quartiles are given by exp. $[\mu \pm 0.67\sigma]$. It follows, of course, that μ is the logarithm of geometric mean, and the quartile relationship is $\log P_i = \mu + x_i\sigma$.

The variance σ^2 increases with the skewness of the distribution, and in fact that standard deviation is directly related to the Lorenz coefficient of concentration L, viz:

σ	0.0	0.25	0.50	0.75	1.00	1.50	2.00	2.50
L	0.0	0.1405	0.2727	0.4309	0.5204	0.7108	0.8415	0.9233
σ		3.00	4.00					
L		0.9660	0.9953					

When L is zero, urban population is evenly distributed over the size strata of cities. A value of unity indicates complete concentration in the largest size category. The data in Table 1 therefore suggest increasing σ with increasing modernization.

In turn, the Lorenz coefficient is directly related to Gini's coefficient of mean difference G, sometimes also used as a concentration coefficient:

$$G = 2\bar{P}L \tag{E4}$$

There is a further relationship that may be noted between L and a measure of entropy. If we write the individual probabilities of the urban population being found in each size stratum of cities as $p_1, \ldots, p_s$ for s strata (the p's sum to unity), the entropy of the distribution is, according to Theil:

$$H(p) = \Sigma_i p_i \log(1/p_i) \tag{E5}$$

$H(p)$ is zero when one p is unity and the rest are zero, and it assumes a maximum value of $\log s$ when all the p's are $1/s$. Now clearly, the former is true when there is complete concentration of the urban population in one size stratum (L approaches 1.0 and σ increases beyond 4.0), whereas the latter reflects complete evenness ($\sigma = L = 0.0$). Therefore, an increasingly skewed distribution is one in which order (the inverse of equiprobable entropy) is increasing, and a measure of order or concentration C is

$$C = 1 - [H_{\text{obs}}/H_{\text{max}}] \tag{E6}$$

where H_{obs} is the entropy of a given system and H_{max} is the maximum attainable degree of entropy in that system.

REFERENCES

ALLEN, G.R., "The 'Courbe des Population': A Further Analysis." *Bulletin of the Oxford University Institute of Statistics,* Vol. 16, 1954, pp. 179–189.

AITCHISON, JOHN AND J.A.C. BROWN, *The Lognormal Distribution.* Cambridge, England: Cambridge University Press, Department of Applied Economics, Monograph No. 5, 1957.

ALONSO, WILLIAM, "Urban and Regional Imbalances in Economic Development." Unpublished manuscript, Harvard Regional Development Program, 1967.

ALVES MARTIN, C.M., *Una Nova Concepçao de Ecologia Umana.* Lisbon: Instituto de Alta Cultura, 1952.

———, "A Sociometric Approach to Human Ecology." Lisbon: University of Lisbon, 1955.

AUERBACH, FELIX, "Das Gesetz der Bevölkerungskonzentration." *Petermann's Mitteilungen,* Vol. 59, 1913, pp. 74 ff.

BECKMANN, MARTIN J., "City Hierarchies and the Distribution of City Size." *Economic Development and Cultural Change,* Vol. 6, 1958, pp. 243–248.

———, *Location Theory.* New York: Random House, 1968.

BELL, GWEN, "Change in City Size Distribution in Israel." *Ekistics,* Vol. 13, 1962, p. 103.

BERRY, BRIAN J.L., "City Size Distribution Development." *Economic Development and Cultural Change,* Vol. 9, 1961, pp. 673–87.

———, "Urbanization and Basic Patterns of Development," Forrest R. Pitts (ed.), *Urban Systems and Economic Development.* Eugene, Oregon: School of Business Administration, University of Oregon, 1962.

———, "Cities as Systems Within Systems of Cities," in John Friedmann and William Alonso (eds.), *Regional Development and Planning.* Cambridge, Mass.: M.I.T. Press, 1964, pp. 116–137.

———, *Geography of Market Centers and Retail Distribution.* Englewood Cliffs: Prentice-Hall, 1967.

———, "Policy Implications of an Urban Location Model for the Kanpur Region," in *Regional Perspective of Industrial and Urban Growth.* Bombay: Macmillan, 1969, pp. 203–219.

———, "Interdependency of Spatial Structure and Spatial Behavior: A General Field Theory Formulation." *Papers of the Regional Science Association,* Vol. 21, 1968, pp. 205–227.

BERRY, BRIAN J.L. and WILLIAM L. GARRISON, "Alternate Explanations of Urban Rank-Size Relationships." *Annals of the Association of American Geographers,* Vol. 48, 1958, pp. 83–91.

BERRY, BRIAN J.L. and ALLEN PRED, *Central Place Studies: A Bibliography of Theory and Applications.* Philadelphia: Regional Science Research Institute, Bibliography Series, No. 1, 1961.

BOAL, F.W. and D.B. JOHNSON, "The Rank Size Curve: A Diagnostic Tool?" *The Professional Geographer,* Vol. 17, 1965, pp. 21–23.

BROWNING, CLYDE E., "Primate Cities and Related Concepts," in Forrest R. Pitts (ed.), *Urban Systems and Economic Development*, Eugene, Oregon: School of Business Administration, University of Oregon, 1962, pp. 16–27.

BROWNING, H.L., "Recent Trends in Latin American Urbanization." *The Annals of the American Association of Political and Social Science*, Vol. 316, 1958, pp. 111–120.

BRUSH, J.E., "The Urban Hierarchy in Europe." *Geographical Review*, Vol. 43, 1953.

——, "The Hierarchy of Central Places in Southwestern Wisconsin. *Geographical Review*, Vol. 43, 1953, pp. 380–402.

CARROLL, JOHN B., "Zipf's Law of Urban Concentration." *Science*, 94 (December 26, 1941), p. 609.

CHERUKUPALLE, NIRMALADEVI, "Demographic Correlates of Urban Size in India." Ph.D. dissertation, Harvard University, 1967.

CLARK, COLIN, *Population Growth and Land Use*. New York: St. Martin's Press, 1967.

CURRY, LESLIE, "The Random Economy: An Exploration in Settlement Theory." *Annals of the Association of American Geographers*, Vol. 54, 1964.

DAVIS, KINGSLEY and HILDA HERTZ GOLDEN, "Urbanization and the Development of Preindustrial Areas." *Economic Development and Cultural Change*, Vol. 3, 1955, pp. 6–26.

DUNCAN, OTIS DUDLEY, "Population Distribution and Community Structure." *Cold Springs Harbor Symposia on Quantitative Biology*, Vol. 22, 1957, pp. 357–71.

DZIEWONSKI, K., "Urbanization in Contemporary Poland." *Geographia Polonica*, Vol. 3, 1964, pp. 37–56.

EL SHAKS, S., "Development, Primacy and the Structure of Cities." Ph.D. dissertation, Harvard University, 1965.

FANO, PIETRO L., "Organization, City Size Distributions and Central Places." Centro di Studi e Documentazione per la Pianificazione Territoriale, Universita' di Roma, 1968.

FISHER, JACK C., *Yugoslavia: A Multinational State: Regional Variation and Political Response*. San Francisco: Chandler, 1966.

FRIEDMANN, JOHN and THOMAS LACKINGTON, "Hyperurbanization and National Development in Chile: Some Hypotheses." *Urban Affairs Quarterly*, Vol. 2 (June, 1967), pp. 3–29.

FRIEDMANN, JOHN, *Regional Development Policy: A Case Study in Venezuela*. Cambridge, Mass.: M.I.T. Press, 1966.

FRYER, D.W., "The Million City in Southeast Asia." *Geographical Review*, Vol. 43, 1953, pp. 474–494.

GIBBS, JACK P. and HARLEY BROWNING, "Systems of Cities," in Gibbs (ed.), *Urban Research Methods*. Princeton: Van Nostrand, 1961, pp. 436–61.

GIBRAT, R., *Les inégalités économiques*. Paris: 1931, pp. 250–252.

GINSBURG, NORTON S., "The Great City in Southeast Asia:" *American Journal of Sociology*, Vol. 60, 1955, pp. 455–462.

——, *Atlas of Economic Development*. Chicago: University of Chicago Press, 1961.

GOODRICH, ERNEST, "The Statistical Relationship Between Population and the City Plan," in E.W. Burgess (ed.), *The Urban Community*. Chicago: 1926, pp. 144–150.

——, "Mass and Density of Building," in *International Town Planning Conference*. Paris, 1928, Papers, Part I.

HARRIS, BRITTON, "Urbanization Policy in India," in *Papers and Proceedings of the Regional Science Association*, Vol. 5, 1959, p. 192.

HAUSER, PHILIP M. (ed.), *Urbanization in Asia and the Far East*. Calcutta: UNESCO, 1957.

——, *Urbanization in Latin America*. New York: UNESCO, 1961.

HIRSCHMAN, A.O., *The Strategy of Economic Development*. New Haven: Yale University Press, 1958.

HOOVER, EDGAR M., "The Concept of a System of Cities: A Comment on Rutledge Vining's Paper." *Economic Development and Cultural Change*, Vol. 3, 1954, pp. 196–198.

HOSELITZ, BERT F., "Generative and Parasitic Cities." *Economic Development and Cultural Change*, Vol. 3, 1955, pp. 276–294.

——, "Urbanization: International Comparisons," in Roy Turner (ed.), *India's Urban Future*. Berkeley: University of California Press, 1962, pp. 157–181.

IJIRI, YUJI and HERBERT A. SIMON, "A Model of Business Firm Growth." *Econometrica*, Vol. 35 (April, 1967), pp. 348–355.

INTERNATIONAL URBAN RESEARCH, *The World's Metropolitan Areas*. Berkeley and Los Angeles: University of California Press, 1959.

ISARD, W., *Location and Space Economy*. New York: John Wiley, 1956, p. 56.

JEFFERSON, MARK, "The Distribution of People in South America." *Bulletin of The Geographical Society of Philadelphia*, Vol. 5, 1907.

——, "The Anthropography of Some Great Cities." *Bulletin of the American Geographical Society*, Vol. 41, 1909, pp. 537–566.

——, "New York and the Four Next Largest American Cities in 1910." *Bulletin of the American Geographical Society*, Vol. 42, 1910, pp. 778–779.

——, "How American Cities Grow." *Bulletin of the American Geographical Society*, Vol. 47, 1915, pp. 19–37.

——, "The Distribution of British Cities and the Empire." *Geographical Review*, Vol. 4, 1917, pp. 387–394.

——, "Great Cities in the United States, 1920." *Geographical Review*, Vol. 11, 1921, pp. 437–441.

——, "Distribution of the World's City Folk." *Geographical Review*, Vol. 21, 1931, pp. 446–465.

——, "Great Cities of the United States in 1930 with a Comparison of New York and London." *Geographical Review*, Vol. 23, 1933, pp. 90–100.

——, "The Law of the Primate City." *Geographical Review*, Vol. 29, 1939, pp. 226–232.

LAMPARD, ERIC E., "The History of Cities in the Economically Advanced Areas." *Economic Development and Cultural Change*, Vol. 3, 1955, pp. 81–136.

LASUEN, A.V., J.R. and A. LORCA J. ORIA, "City Size Distribution and Economic Growth." Unpublished manuscript, Universidad de Barcelona, 1966.

LINSKY, ARNOLD S., "Some Generalizations Concerning Primate Cities." *Annals of the Association of American Geographers,* Vol. 55, 1965, pp. 506–513.

LORENZ, M.O., "Methods of Measuring the Concentration of Wealth." *Publications of the American Statistical Association,* New Series, Vol. 9, 1904–05, pp. 209–210.

LOTKA, ALFRED J., *Elements of Physical Biology.* Baltimore: 1924, pp. 306–307.

——, "Frequency Distribution of Productivity." *Journal of the Washington Academy of Sciences,* Vol. 16, 1926, p. 323.

——, "The Law of Urban Concentration." *Science,* Vol. 94, 1941, p. 164.

MACGREGOR, D.N., "Pareto's Law." *The Economic Journal,* Vol. 46, 1936, pp. 80–88.

MADDEN, CARL R., "Some Indications of Stability in the Growth of Cities in the United States." *Economic Development and Cultural Change,* Vol. 4, 1956, pp. 236–252.

MALIZIA, EMIL E., "Observed City Size Distribution and Regional Economic Growth and Development in the Northeast." Unpublished M.R.P. Thesis, Cornell University, 1966.

MARTIN, GEFFREY J., "The Law of the Primate Cities' Re-Examined." *Journal of Geography,* Vol. 60, 1960, pp. 165–172.

MCGEE, T.G., *The Southeast Asian City: A Social Geography of the Primate Cities of Southeast Asia.* London: G. Bell, 1967.

MEHTA, SURENDER K., "Some Demographic and Economic Correlates of Primate Cities: A Case for Re-Evaluation." *Demography,* Vol. 1, 1964, pp. 136–147.

MOORE, FREDERICK T., "A Note on City Size Distribution." *Economic Development and Cultural Change,* Vol. 7, 1958–59, pp. 465–466.

MORRILL, RICHARD L., "Simulation of Central Place Patterns over Time." *Lund Studies in Geography,* No. 24, 1962, pp. 109–120.

MURPHEY, RHOADS, "New Capitals of Asia." *Economic Development and Cultural Change,* Vol. 5, 1957, pp. 216–243.

MYRDAL, G., *Rich Lands and Poor.* New York: Harper, 1957.

NORDBECK, STIG, "The Law of Allometric Growth." *Michigan Inter-University Community of Mathematical Geographers,* Discussion Paper 7, 1965, pp. 1–28.

PEDERSEN, POUL OVE, "Innovation Diffusion in a National Urban System." Unpublished manuscript, 1968.

PERROUX, F., "Consideraciones en torno a la noción de polo de crecimiento." *Cuadernos de la Sociedad Venezolana de Planificación,* Vol. 2 (June–July 1963), Nos. 2–3.

RASHEVSKY, NICOLAS, "Development of Mathematical Biophysics in the U.S.A. from 1939–1945 Inclusive." *Relationes De Avetis Scientus Tempere Belli,* Pontificia Academia Scientiarum, Vol. 14, 1946, pp. 1–28.

——, *Mathematical Theory of Human Relations.* Bloomington, Ind.: Mathematics Biophysics Monograph Ser., No. 2, 1947.

——, "Mathematical Theory of Human Relations: VII, Outline of a Mathematical Theory of the Sizes of Cities," *Psychometrica,* Vol. 8, 1953, pp. 87–90.

——, *Mathematical Biology of Social Behavior.* Chicago: University of Chicago Press, 1959, pp. 148–156.

ROSING, KENNETH E., "A Rejection of the Zipf Model (Rank Size Rule) in Relation to City Size." *The Professional Geographer,* Vol. 18 (March, 1966), pp. 75–82.

SENDUT, HAMZAH, "City Size Distribution in Southeast Asia." *Journal of Asian Studies,* Vol. 4 (August, 1966), pp. 268–280.

SIMON, HERBERT A., "On a Class of Skew Distribution Functions." *Biometrika,* Vol. 42, 1955, pp. 425–440.

——, *Models of Man.* New York: John Wiley, 1957, pp. 145–164.

——, "The Size Distribution of Business Firms." *American Economic Review,* Vol. 48, 1958, pp. 607–617.

SINGER, H.W., "Courbes des Population: A Parallel to Pareto's Law." *Economic Journal,* Vol. 46, 1936, pp. 254–263.

SOVANI, N.V., "The Analysis of Over-Urbanization." *Economic Development and Cultural Change,* Vol. 12 (January, 1964), pp. 113–122.

STANFORD RESEARCH INSTITUTE, *Costs of Urban Infrastructure as Related to City Size in Developing Countries.* Menlo Park, Calif.: Stanford Research Institute, 1968.

STEWART, CHARLES T. JR., "The Size and Spacing of Cities." *Geographical Review,* Vol. 48, 1958, pp. 222–245.

STEWART, JOHN Q., "Empirical Mathematical Rules Concerning the Distribution and Equilibrium of Population." *Geographical Review,* Vol. 37, 1947, pp. 461–485.

——, "A Basis for Social Physics." *Impact,* Vol. 3, 1952 (UNESCO), p. 117.

STOLPER, WOLFGANG, "Spatial Order and the Economic Growth of Cities: A Comment on Eric Lampard's Paper." *Economic Development and Cultural Change,* Vol. 3, 1955, pp. 137–146.

SUNDRUM, R.M., "Urbanization: The Burmese Experience." Rangoon: Departments of Statistics and Commerce, University of Rangoon, Economic Paper No. 16, Economic Research Project, 1957.

TANGRI, SHANTI, "Urbanization, Political Stability and Economic Growth," in Roy Turner (ed.), *India's Urban Future.* Berkeley: University of California Press, 1962, pp. 192–212.

THEIL, HENRI, *Economics and Information Theory.* Amsterdam: North-Holland, 1967.

THOMAS, EDWIN N., "The Stability of Distance Population-Size Relationships for Iowa from 1900–1950." *Lund Studies in Geography,* No. 24, 1962, pp. 13–29.

——, "Additional Comments on Population-Size Relationships for Sets of Cities," in *Quantitative Geography.* Evanston: Northwestern University, Studies in Geography, 1967, pp. 167–190.

THORNDIKE, E.L., " 'National Unity and Disunity' A Review." *Science,* Vol. 94, 1941, p. 19.

——, [Comment following Lotka article which replied to his review of *National Unity and Disunity, op. cit.*]. *Science,* Vol. 94, 1941, p. 164.

ULLMAN, L., "A Theory of Location for Cities." *American Journal of Sociology,* Vol. 46, 1941, pp. 853–864.

ULLMAN, EDWARD L. "Regional Development and the Geography of Concentration." *Papers and Proceedings of the Regional Science Association,* Vol. 4, 1958, pp. 179–198.

UNITED NATIONS, *Report on the World Social Situation Including Studies of Urbanization in Underdeveloped Areas.* New York: 1957, Part II.

UNESCO, *Report by the Director-General on the Joint UN/UNESCO Seminar on Urbanization in the ECAFE Region.* Paris: 1956.

VAPNARSKY, CESAR A., *Rank Size Distribution of Cities in Argentina.* Unpublished M.A. Thesis, Cornell University, 1966.

VINING, RUTLEDGE, "A Description of Certain Spatial Aspects of an Economic System." *Economic Development and Cultural Change,* Vol. 3, 1954, pp. 147–195.

WARD, BENJAMIN, "City Structure and Interdependence." *Papers of the Regional Science Association,* Vol. 10, 1963, pp. 207–221.

WILLIAMSON, G.G., "Regional Inequality and the Process of National Development: A Description of the Patterns." *Economic Development and Cultural Change,* Vol. 13 (July, 1965).

WINGO, LOWDON, "Latin American Urbanization: Plan or Process?" in Bernard J. Frieden and William Nash (eds.), *Shaping an Urban Future.* Cambridge, Mass.: M.I.T. Press, 1969.

WURSTER, CATHERINE BAUER, "Urban Living Conditions, Overhead Costs, and the Development Pattern," in Roy Turner (ed.), *India's Urban Future.* Berkeley: University of California Press, 1962, pp. 281–287.

ZIPF, G.K., *National Unity and Disunity.* Bloomington, Ind.: Principia Press, 1941.

——, *Human Behavior and the Principle of Least Effort,* Cambridge: Addison Wesley, 1949.

CHAPTER 6

T.G. McGee

CATALYSTS OR CANCERS?
The Role of Cities in Asian Society

"For cities exist as points of economic articulation on the landscape, and their *raison d'etre* can only be understood in terms of the economic system they grew up to serve."

—Akin L. Mabogunje [1]

Introduction

The evaluation of the role of cities in such a vast region as South and Southeast Asia [2] poses many difficulties, not only because of the immense cultural diversity of these societies, but also because of the varying historical urban traditions and the variety of functional types of urban center. It is thus important to establish at the initial stage of this paper that it is possible to make valid regional-wide models concerning the urban processes in this region. Hoselitz is rather pessimistic concerning the prospects of constructing models which will allow cross-cultural generalization.[3]

"The difficulty in constructing even an ideal model of urban culture is due to the fact that its outstanding characteristic is its heterogeneity and that, therefore, sets of culture traits found in the urban centers of one country need not be repeated in those of another country."

The accuracy of Hoselitz's statement is hard to debate, for even the most casual observer can attest to the cultural diversity of these cities. The existence of this cultural heterogeneity has not, however, prevented researchers from attempting to devise model and classificatory schemes which attempt to group cities on the basis of such criteria as size, economic and political function, or even cultural roles.[4] The principal drawback of such efforts is that these models and classificatory systems tend to be based on one variable rather than on an attempt to build up a holistic model of the city. In earlier studies the author has attempted to move towards such holistic models, by following Sjoberg's twofold division of cities into pre-industrial and industrial states, and introducing a third model called the "colonial city" to explain the type of city which emerges in many of the Asian countries where the features of Sjoberg's two conceptual models appear to be juxtaposed.[5]

Not surprisingly these models have not remained uncriticized. Attention has been drawn to the fact that they do not successfully incorporate historical change; that they rely heavily on the technological variable; that they are heuristic rather than predictive; and finally that they do not take sufficient account of the processes of urbanization, in particular the relationship between cities and the nation or territory of which they form a part. From the point of view of this paper, which attempts to analyze the roles of cities in contemporary Asia, the last criticism is most telling, for the role of the city cannot be separated from the territory to which it is linked. Nor for that matter can a city's role be understood without clearly delineating the processes which are affecting the cities and the nation at large. In some cases such as Singapore and Hong Kong the City and State are synonymous, but in the majority of Asian nations the cities are only part of a much larger nation and in many cases represent only a small portion of a country in demographic and territorial terms. To reiterate the argument: to understand the role of cities it is first necessary to understand the processes operating within the country and internationally; and secondly, to build some conceptual model of the city incorporating its relationship to the countryside of these Asian societies which transcends the dichotomy between city and the nation at large. In later sections of the paper an effort will be made to come to terms with these necessary prerequisites for analysis.

Before this is attempted, however, some effort must be made to investigate the meaning and use of the term "role" when applied to cities. Superficially it would seem that any discussion of the roles of cities might be approached from two different viewpoints. The first approach, basically *descriptive,* would list the various *ideal* roles of cities and contrast them with the *actual* roles of cities in the Asian region. Thus, for instance, the economic, political, social, and cultural roles of cities would be considered. A second approach would be more analytical; that is to say, not only are the roles of cities described but a theoretical model is developed as a basis for investigating the question of *why* cities play certain roles.

This paper adopts the analytical approach for three reasons. First, writers utilizing the descriptive approach seem to have constantly misinterpreted the meaning of the concept of role. In itself this is not surprising, since role is a concept which (along with many other terms in the social sciences) lacks specificity and thus is open to many interpretations. The author would argue, however, that the principle weakness in its use by the writers who have utilized the descriptive approach is the fact that it has been used interchangeably with the concept of function—which has led to a misinterpretation of the roles of cities. In sociology, a common definition of "role" is given by Smelser as ". . . selected aspects of inter-action among persons." [6] In rather more general terms it would seem that role can be also used to describe the "identity" that a person carries, such as businessman or husband, etc.; in other words how he is identified, or identifies himself within a society. Whether one may extend this concept of role to such a complex sociological phenomenon as a city is open to question, but there seems little doubt that people both within and outside cities do identify cities in just these terms. For the Indian rural migrant about to move to Calcutta, the city is identified as a city of "hope" for the future. Why else would he move?

On the other hand, a "function" in Riggs' words is the ". . . conse-quences of a structure insofar as they affect other structures or the total system of which they are a part." [7] and thus it is possible to classify social structures in terms of the functions they offer or perform. Thus a city, which is a social structure, performs a variety of economic, political, and social functions. In the light of these two definitions there would seem to be a marked difference in the meaning of these two concepts. Not only is a role an aspect of interaction, it is also an expected form of interaction, with certain predetermined qualities. Thus, if an individual is a business-man he will be expected to act like a businessman. In a similar fashion, a city will be expected to perform the role or roles of a city. The actual functions of a city will reflect the structures of that city and may be very different from the identifying role of that city. Thus, to confuse role and function when discussing cities is an unfortunate mistake. Yet it is all too common in the work of the descriptive approach. For example, consider Breese's discussion of the role of cities in the developing countries in which he uses the concepts "role" and "function" interchangeably: "A third existing function or role of urban centers in newly developing coun-tries is as magnets for the national population." [8] In this particular context how can role and function be used interchangeably? In terms of *role* the city is a *mecca*—an attraction point for perceived economic and social opportunity—this is the identity of the city in the minds of the national population. The city *functions,* however, as a *magnet.* Two distinct actions are involved, perception and movement, and it is important to keep them apart.

A second reason for discarding the descriptive approach is that, even when writers of the descriptive approach utilize the concept of role cor-rectly, they tend to identify roles in terms of their experience of roles of cities in Western Europe and the United States.[9] As a logical technique of the social sciences there is nothing wrong with this approach, but it does have a built-in assumption that the roles and processes creating cities in Asia are the same as those which characterized Western Europe and the United States. For instance Reissman states: [10]

"The urban process has begun in the rural and undeveloped countries of
the world. From what is now known about it, that process of urbanization is strikingly close to the lines followed in the West a century and
half before."

For various reasons which will be elaborated in greater detail in the next
section I believe this assumption to be unacceptable in the majority of
Asian countries and thus inadequate as a basis for the investigation of
the roles of cities.

Finally, I have discarded the descriptive approach because it has already been fully exploited in a series of papers.[11] To repeat it would place
nothing new before the reader.

This paper then adopts an analytical approach. In utilizing this approach there are built-in assumptions as well. The first assumption is that
cities are only parts of a total social and economic system which is not
only national but also international. Thus the roles of cities as seen by
people within the nation may be different from those viewed by people
living outside the nation. In a like fashion the functions of cities are a
consequence not only of their structure but of the total system of which
the city is a part. In Takeo Yazaki's words "We view the city as a sub-
system and grasp its structural relationships to the total social system." [12]
It is only through the understanding of this facet that the relative importance of city roles and functions can be assessed. A second assumption of
this approach is that emphasis is placed on function rather than role. (In
a sense then the title of this paper is a misnomer!) Again, it would be
useful to analyze the roles of cities in Asia in terms of their expected roles
in the light of the policies within these countries. This would be most
interesting in countries such as India and Malaysia, where government
policy-makers have an exceptionally ambivalent attitude with respect to
the role of cities—not unjustified, it would seem, in the case of Malaysia
—but in view of our attempt to build a theoretical model which will have
some predictive powers it seems more important to concentrate on the
functions of these cities.

Since it is basic to our argument that the structure of Asian cities
should be understood as a basis for model building, and since the socio-
economic structures of these cities are a consequence of the processes
operating at a national level, it is necessary to delineate the features of the
urbanization process in the Asian region.

Characteristics of the Urbanization
Process in Southeast Asia [13]

There is by now common agreement that the term "urbanization
process" is a concept which urbanologists may manipulate in many ways
to incorporate their particular viewpoints. The "urbanization process,"

for instance, may be simply defined as "the growth of cities" or more grandly designated as a process of "societal change." Lampard's definition of the urbanization process as ". . . a way of ordering a population to attain a certain level of subsistence and security in a given environment," [14] is perhaps a little broad, but it does carry the main conceptual meaning of a spatial reorganization of society.

Whatever the debates on the definition of the urbanization process, there is general agreement on the measure of its end product—the urbanization level—which is generally assumed to be the proportion of the total population of a country (or any designated statistical unit) resident in urban places.[15] Thus an increase in the level of urbanization is the symptom of the successful operation of the urbanization process.

It is tempting, after a superficial evaluation of the end product of the urbanization process in Southeast Asia, to suggest that the Western experience of urbanization is being repeated. Since 1945 the rate of urbanization in many countries has been almost double that of the total population, although in the case of some countries (notably the Philippines and Indonesia) this rate of urbanization has not led to a rapid increase in the level of urbanization.[16]

It cannot be denied, however, that individual cities are undergoing booming rates of increase. What is more, the largest metropolitan areas are growing at rates greatly in excess of the smaller cities. Thus if the rates of growth of the first and second largest cities in selected Southeast Asian countries are compared, it is obvious that in most cases the larger city is growing faster than the second city.

Exceptions to this generalization occur when the larger colonial political units are split up due to independence. Thus Kuala Lumpur has grown at almost double the rate of Singapore. Similar patterns characterized Phnom-Penh and Saigon, the former trading capital of French Indo-China before the Viet Nam War was accelerated, turning it into a gigantic refugee camp.

Thus it would appear that Southeast Asia, statistically at least, is conforming to the Western experience. It would not appear unreasonable then to analyze the Southeast Asian urbanization process in terms of Western-based theory. But when we turn to a closer analysis of the facets of the urbanization process (not the end product) considerable doubt arises as to their similarity to the West. First, consider the demographic components of the process. These are difficult to establish with any degree of accuracy in the Southeast Asian context, particularly since the quality of demographic data varies markedly from country to country. Isolated studies would appear to indicate that boundary expansion of existing cities and the reclassification of rural settlements as urban is of some consequence, inflating the figures of urban growth. In the case of large metropolitan centers such as Kuala Lumpur, where such boundary extension incorporated genuine suburban development, it is a valid process of urban growth; [17] but in other cases such as the Chartered Cities of the Philippines (which often include large rural populations),[18] this represents no real increase of urban population. The basic difficulty involved in analyzing this part of the demographic process is that of estimating the relative contribution of boundary expansion, or new urban creation to overall urban population growth, and on this point census authorities are notably reticent and ill-equipped to aid the researcher.

The contribution of natural increase is even more difficult to ascertain, for adequate figures on rural-urban differentials are scarce. With respect to births, a similar occurrence of declining rates, induced by city residence, is said to be occurring in Southeast Asia.[19] There is evidence to support this statement in the case of Singapore.[20] Milone suggests that this may also be true for Indonesia: [21]

"To date, though all census tables have not been completed, there is strong evidence (based on a 1% sample of age data) that there is a lower birth rate in the urban than in the rural areas for the whole of Indonesia (urban 38.5 per 1,000; rural 43.8 per 1,000) . . . there is also a firm indication from certain unpublished Greater Djakarta statistics of a lower degree of fertility among the more educated and urbanised population of the cities."

Similar evidence on the last point has been presented by Amos Hawley in the case of the dominantly Catholic Philippines.[22] There thus appears some evidence to support the argument of urban-induced decline in fertility.

It is, however, the question of rural-urban differentials in mortality which is of more significance since, as I have already pointed out, the high mortality rates of the Western city populations during the early phases of urban growth was an important factor allowing rural migrants to move into the city and find jobs. Here there is little doubt that the advent of the medical revolution has led to a substantial drop in crude death rates and infant mortality rates in the cities of Southeast Asia. For instance, in Singapore the crude death rate fell from 24.2 per 1,000 population in 1931 to 6.4 in 1959. The decline in the infant mortality rate was even greater— from 191.3 per 1,000 live births in 1931 to 36 per 1,000 in 1959. In general the urban death rates are much lower than the rural areas throughout the region. Even in a region such as Java where urban areas appear superficially to be little more than congested extensions of the rural areas, Milone reports that the urban death rate at 15.9 per 1,000 ". . . is appreciably lower than that of the Javanese rural areas (22.6)." [23] The most important conclusion of these vital statistics is that mortality rates lower in the cities than in the countryside, and only slightly lower fertility rates, means that there may be higher rates of natural increase in the cities than in the rural areas. This is certainly the case in Java. If this situation pertains in other

Southeast Asian countries then one vital variable in the demographic component of the urban process is very different from the experience of the West, for natural increase is a far more important contributor to city population growth than it was in the "industrial urban revolution." As Davis has commented: [24]

> "Today the underdeveloped nations—already densely settled, tragically impoverished and with gloomy economic prospects—are multiplying their people by sheer biological increase at a rate which is unprecedented. It is the population boom that is overwhelmingly responsible for the rapid inflation of city populations in such countries. Contrary to popular opinion both inside and outside those countries, the main factor is not rural-urban migration."

While one may be critical of the generalities of Davis' statement, particularly as it applies to some Southeast Asian nations (particularly Malaysia), its analytical significance cannot be denied, for it raises a real question as to whether the spatial redistribution of population from rural to urban areas which occurred in the West will occur at the same fast rate in every Southeast Asian country as it did in the West.

The implications of this situation are interesting, for in the heavily populated parts of Asia (i.e. Java and India), what happens if the spatial redistribution of population which occurred in the West does not occur? Will the populations of rural areas continue to grow to a point where they absorb the food supply of the urban areas completely? Or, can food production be increased to permit the continued feeding of the cities? Do we have to envisage a situation in which a permanently lower level of urbanization will be maintained in Asia than in the West? These questions are of course speculative but they do challenge the acceptance of the inevitability of the "urban revolution." [25]

But at the present there seems to be no doubt that every country in Southeast Asia is experiencing rural-urban movement, although it is difficult to establish its contribution to urban growth. An analysis of the censuses of various countries indicates that while there are considerable variations from country to country, in many cases up to a third of the urban population have been born outside their present place of residence—in most cases in rural areas.[26]

It has been customary for researchers to analyze the motivations for rural-urban migration within a "push-pull" framework. Their conclusions are generally that the "push" factors have been more significant in forcing migrants into the city as opposed to the "pull" of migrants in the West. For instance, Hauser reports: [27]

> "It gives a clue to one of the most important features of Asian rural-urban migration; namely, the *push* of people from the country-side to the cities rather than the pull of industrial and employment opportunities in the urban areas."

This experience appears to be in direct contrast to that of the Industrial Revolution in Western Europe where there was a close connection between the demands for labor exerted by the rapidly growing urban industry and the growth of cities. In reality, it would seem that this push-pull model is over-simplified. While there is ample evidence in Southeast Asia of people being pushed to the cities—the most horrifying contemporary example being the mass exodus of the South Vietnamese rural population to the cities [28]—there is evidence that the cities in Southeast Asia are also attracting migrants because of suspected availability of jobs, the concentration of educational institutions, and even the glitter of modern life. I have conducted research among Malay migrants in Kuala Lumpur City which has led me to question the validity of the push-pull dichotomy, for it would appear that often there is a combination of motivations operating which cannot simply be classified as either "push" or "pull." [29] Unquestionably there is need for much more detailed research on this subject in the Southeast Asian context.

While I have dealt in rather general terms with the demographic components, and realize that one must treat the available statistics with caution, there seems at least enough evidence to suggest that the demographic components of the urbanization process in Southeast Asia are not identical to those of the West.[30]

The second facet of the urbanization process in the Western world was the favorable association of urban growth with economic development. Here, once again, the analysis of the economic features of the urbanization process in Southeast Asia indicates substantial differences from the experience of the West. The first of these differences is the fact that many countries in the region are characterized by a level of organization high in relation to the level of economic development. The statement of the ECAFE Bulletin summarizes the situation for the wider region of South Asia.[31]

> "When the major industrialised countries of Europe and North America were at a comparable level of urbanisation they were far more developed as is shown by the fact that approximately 55 percent of their labor force was engaged in non-agricultural occupations as against the present figure (for South Asia) of 30 percent."

While there are substantial variations within the region as to the validity of this generalization, it certainly is least valid in Malaysia, and most valid in Indonesia. Its inference that Southeast Asia is "over-urbanized" in relation to its level of economic development has been widely accepted.[32] What is more, the empirical evidence for the dominance of service occupations is strong. Indeed, the pattern throughout much of the underdeveloped world including Southeast Asia appears to be ". . . that urbanisation is proceeding at a more rapid pace than the expansion of manufacturing employment, resulting in a direct shift out of agriculture into services." [33] This is not to deny that industrial output has increased in some countries of the region, but this has not always led to large increases in the manufacturing work-force; principally because these industries tend to be capital-intensive.[34]

The consequences of this pattern have been a shift of population from agricultural occupations to the low productivity occupations in the city particularly in the service sector.

This traditional division according to occupations does not, however, do justice to the true economic structure of the Asian cities which are in many cases divided into two economic sectors, consisting of a *bazaar sector* which is highly labor intensive and a *modern sector* which is capital intensive. Since the existence of this dualistic structure is basic to the model developed later, it is not treated in great detail at this point. But it should be made clear that in the writer's opinion the persistence of this dualistic structure, basically a symptom of economic underdevelopment, is the most important variable affecting the functions of contemporary Asian cities.[35]

The problem of the dual economic structure is further aggravated in the Southeast Asian cities by the fact that the alien Asian communities are frequently unevenly distributed throughout the occupational structure. Thus the alien Asian communities, particularly the Chinese, have tended to dominate the commercial, financial, industrial, and artisan occupations, while the indigenous group are polarized at the extreme ends of the occupational scale—from the government services to unskilled and domestic occupations. This situation has been an important source of conflict between the politically powerful indigenous groups and the economically powerful alien communities. In some cases, as in Indonesia and Burma, it has resulted in the drastic measures taken against these communities. Whether or not government interference will improve this situation remains uncertain. It is sufficient to say that the plural societies remain one of the most critical problems of these cities.[36]

Finally, the question of social change in the Southeast Asian city may be considered. Here once again the paucity of empirical studies and the wide diversity of cultures in the region make generalization hazardous. In general one may argue that the persistence of pluralistic ethnic structures and a large bazaar sector have inhibited the kind of social changes said to have occurred in Western cities. Bruner's claim that ". . . the social concomitants of the transition from rural to urban life are not the same in Southeast Asia as in Western society" [37] is certainly true. Hauser has explained the reasons for the different patterns of social change as stemming primarily from the persistence of "folk" conditions within the cities. In many of the cities, in fact, the creation of squatter colonies has allowed the retention of basic village forms of political and social organization.[38]

What is more, the people living in these folk areas often have persistent and interlocking contacts with kinfolk in rural areas, both in the form of social contacts and economic remittances. Thus among migrants principally engaged in the bazaar sector, I would argue social change is not great. In most Asian cities there are, however, an elite and growing middle class who cannot be identified with the former group. It is tempting to argue that this group is changing its behavior towards fitting the model of the Western "urban way of life," for the most observable feature of their behavior (their consumer spending) seems to be the same as the wealthier inhabitants of the West. But in other areas of social behavior, I am not so certain that social change is occurring.[39]

To sum up, social change is not occurring in Asia at the pace of the urban revolution in the West. This is basically because the dimensions of the urbanization process are so different. Hauser identifies the major reason: [40]

"In large measure, the problems—social and personal—in the great cities of Asia, derive not so much from 'urbanism as a way of life,' but reflect rather the problems of the nation at large, problems arising from low productivity and mass poverty."

A Socio-Economic Model of the Asian City

The discussion of the processes of urbanization occurring in Southeast Asia suggests that in the present period they are producing a rather different socio-economic structure in these cities than that which has emerged in the developed Western world. Since the analytical approach advocated by this paper argues that an understanding of the socio-economic structure of these cities is basic to the delineating of their functions, it is necessary to elaborate on the features of the distinct structures. In developing a model of the Asian city, I have chosen to emphasize the economic basis of the city—arguing that in virtually all Asian cities there exist two distinct systems of economic production which affect their economic, political, and social functions. This is not to argue that these systems of production are the only sectors of the social system which bring about certain functions, for this would be to deny the important functions of power and culture in influencing the operations of a city.[41] But for the purpose of this analysis I will place the emphasis upon these models of production as the most important elements in the city sub-structure.

In order to construct a model I have drawn heavily upon the work of two writers, Franklin and Geertz. First, following Franklin,[42] I have accepted his threefold division of systems of production into peasant, capitalist, and socialist in which the ". . . fundamental differentiator is the labor commitment of the enterprise." [43] This division provides a conceptual framework which avoids the false dualism between urban and rural, and allows a city or system of cities to be seen in relation to the total nation. Secondly, I have drawn heavily upon the work of Geertz in Indonesia, particularly his study of Modjokuto,[44] to delineate the socio-economic structure of the city, particularly with regard to his division between the firm-type and bazaar-type sectors.

In utilizing Geertz's study as a basis for a model of other cities in Asia, it may be argued that Modjokuto is an aberrant example with little relevance to the rest of Asia. This charge can be refuted, since the model does not argue that there will be other cities in Asia with the same characteristics of this Javanese town; indeed it would be highly unlikely that this would be the case. But in terms of the town's structural make-up— that is to say the division between the two systems of production—other cities of Asia will contain this feature in varying degrees. The relative importance of the two systems in terms of employment or contribution to the gross national product or any other factor will be affected by several elements, particularly the economic base of the city. Thus single-function centers such as exploitative oil centers (e.g. Serai in Brunei), will be largely within the orbit of the capitalist system of production. In other cases (e.g. Kudus in Java, largely concerned with the manufacture of cigarettes [45]), the town is almost entirely within the peasant mode of production as defined by Franklin. Despite such differences it is the assumption of this paper that the model is valid for the majority of Asian cities, and particularly the large primate metropoli which dominate the urban hierarchy of many countries.

Geertz points out that in Modjokuto the economic structure of the town is divided into two parts: (a) a firm-centered economy [46]

> ". . . where trade and industry occur through a set of impersonally-defined social institutions which organize a variety of specialized occupations with respect to some particular productive or distributive end."

and (b) a bazaar economy which is based on

> ". . . the independent activities of a set of highly competitive commodity traders who relate to one another by means of an incredible volume of *ad hoc* acts of exchange."

It is clear that Geertz would also include in this latter system cottage industries organized on a family, or kinship, basis.

Superficially this distinction between the two-part city economy might be said to approximate the model of the dual economy. Certainly it has affinities with the model of technological and economic dualism put forward by Higgins,[47] to the extent that the firm-centered economy is largely capital-intensive and often technologically sophisticated, whereas the bazaar economy is labor intensive, often using limited technology. It must be stressed, however, that this dualistic view of the city must not inhibit the view of it as a whole, for while these two systems conceptually fall into "separate" boxes they should be regarded more accurately as being interlaced and interlocked. The manufacture and flow of goods and services are generated in each of these sectors and flow between them (see Diagram 1). There are of course certain categories of employment in the city which do not fit neatly into these two divisions. Where for instance are government employees, who often form a large proportion of the Asian city work force? In some countries, most government employment would appear to fall into the firm-centered sector; but, in general, observations of the pattern of government employment in underdeveloped countries suggest that burgeoning bureaucracy is more often employed with bazaar economy principles. Excellent examples may be found in the Philippines, India, and Indonesia.

A closer assessment of Diagram 1 shows that the model introduces the reader to a broader structural model of the large Asian city in (a) its relationship to both the developed economies and its hinterland (in most cases its own nation),[48] and (b) to a broader application of the systems of production prevailing throughout the country. It is necessary to examine these two facets in greater detail since they are basic to the predictive model of city functions.

First, it is clear from the conceptual terms used in the model that concepts of firm and bazaar types have been extended beyond their strict meaning as related to the tertiary sector in the city (as expounded by Geertz) and included within a broader conceptual definition of the mode of production. If one accepts, following Franklin, the threefold division of peasant, capitalist, and socialist systems of production in which the ". . . fundamental differentiator is the labour commitment of the enterprise," a broader and more useful conceptual framework is provided for the purposes of our analysis. The peasant economy is characterized by the commitment of the *chef d'enterprise* to the utilization of his family (kin); the capitalist and socialist systems of production are characterized by the fact that ". . . labour becomes a commodity to be hired and dismissed by the enterprise."[49] This is a conceptual model which, as Franklin has pointed out. is not impinged upon by any urban-rural division or agricultural-industrial division.

It would be possible to elaborate the features of these two systems in greater detail, particularly in terms of the varieties of economic organization and different values which pertain in the two sectors, but I shall be content in this paper to leave the model sketched in these broad outlines.[50] Of more importance is the elaboration of the relationship between the primate Asian city and the developed country in which the capitalist economy dominates and the predominantly peasant sector of the hinterland. The first point to be made is that the majority of Asian cities are part of what Frank has labeled a "world metropolis—national metropolis-

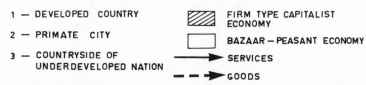

A SOCIO-ECONOMIC MODEL OF
THE ASIAN PRIMATE CITY

1 — DEVELOPED COUNTRY

2 — PRIMATE CITY

3 — COUNTRYSIDE OF
    UNDERDEVELOPED NATION

FIRM TYPE CAPITALIST ECONOMY

BAZAAR — PEASANT ECONOMY

SERVICES

GOODS

Diagram 1.

satelite relationship" [51] in which the national centers act as centers for the expropriation of surplus wealth to be transferred to the world metropolis. In delineating this process in Latin Ameria, Frank traces almost the same process as has occurred and is occurring in Asia. He comments: [52]

> "The colonial city became the dominant domestic metropolitan center and the countryside the dependent peripheral satellite. At the same time, the domination and capacity for the economic development of the Latin American city was limited from the very beginning—not by the hinterland or any supposed feudal structure . . . but rather by its own satellite status in relation to the world metropolis overseas. Nor in some four hundred years has any Latin American metropolis overcome this structural limitation to its economic development."

In this interpretation, then, the role of the city as the leading sector in economic growth is disparaged; its function in terms of economic development is doubtful. The effect of its relationship to the world metropolis is to keep the bazaar and capitalist systems of production relatively balanced in the city for long periods. Perhaps this is what Hirschman means when he says: [53]

> "It is often said that the underdeveloped but developing countries are apt to pass from the mule to airplane in one generation. But a closer look at most countries reveals that they are, and appear to remain for a long time, in a situation where both *airplane and mule* fulfill essential economic functions."

But in fact no such balance between bazaar and capitalist systems of production persists in quite these terms in the Asian cities. The penetration of the capitalist production system is a dynamic fact although varying in degree and speed from country to country. Since the majority of Asian countries are still dominated by peasant bazaar systems of production, a dynamic model of this process would emerge as in Diagram 2. This process of capitalist penetration into the peasant bazaar system does not always lead to immediate collapse of this sector; the creation of new-felt needs—shoes, bicycles, etc.—can gradually cause the decline of uncompetitive cottage industries in the city, or it can motivate a young peasant to move to the city to acquire these possessions. This is a dynamic process constantly changing the structure of the city and therefore affecting both its role and its function, and becomes a basic constituent for our model.

The Model and Asian Reality

What are the implications of this model's effectiveness in defining the roles and functions of Asian cities? If role and function are indeed a reflection of the structural setting (a phrase which I hope conveys the meaning of structure and relationship as discussed in the previous section)

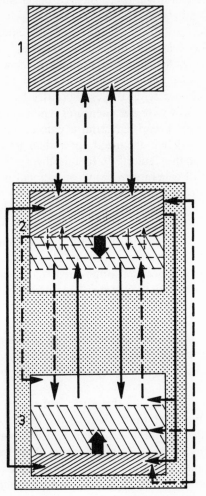

A SOCIO-ECONOMIC MODEL OF
THE ASIAN PRIMATE CITY
(DYNAMIC MODEL)

1 — DEVELOPED COUNTRY

2 — PRIMATE CITY

3 — COUNTRYSIDE OF
UNDERDEVELOPED NATION

CAPITALIST PENETRATION OF
BAZAAR–PEASANT SECTOR

FIRM TYPE CAPITALIST
ECONOMY

BAZAAR–PEASANT ECONOMY

RESULT OF CAPITALIST
PENETRATION (NEW AREAS OF
CAPITALIST PRODUCTION)

GOODS

SERVICES

Diagram 2.

then it should be possible to consider these aspects with respect to the relative balance between the two systems of production and the relationship of the city to the countryside and to the developed countries. The aspects of role and function will be considered under three broad headings: economic, political, and social.

First, with respect to the economic roles of cities, there is divided opinion (both within and outside Asian countries) as to whether the Asian cities play a beneficial role in promoting economic growth. Some writers have argued that the economic advantages of the central city in these developing countries require these societies ". . . to phase the investment process carefully over a sequence of regions concentrating initially upon the points of rapid urban-industrial expansion and moving outwards towards the periphery when the need for public investment declines in relative importance at the center." [54] In this theory the city is seen as a growth pole which will generate the kind of national wealth which can be eventually redistributed to the peripheral regions. In point of fact many Asian governments do not consider the city's role in this light. In some a Gandhian fear of the city and a realization in Nehru's words that there is a need for ". . . a distribution of wealth for the public good," [55] means that in terms of public investment the city is not always given this kind of priority. Emphasis is placed on other areas of government investment, for instance, education, social welfare, and industrialization. Private investment is often encouraged in the city, as in Malaysia or Thailand, where the private investor is most willing to invest because of available cheap labor, economies of scale, and infrastructural advantages. Whatever the identifying role of the city, good or bad, this does not of course mean that the Asian primate city will necessarily function in the same manner as its perceived role.

What are the implications of this structural model to the actual economic functions of the city? First, there is the highly important question of the function of the city as a center in which labor opportunities are created. If as in many of these Asian countries there is a constant stream of migrants moving into the city, as well as high rates of natural increase, then there will have to be large numbers of employment opportunities made available. The important function of the city as a training ground for such labor forces has often been commented upon, but an even more important function is in terms of the creation of employment opportunities. The point here is that despite increases in industrial production the number of jobs has not increased at a fast enough rate to absorb all the available labor force. Hoselitz makes this point in discussing the changes in the composition of the labor force of India between 1961 and 1965: [56]

"But if the same number of workers employed in an industry is approximately the same as five years ago, this means that there has been no additional employment opportunities. And in India the growth of new industry has not provided the jobs as expected."

The principal reason for this is that much of the growing industry is concentrated within the capitalist sector and tends to be capital intensive. On the other hand the city populations continue to grow and employment opportunities tend to be created either in the governmental or bazaar economy sectors. Hoselitz, commenting on the Indian labor force, says that "only bodies of government, including the State government, the quasi government and the local government, gained in the number of workers, from 5,157,000 in 1961 to 6,650,000 in 1965." [57] This proliferating bureaucracy frequently coincides with the growth in numbers of the armed forces of these societies and certainly is one reason for the power which the armed forces have in the political structure.

It is, however, the bazaar sector which frequently acts as the chief absorber of the labor forces of these Asian cities, and this is reflected in the very high proportions of city population employed in the service sectors as compared with comparable periods of urban growth in the Western world. What are the reasons for this? First, the institutional basis of the enterprise in the bazaar economy is the family or some kin association, and every effort is made to ensure some type of employment for kinfolk. Secondly, the bazaar system has a certain self-inflationary quality, and thus the more people who enter the system the larger the market. The simplest example is the proliferation of food vendors whose number increases as the population of the city increases. The third factor is the relationship of the bazaar sector to the peasant economy of the countryside. Population movement between the two sectors is common, since kin ties are strong and there is little reason for economic disruption in the shift from the countryside to the city or vice-versa. Finally there is the relationship of the bazaar sector to the capital-intensive sector. Here, there is a siphoning downward of the greater wealth earned in the sector, whether it be in wages earned by servants, employment in prestige projects, or in forms of welfare.[58] In economic terms the existence of large bazaar sectors in these cities, it can be argued, is inhibitive to economic growth, for they are characterized by under-employment and low productivity, but in social and political terms they perform important functions as absorbers of labor which would otherwise be unemployed and a much greater burden—one would be tempted to say *threat*—to the State.

When the city is looked at in these structural terms it can be seen that its economic functions are exceptionally diverse. The economic advantages accruing from a favorable infrastructure, the economies of scale, and other such factors should certainly enable the Asian city to function as a center of investment and industrial development (particularly in consumer industries) which will be a catalytic agent in economic growth. But whether this desirable function occurs will depend upon the relationships between the various parts of the structural model. For instance, what are the implications if the firm-type sector is largely controlled or at least heavily invested in by foreign companies? [59] Not only do they desire to export the surplus of their profits to their own country but they combine with indigenous elites interested in keeping the economic situation in terms of the status quo. If this occurs, the balance between bazaar and firm-type sectors can be continued through such interests. Thus in terms of the economic roles of these cities it can be observed that while they may be identified as playing developmental roles in economic change they may not necessarily function as such because of their structural setting.

Secondly, I want to consider the political roles and functions of Asian cities. There is a sharp division of opinion concerning the political roles of cities in Asia. In an earlier study I have suggested that, because many of the Asian nations were still (at a period in which they were) concerned with establishing custody of their countries, through a strident nationalism propagated by city elites in many of the Asian countries, the political role of the city seemed supreme. This concentration of the nation's political energies and national ethos in one city (generally the largest) is not without its dangers since this nationalism requires investment, or as Ginsberg has labeled it "pseudo investment" in the costly trappings of nationalism,[60] while investment in other areas might bring more useful rewards in terms of economic development. An even more important consequence of this rampant nationalism propagated by the city elites is that increasingly it is used as an excuse to avoid more intractable problems of economic and social change both in the countryside and city. In such a situation the country's leader assumes some of the charisma of the god-king of the pre-industrial city; the cities play the role of "cult-centers" as they did in the pre-industrial era.[61] Any traveler to Djakarta during the Sukarno era would not find it difficult to identify that city as a cult center disinterred from the past.

This interpretation of the political role of contemporary Asian cities, particularly the capitals of these countries, has not gone without criticism. Indeed Lacquian in a provocative paper has argued that urbanization plays a beneficial role in the emergence of new nations. He writes: [62]

> "It is questionable to assume, as the doctrine does, that because cities in Asia are reverting to their role as 'cult-centres' their influence on development is negative. It must not be forgotten that many of the nation-states in Asia flourished for centuries with the help of such centres. In a region of many agricultural villages, the great city served as a symbol for nationhood. That it continues to enhance unity and stability, even in the present age, may be considered developmental."

He goes on to argue that there is an important difference between the urban elites of the pre-industrial "cult-center" and the contemporary Asian cities in terms of their power, since the latter rests ". . . on reciprocity and the legitimating acknowledgement of the masses," [63] not the absolute power of the former. He states that, "to the extent that their policies [i.e. of contemporary elites] are made in the light of this need for legitimacy they cannot help but be developmental." [64]

Lacquian's argument may well be valid in the context of the society with which he is most familiar, the Philippines. But in the majority of Asian countries where military rule prevails, one cannot but agree with De Briey that the legitimacy of the ruler's claim to power remains in doubt [65] and that the majority of policies are performed in as an authoritarian a manner as the rulers of the pre-industrial city. Thus the validity of the argument of the political role and dangers of these "cult-centers" is not totally inadequate. It is clear, however, that there is divided opinion on the political roles of these Asian cities.

What of the question of the political functioning of these cities? If the social and political structure of the cities is related to the model elaborated earlier, it is possible to suggest a three-tiered model of the social structure of these cities which reflects both political power and social status. At the top of the social structure are a small elite consisting of military leaders, governmental bureaucrats, wealthy industrialists, or commercial entrepreneurs. Their political power not only rests with their control of the government but also with the links they have with the capital-intensive sector, a major source of wealth and productivity in the city. Secondly, there is a growing middle class consisting of white collar government workers, intellectuals, journalists and office workers who are also largely engaged in the capitalist sector. Finally there is the major part of the city population (of which a small proportion may be engaged in modern industry) working in the bazaar sector where their productivity is low and under-employment is characteristic. The political functions of the elites are peculiarly ambivalent. On one hand they are committed to programs of modernization which improve the conditions of their nation's population—the majority of whom are located in the peasant-bazaar sector —while on the other hand they must be careful not to lose their links and sources of wealth in the capitalist sector. Thus the elites are forced into a situation where they must keep a balance between the two sectors, often meaning that some of the important structural changes such as land reform, are not put into practice. Few researchers would deny that urbanization can and does play an important function in nation building; but an understanding of the existing structures of these Asian cities does not indicate that they necessarily provide the environment which invariably leads to the situation in which ". . . major clusters of old social, economic and psychological commitments are eroded and broken," [66] leading to the creation of large political communities. The continuing existence of the large bazaar sectors within these cities in which particularistic and subnational values persist inhibits this function, as the current Malay-Chinese conflict in Malaysia shows only too well.

Finally, there is the question of the role of the city in social change. Here, too, there is divided opinion. Many writers, both within Asian societies and outside, stress the beneficial and positive role of the city in social change. The city is seen in this interpretation as providing the environment in which the peasant undergoes the necessary psycho-social transformation to become an urbanite. This presumably creates a situation in which the urbanite is more motivated towards modernization and development which will be beneficial to the country. Such writers find a new and distinct mode of urban life emerging which they sometimes greet with an enthusiasm approaching an "urban Rousseauism." [67] Other writers have not seen the city's role in social change in such a beneficial light. They argue that shift of rural people to the city is a psychologically disruptive process which creates mentally disturbed and socially dislocated urban populations, who find great difficulty adapting to the urban situation. Neither of these interpretations of the role of the Asian city in social change seems correct, in part because of the theoretical assumptions underlying the viewpoint; in part because of the identification of the city as a mono-entity and the failure to acknowledge that different processes of social change are occurring in different parts of the structure of the city.

If this point is explored further it can be argued that the structural model of the city in relation to its hinterland again aids the delineating of the function of the city in social change. In research that I carried out on the movement of Malays to Kuala Lumpur City in the early 1960s I found this model most useful in explaining social adaptation in the city. For instance, I found that it was important to distinguish between migrants who were leaving the countryside (used in the broadest sense to include small country towns) as a result of processive forces operating at a national level (such as education, the penetration of capitalist production systems, and political dislocation) which had already dislocated them from the peasant production system before they entered the city, and those migrants who had remained within the peasant system of production but had left the countryside because of economic or social pressures. The first group had, in effect, been urbanized before they left the countryside, and the social changes they underwent in the city were slight. The second group moved largely into the bazaar economy where their social relations were little different from the countryside, and they too had little problem adapting to the city. Both groups underwent certain changes in terms of what may be labeled the city situation, in that work associations and residential associations assumed increased importance, as did the realization of their socio-economic position vis-a-vis the majority group of the city (the Chinese), but it did not seem that they suffered major problems of adaptation at the personal level. [68]

The point I seek to make here is that the identification of the city as playing a major role in social change, whether that change be interpreted as beneficial or destructive, does not relate to the function of the city in social change where it is essentially subordinate to the processive changes occurring at a national level. The function of an Asian city in social change will essentially be related to the relative dominance or size of the bazaar sectors in the city and the effectiveness of processive changes at a national level. For instance, in a city such as Djakarta or Vientianne, in which the bazaar sector is prevalent, we need not expect a great degree of social change. On the other hand, in a City State such as Singapore, which is rapidly trying to develop its capitalist sector, we may argue that social change is likely to be occurring at a much faster rate.

At this point it is useful to pull together the threads of the argument of this section. It would be possible to apply the model to specific Asian examples, arguing that some form of continuum of cities and nations might be constructed, running from cities such as Djakarta and Vientianne, where bazaar systems dominate, to some intermediary type where bazaar and capitalist systems are relatively balanced, as in Kuala Lumpur in West Malaysia, to Singapore where, although the bazaar system still is important, there is a vigorous effort to encourage the growth of the capitalist sector. One might postulate that in relation to this continuum cities will play certain distinct political and economic and social functions. It might also be suggested that a less preconceived description of the likely roles of these cities might be developed. This, at least, is a preliminary recommendation.

Conclusion

The most obvious conclusion to emerge from the analysis of this paper is that there is a very large gap between the preconceived roles of cities and their actual functions. This is a fact which I am certain would emerge if a similar analysis of Western cities at comparable periods in their process of urbanization were carried out. Of more importance is the fact that the different dimensions of the urbanization process in the Asian countries, as compared to the West, combined with an understanding of the structural setting of these cities, allows a more accurate identification of the roles and particularly the functions of the cities. There is no assumption that the roles and functions of the cities will be the same as those of the West. Clearly at this stage of Asian development one cannot argue that all Asian cities are contributing to national development; some cities contribute more, depending on their structural setting. Nor can one argue along with Lacquian that ". . . all cities are contributing to nation-building because of the liberating influence of urbanism." [69] There are as many forms of "urbanism" as there are variations in the structural settings of the Asian city.

The implications of this conclusion and indeed of the paper are that researchers, planners, and politicians, indeed every person concerned with attempting to unravel the "mystery of the city" in Asia, would do well to look carefully at the empirical circumstances and structure of Asian societies rather than assume that these cities will play identical roles and functions to those of the West.

The title of this paper posed the question: Asian cities—cancers or catalysts? In asking this question the author was assuming the roles of Asian cities to be either destructive or beneficial. The truth is that at this stage of Asian history, the cities are both catalysts and cancers, aiding and hindering the development process. Whether these cities become catalysts or cancers depends on the success of the Asian countries' efforts at social and economic development. These are processes mounted at a national level in which the city plays an important but not necessarily a central role.

NOTES

1. Akin L. Mabogunje, "Urbanisation in Nigeria—A Constraint on Economic Development." *Economic Development and Cultural Change,* Vol. 13, No. 4, 1965, p. 420.

2. Much of the discussion in this paper relates to a region of Southeast Asia with which the author is most familiar. This does not, however, in the author's opinion, limit the application of the arguments to countries such as India and Pakistan.

3. Bert F. Hoselitz, *Sociological Aspects of Economic Growth.* Glencoe: 1960, p. 177.

4. For instance see Chauncy Harris, "A Functional Classification of Cities in the United States." *Geographical Review,* Vol. 33, January 1943, pp. 86–99, and Robert Redfield and Milton B. Singer, "The Cultural Role of Cities." *Economic Development and Cultural Change,* Vol. III, No. 1, 1954, pp. 53–73. The author has attempted to apply Redfield and Singer's division between the orthogenetic and heterogenetic cultural role of cities to the case of Kuala Lumpur. See T.G. McGee, "The Cultural Role of Cities: A Case Study of Kuala Lumpur." *The Journal of Tropical Geography,* No. 17, 1963, pp. 178–196.

5. See Gideon Sjoberg, *The Preindustrial City,* Glencoe: 1960, and T.G. McGee, "The Rural-Urban Continuum Debate, The Preindustrial City and Rural-Urban Migration." *Pacific Viewpoint,* Vol. 5, No. 2, 1964, pp. 159–181, and McGee, *The Southeast Asian City.* London: 1967, pp. 52–75.

6. Neil J. Smelser, *The Sociology of Economic Life.* Englewood Cliffs: 1963, p. 27.

7. Fred W. Riggs, *Administration in Developing Countries.* Bosyon, 1964, p. 20.

8. Gerald Breese, *Urbanization in Developing Countries.* Englewood Cliffs, N.J.: 1960, p. 43.

9. See *e.g.* Bert F. Hoselitz, "The Role of Cities in the Economic Growth of Underdeveloped Countries." *The Journal of Political Economy,* Vol. LXI, No. 3, 1953, pp. 195–208.

10. Leonard Reissman, *The Urban Process: Cities in Industrial Society.* New York: 1964, p. 153.

11. See Bert F. Hoselitz, "Generative and Parasitic Cities." *Economic Development and Cultural Change*, Vol. III, No. 3, 1955, pp. 278–294 and by the same author, "The Role of Urbanization in Economic Development: Some International Comparisons," in Roy Turner (ed.), *India's Urban Future*. Berkeley and Los Angeles: 1962. See also Pierre De Briey, "Urbanisation and Under-Development." *Civilisations*, Vol. XV, No. 4, 1965, pp. 2–14, and De Briey, "Urban Agglomerations and the Modernisation of the Developing States." *Civilisations*, Vol. XVI, No. 1, 1964, pp. 3–25. Also Norton S. Ginsburg, "The Great City in Southeast Asia." *The American Journal of Sociology*, Vol. 60, No. 5, 1955, pp. 455–462. Finally three articles by Rhoads Murphey: "The City as a Center of Change: Western Europe and China." *Annals of the Association of American Geographers*, Vol. XLIV, pp. 349–362; "New Capitals of Asia." *Economic Development and Cultural Change*, Vol. V, No. 3, 1957, pp. 216–243, and "Urbanization in Asia," in Gerald Breese (ed.), *The City in Newly Developing Countries*. Englewood Cliffs, N.J.: 1969, pp. 58–75.

12. Takeo Yazaki, *The Japanese City: A Sociological Analysis*. Tokyo: 1963, p. 32.

13. I have limited the discussion of the processes of urbanization to the region known as Southeast Asia. Much the same patterns are exhibited in India and Pakistan.

14. Eric E. Lampard, "Historical Aspects of Urbanization," in P.M. Hauser and L.F. Schnore (eds.), *The Study of Urbanization*. New York: 1965, p. 521.

15. It should be noted that there is considerable disagreement over the definition of "urban places." There is a wide variety in country-to-country definitions, urban places sometimes appearing to be defined on the basis of "political whimsy" rather that reality. Many governmental authorities prompted by the United Nations are moving towards an acceptable upper limit of urban concentrations of 100,000 population, or more, as a basis for an urban definition. See Kingsley Davis (ed.), *The World Metropolitan Areas*. Berkeley and Los Angeles: 1959.

16. This point is difficult to establish empirically because of the wide variety of statistical definitions of "urban areas" in the Asian region. See my discussion of this aspect in T.G. McGee, *The Southeast Asian City, op. cit.*, pp. 76–78.

17. See T.G. McGee, "Malays in Kuala Lumpur City: A Geographical Study of the Process of Urbanisation." Unpublished Ph.D. Thesis, Victoria University of Wellington, New Zealand, 1969, pp. 379–383, and Lance Castles, "The Ethnic Profile of Djakarta." *Indonesia*, Vol. II, 1967, pp. 153–204.

18. See J.E. Spencer, "Cities of the Philippines." *Journal of Geography*, September 1958, pp. 288–294, and A.A. Lacquian, *The City in Nation-Building*. Manila: 1966.

19. Gunnar Myrdal, *Asian Drama: An Inquiry into the Poverty of Nations*. 1968, p. 470.

20. See K. Barnett, "Mooncake for the Millions?" *Far Eastern Economic Review*, Vol. LXIII, No. 50, 1968, p. 603.

21. P.D. Milone, *Urban Areas In Indonesia: Administrative and Census Concepts*. Berkeley: 1966, p. 95.

22. See A.H. Hawley, *Papers in Demography and Public Administration*. Manila: 1954, pp. 27–42.

23. P.D. Milone, *op. cit.*, p. 96.

24. Kingsley Davis, "The Urbanization of the Human Population," in *Scientific American* (ed.), *Cities*. New York, 1967, p. 227.

25. See Nathan Keyfitz, "Political-Economic Aspects of Urbanization in Southeast Asia," in P.M. Hauser and L.F. Schnore (eds.), *The Study of Urbanization, op. cit.*, pp. 265–309, for an excellent discussion of this point and its relevance to Southeast Asia.

26. I have presented evidence for these statements in another paper. See T.G. McGee, "An Aspect of the Urbanisation of Southeast Asia: The Process of Cityward Migration." *Proceedings of the Fourth New Zealand Geographical Conference,* Dunedin: 1965, pp. 207–218.

27. P.M. Hauser (ed.), *Urbanization in Asia and the Far East.* Calcutta: 1957, p. 133.

28. A recent *Newsweek* article cited a U.S. Senior Civilian Advisor who claimed that over 3 million Vietnamese farmers had moved to the cities in the last three years. Even assuming that he meant the farmers and their families, this is a remarkably high figure. *Newsweek,* January 20, 1969, p. 20.

29. See T.G. McGee, "Malays in Kuala Lumpur City," *op. cit.,* pp. 668–681.

30. I have not dealt with the important factor of ethnic heterogeneity in these cities, but the presence of large alien communities within these Asian cities poses further difficulties in accurate demographic interpretation. See T.G. McGee, *The Southeast Asian City, op. cit.,* p. 118.

31. U.N.O., ECAFE, *Economic Bulletin,* Vol. X, No. 1, 1959, p. 19.

32. Most recently by Gunnar Myrdal, *op. cit.* A major critique of the concept of "over-urbanization" has been written by N.V. Sovani, *Urbanisation and Urban India.* London: 1966.

33. W.E. Moore, "Changes in Occupational Structures," in N.J. Smelser and S.M. Lipset (eds.), *Social Structure and Mobility in Economic Development.* London: 1966, p. 203.

34. Sir Arthur Lewis has discussed the economic reasons for this emphasis on capital-intensive industry in "Unemployment in Developing Countries." *The World Today,* No. 1, 1967, pp. 13–22.

35. This is a point developed in much greater detail in the later sections of the paper. Such dual economies also characterize some cities in underdeveloped regions of the developed world; Naples is an excellent example.

36. The current communal rioting in the capital of Malaysia, Kuala Lumpur, whatever its immediate causes, is basically brought about because of the government's failure to solve the problems of Malaysia's plural society.

37. E.M. Bruner, "Urbanisation and Ethnic Identity in North Sumatra." *American Anthropologist,* Vol. 63, No. 3, 1961, p. 508.

38. See A.A. Lacquian, *Slums are for People.* Manila: 1969, and T.G. McGee, "Malays in Kuala Lumpur City," *op. cit.*

39. See T.G. McGee and W.D. McTaggart, *Petaling Jaya: A Socio-Economic Survey of a New Town in Selanger, Malaysia.* Wellington, New Zealand: 1967.

40. P.M. Hauser, *op. cit.,* p. 88.

41. See Takeo Yazaki, *op. cit.,* for a full discussion of aspects of political power and its relationship to the cultural characteristics of cities.

42. See S.H. Franklin, "Systems of Production, Systems of Appropriation." *Pacific Viewpoint,* Vol. 6, No. 2, 1965, pp. 145–166.

43. *Ibid.,* p. 148.

44. See C. Geertz, *Peddlers and Princes: Social Change and Economic Modernisation in Two Indonesian Towns.* Chicago: 1963.

45. See L. Castles, *Religion, Politics and Economic Behaviour in Java: The Kudus Cigarette Industry.* New London: 1967.

46. C. Geertz, *op. cit.,* p. 28.

47. See J.H. Boeke, *Economics and Economic Policy of Dual Societies as Exemplified by Indonesia.* New York: 1953, and B. Higgins, "The 'Dualistic Theory' of Underdeveloped Areas." *Economic Development and Cultural Change,* Vol. 4, No. 2, 1956, pp. 99–115.

48. It is, of course, true that in the larger countries of Asia, such as India,

Pakistan, and Indonesia, that this model to some extent falsifies the position. There are in fact a number of exceptionally large regional urban centers, such as Calcutta, Bombay, and Madras in India or Medan and Djakarta in Indonesia, which share the national hinterland. Thus a more accurate diagrammatic representation of the model would show a number of cities within each of these nations each related to its own rural hinterland. This point does not, however, invalidate the broad structural relationships which are shown in the model.

49. S.H. Franklin, *op. cit.*, p. 148.

50. A brief comment should be made upon the principal criticism of the model; namely that it does not allow for the incorporation of intermediate forms of economic organization. This is particularly true of some immigrant organizations in the Asian Cities. It may be that Rigg's bazaar-canteen model of the economic structure of these cities is more accurate, in that it portrays a more complex mixture of economic behavior and organization more accurately. See F.W. Riggs, *op. cit.*, pp. 100–121.

51. See A.G. Frank, *Capitalism and Underdevelopment in Latin America.* New York: 1967.

52. *Ibid.*, p. 25.

53. A.O. Hirschman, *The Strategy of Economic Development.* New Haven: 1958, pp. 125–126.

54. John Friedmann, *Regional Development Policy: A Case Study of Venezuela.* Cambridge, Mass.: 1966, p. 12.

55. Cited by G. Rosen, *Democracy and Economic Change in India.* Berkeley and Los Angeles: 1966, p. 123.

56. Bert F. Hoselitz, "Urbanisation and Economic Growth." Unpublished paper presented to the Conference on "The City As A Centre Of Change In Asia," University of Hong Kong, June 1969, pp. 10–11.

57. *Ibid.*, p. 11.

58. For a wider discussion of this aspect, see W.R. Armstrong and T.G. McGee, "Revolutionary Change and the Third World City: A theory of urban involution." *Civilisations,* Vol. XVIII, No. 3, 1968, pp. 353–378.

59. In the contemporary period many Asian countries are attempting to limit foreign control of companies in their countries through a variety of measures.

60. N.S. Ginsburg, "Cities without Nations?" *Pacific Viewpoint,* Vol. 9, No. 2, 1968, p. 209.

61. See T.G. McGee, *The Southeast Asian City, op. cit.,* pp. 19, 171–175, for an elaboration of this argument.

62. A.A. Lacquian, "The Asian City and the Political Process." Unpublished paper submitted to the Conference on "The City As a Centre Of Change In Asia," University of Hong Kong, June 1969, p. 11.

63. *Ibid.*, p. 12.

64. *Ibid.*, p. 12.

65. See P. De Briey, "Urbanisation and Under-Development," *op. cit.*, p. 4.

66. A.A. Lacquian, "The Asian City and the Political Process," *op. cit.*, p. 7.

67. For instance see M.R. Hollinsteiner, "Becoming an Urbanite: the Neighbourhood as a Learning Environment." Unpublished paper presented to the Conference on "The City As A Centre Of Change In Asia," University of Hong Kong, June 1969, for an excellent example of this glorification of urban life.

68. See T.G. McGee, "Malays in Kuala Lumpur City," *op. cit.*

69. A.A. Lacquian, "The Asian City and the Political Process," *op. cit.*, p. 21.

CHAPTER 7

Aprodicio A. Laquian
SLUMS AND SQUATTERS IN SOUTH AND SOUTHEAST ASIA

Introduction

Many a traveler's account of the perils of the Asian city is usually spiced with tales of Calcutta's pavement dwellers, Hong Kong's packed tenements, Manila's stilt-shanty slums, or Saigon's refugee settlements. The slums and squatter colonies of South and Southeast Asian cities are indeed most visible. This is especially so because they provide such a sharp contrast to the high-rise concrete and glass buildings of commercial centers and the pseudo-classical facades of government offices dating from a colonial past.

In spite of the marked visibility of slums and squatter areas, however, there is some confusion about what they really are, what they do, and what they mean, not only for the development of the city but of the nation-state as well. Many of the questions raised by McGee in considering whether the cities in Asia are "catalysts or cancers" are directly linked to the role and/or function of the slum and squatter area in development. Much of the debate on whether urbanization in Asia is "true" or "false" and the question of the "over-urbanization" of Asian cities also revolves around the slum and squatter population of such cities.

It has been proposed by some students of Asian urbanization that the link between urban and economic and social development does not contribute to positive overall development anymore. Much of the blame for this state of affairs is placed on the "population explosion," which has resulted in huge population concentrations in urban centers. Economic and social development, it is argued, become extremely difficult to achieve with so many mouths to feed. In the cities, the increasingly high demands for services, in the light of limited resources, create frustration and unrest, which often make the very viability of the countries themselves questionable.

The most commonly used symbol for the inability of South and South-east Asian cities to serve as catalysts for development is the slum. All the cities in the region seem to be plagued with squatters and slum dwellers. Some 25 percent of the populations of Djakarta and Kuala Lumpur as well as 26 percent of Singapore's population are squatters or slum dwellers. In Manila, about one-third of the population live in slums or squatter shanties. In almost all cases, squatters and slum dwellers are predominantly composed of migrants from the countryside or lesser cities. As a group, squatters and slum dwellers are usually at the bottom of the socio-economic ladder. They are more commonly found in jobs that require little or no skill, especially in the service fields.

Because urbanization in South and Southeast Asia is mainly characterized by massive population shifts rather than technological and industrial developments, it has been argued that it is not true urbanization. Because the cities' populations are so large in relation to their level of technological and industrial development, they are called "over-urbanized." It is claimed that the modernizing influence of the city is lessened by the dominance of still largely rural migrants in urban life. Further development in the countryside only brings "creeping urbanism," attracting more migrants to the urban areas and swelling the city's slum and squatter colonies. Thus, the city ceases to be a developmental catalyst because its abilities and energies are suffused and swamped by the rural element it is supposed to change.

It is possible that the hypotheses mentioned above mainly revolve around questions of fact. For one of the most interesting aspects about the whole question of Asian urbanization is the lack of studies and even basic data. In spite of the serious implications of the questions posed, there are still very few empirical studies available. This is especially true in regard to squatters and slum dwellers, where the few studies available are mainly concerned with specific squatter and slum communities. A truly comparative study of slums and squatters in the cities of South and Southeast Asia is still to be made.[1]

In view of the lack of data mentioned above, this article will rely on a review of what is available not only in terms of slums and squatters in Asia but in other developing countries as well. Its main focus is on the question of whether slums and squatters are developmental or not. More precisely, this question is divided into what slums and squatters mean in terms of economic and social (including political) development.

As far as economic development is concerned, we take its conventional meaning of increase in goods and services produced and its just allocation among the members of the community. We therefore concentrate on a few economically relevant questions which include: (a) do squatters and slum dwellers contribute to production of goods and services; (b) are they integrated with the urban and/or national economic system; (c) do they contribute to capital formation; and (d) are squatters and slum dwellers upwardly mobile economically? These questions will be considered in the light of short and long term economic development trends in the areas covered.

For social development, relatively simple indices are used in this study. These include: (a) social stability or the absence of disruptive events such as revolutions, riots, and other disturbances in the urban areas; (b) social and political participation of squatters and slum dwellers in community, city, and national affairs; (c) integration of squatters and slum dwellers with the larger society; and (d) openness of squatters and slum dwellers to external influences and values. These variables, of course, reflect psycho-social rather than mainly sociological dimensions found in urban slum life. They also reflect values which the nations of South and Southeast Asia are committed to—at least formally.

Types of Slums and Squatter Areas

The role and function of slum and squatter areas in economic and social development will depend, of course, on the nature of such areas. In South and Southeast Asian cities there are a great variety of urban settlements that may be called slums or squatters'. At first glance, there seems to be little similarity between the rural village-like *kampongs* of Kuala Lumpur and Djakarta and the tin and old wood *barung-barongs* of Manila and Cebu. The "Chinatowns" of Singapore, Kuala Lumpur, Bangkok, and Manila similarly seem to have little in common with the peripheral area *bustees* of Calcutta or even the low-income Indian communities in Rangoon. These seeming disparities, however, have not prevented social scientists from setting up typologies classifying slums and squatter areas.[2] In fact, there is a tendency to lump slum and squatter settlements together as though they represent homogenous entities.

A distinction, however, is usually made between squatter communities and slums. Generally, the former are located on land or buildings which are occupied without the consent of the owner. While ownership serves to distinguish between squatter and non-squatter settlements, however, it does not clearly differentiate between squatter communities and slums. It seems to be more productive, therefore, if variables other than ownership are used in a classificatory scheme.

Some of the more significant variables that may be used in classifying slum and squatter settlements are: (a) spatial location (central city or peripheral area); (b) degree of physical deterioration of dwellings and surroundings; (c) extent of overcrowding and congestion of dwellings and people; (d) relative age of the settlement; (e) type of land occupied (valuable land or marginal land); (f) adequacy of urban services (water, fuel, light, medical, and welfare services); (g) community organization or disorganization; (h) ethnic or class homogeneity or heterogeneity; (i) extent of deviant behavior (crime, juvenile delinquency); (j) apathy and social isolation; (k) disease rates and extent of health and sanitation; etc. These variables are amenable to specific forms of measurement and with their operationalization, similarities and differences among slum and squatter areas may be determined. By empirical observation of the existing settlements in Asian cities, general configurations may emerge, and the developmental meaning of such configurations may then be determined.

Instead of being captives of imprecise terminologies as slums and squatters, therefore, we may be able to describe and analyze the settlements we are interested in by this operational method. For example, it is highly probable that high correlations may be found among such variables as central city location, high degree of physical deterioration, extensive overcrowding and congestion of dwellings and people, old age of areas, high value of land, relatively more adequate services, high community disorganization, greater ethnic and class heterogeneity, and higher degrees of apathy and social isolation. On the other hand, peripheral city areas which have less physical deterioration may be far less congested, relatively younger, built on marginal lands, have inadequate services, be relatively more organized, have greater homogeneity, display less deviant behavior and may be less apathetic and isolated. It does not mean, of course, that a particular settlement has to conform to *all* the variables mentioned to be included in the former or latter configuration. What matters is the predominant character of the area. For greater precision, each variable may be given a particular weight derived from the value system of the particular society wherein the settlement exists and the relationship of that value system to the goal of development. By adding this value dimension, the classification becomes more valid and more relevant to the realities existing in each particular social setting.

Using the very broad configurations mentioned above, it is readily apparent that communities in the first group—a central city location— are less developmental than those in the second. Where the second group— peripheral city areas—predominates, there is a higher probability that cities will have a positive influence on overall national development.

With additional input of data, these configurations may provide greater depth and refinements to the typologies proposed and used for classifying slum and squatter areas. Many of the elements of the first configuration readily correspond to the traditional model of the slum area. They also approximate the features of Stokes' "slum of despair," or Frankenhoff's "dead-end" slum. In South and Southeast Asia, the inner city slums of Calcutta, Bombay, Delhi, and other cities in India, as well as the slums of Karachi, Bangkok, and Manila display these elements.

A peculiar variant of inner city slums—the "Chinatowns" of most Asian cities—alters the meaning of the first configuration somewhat. Close to the core of Kuala Lumpur, Singapore, Penang, Manila, Bangkok, and other Asian cities, are areas inhabited by "Chinese living abroad." These are characterized by high densities, physical deterioration, old age, high land values, and relatively more adequate services. However, the communities are not socially disorganized and the residents do not show apathy and social isolation. What isolation exists is likely to be political rather than economic. This is due to the high ethnic homogeneity of the Chinese and their tendency to maintain their identity.

The developmental role of these Chinatowns, and the Chinese in Southeast Asia in general, is too large a subject to be fully treated in this essay. Considerable debate surrounds the subject. Within the Chinese communities themselves there is ample evidence of economic and social mobility—particularly in business. In earlier times, when Chinese craftsmen and artisans were first imported into Southeast Asia and Chinese entrepreneurs founded their businesses, they obviously played a most developmental role. However, with the growth of Chinatowns and Chinese colonies, and the inward orientation of these enclaves, there is some doubt whether they still play as large a developmental role. Certainly, where they have controlled trade and commerce by their skills and efficiency, they have effectively blocked the growth of an indigenous enterpreneurial class. Veiled or outright discrimination against the Chinese in Asian cities has not slackened their hold over such trade and commerce. On the contrary, it has had undesirable effects in many countries. It has also made political and social integration of the Chinese more difficult.

As far as new and peripheral squatter and slum areas are concerned, considerable variations also exist. The *bustees* and other settlements that gird most Indian cities show greater deterioration than even central city slums, constructed as they are of mud, straws, and odd assortments of wood and board. Unlike the "invasions" in Latin American cities, they do not seem to be well organized. In the face of lack of services, congestion, and an apparent lack of a sense of community, it is extremely difficult to see developmental potentials in these settlements. Perhaps if they are allowed more time to develop a sense of community instead of being perennially harassed by governmental eviction, they may be able to improve. Resettlement to better planned areas may also help.

In Manila and other cities where squatters occupy railroad rights of way, roadsides, the banks of streams, and rivers, the "string development" of the resulting settlements also makes the growth of a community sense difficult. The elongated shape of the settlements does not contribute to closer human interaction—instead increasing the difficulty of creating organizations and initiating group activities. Perhaps one solution to this problem is the relocation of these "string settlements" to better areas.

More developmental potentials are apparent in peripheral city settlements made up of formerly rural communities now engulfed in the city's urban sprawl. In most of these places, the traditional community power structure persists for some time, until it is effectively changed by the participation of new migrants. If the transition is relatively smooth, it has vast developmental significance. The urbanization of the people, old and new, in the peripheral settlement is relatively gradual, thus avoiding disruption and social dislocation. Where the outward settlement is in response to the location of industries in the suburban areas, it also lessens urban transportation problems and decongests the city core. Of course, uncontrolled outward sprawl may entail higher costs in the provision of urban services. However, these costs may be shared with private developers. They may also be assumed by peripheral local government units whose tax rolls are considerably benefited by this outward movement. Finally, area-wide systems of government or regional planning may also help in assuring more rational urban development.

ECONOMIC DEVELOPMENT

Contribution to Production

The first question usually posed regarding squatters and slum dwellers is the extent to which they contribute to the process of production. Mangin, in a review of squatter settlement studies in Latin America, noted that squatters do contribute to production by providing their own housing, filling up jobs in the urban economic system, operating small businesses, and providing social capital such as community organizations, self help, and cooperative activities.[3]

One of the frequently ignored facts which tend to be hidden by the dilapidated and poorly constructed dwellings set up by squatters and slum dwellers is the relative magnitude of investment in this type of housing. In a survey of squatters on the railroad right of way in Metropolitan Manila conducted by the author in 1968, investments ranged from Pesos 2,000 to Pesos 15,000—in an area where tenure was highly uncertain and physical danger to the squatter and his family was great. In areas where there is more security of tenure, such as privately owned lands where the housebuilders pay rent or where the government assures squatters they will not be evicted, investment in housing may be higher. Furthermore, squatters and slum dwellers also tend to make "improvements" on their houses as time goes by.

In view of the acute shortage of housing in almost all of the cities in South and Southeast Asia, the provision of dwellings by squatters and slum dwellers may be a solution rather than a problem.[4] In recognition of this, Van Huyck and Rosser, both of the Ford Foundation, have proposed the following housing policies for India's cities: [5]

1. Existing housing stock even in slums must be preserved wherever possible so that new housing results in the maximum net gain;

2. Major and rapid improvements in the living conditions of slum dwellers can be achieved through well-organized programs of environmental improvements without replacing existing housing units;

3. Programs for the establishment of viable settlements housed in non-permanent, very cheaply constructed shelter but with adequate environmental facilities must be undertaken on a sufficiently large scale to provide a real alternative to the proliferating shanty-slum in urban areas; and

4. The achievement of acceptable environmental standards (to render even slum housing tolerable) depends essentially on the successful introduction of imaginative and efficient management services and on associated programs of community development based on voluntary effort.

Along the same lines, the author has also proposed "planned slums" in Manila not only to conserve considerable investments in housing by squatters and slum dwellers but also to extend basic services to squatter areas which are denied even the minimum of such services because of a strict adherence to a legalistic definition of squatting.[6] To planners who question the advocacy of "planned slums" in "unplanned cities" and who worry about the fact that the land occupied by squatters and slum dwellers may be used more "logically" in an economic sense, an observation by Mangin is very apropos. "This dilemma," Mangin says, "is only a dilemma for planners. People who need housing can't be kept in the pipeline for years as a plan can."[7] An economic good, such as land, is wasted if it is not used. When the time for a more rational use of squatted land comes, the relative costs and benefits of the proposed use, measured in the light of total societal needs, may be applied. If this assessment shows that the squatters have to be evicted, this may be done. At least, the costs incurred by having to evict squatters may be offset by the use of the land when it was idle.

A controversial question related to the contribution of squatters and slum dwellers to production involves their participation in the job market. Dwyer, McGee, Geertz, and Wertheim, among others, have argued that people who live in slums and squatter areas have a negative influence on economic development because they are usually unemployed or underemployed, and when they do work, they tend to be concentrated in the tertiary sector (mainly service occupations) which are not as developmental as manufacturing, industry, and other sectors.[8] According to Dwyer,[9]

"For the masses of immigrants from the countryside who now crowd into the cities, the urban standard of living must remain only marginally above that which they can obtain in their home villages, an abysmally low level, in order to be acceptable."

These observations have the ring of truth in them, but only if they are seen in the light of the premise that the employment patterns in the cities of South and Southeast Asia are different from those that occurred in the

West, and that since the Western model of development is the only true road to progress, the deviation from this model shown by South and Southeast Asian countries makes them incapable of development.

Aside from being quite ethnocentric, the above premise ignores the fact that, as noted by Wittermans, "Western urbanization, far from being the norm from which developing countries seem to deviate, should itself be regarded as a variant—an outgrowth of another remarkable variant development: industrialization." [10]

In a study of the 1961 census figures for Greater Bombay, Zachariah found that one out of every four migrant workers in the city is a casual laborer or a "peon," a cook, a waiter, a hawker, or a petty trader. He admits the likelihood that "the relative number of manual laborers and low level service personnel will be small in Western cities in comparison with Bombay." However, Zachariah's conclusions are worth noting. He said: [11]

> ". . . at the present stage of technological development in India, it is doubtful whether these so-called 'miscellaneous, usually menial, unskilled service' personnel are nonessential. The large number of laborers engaged in loading and unloading trucks, railway cars, and ships, included in the occupational group 'laborers not elsewhere classified,' for example, can in no way be considered nonessential in a port city like Bombay."

Almost all studies of slum and squatter areas in the cities of South and Southeast Asia to date show that these settlements are peopled mainly by migrants from the countryside. There is also ample evidence supporting the belief that it is the educated, ambitious, and highly motivated persons who move to the cities and find themselves in the slum and squatter areas. Sendut notes that "a few years' education is a powerful stimulant to young people to leave their villages and seek a new life in towns," and that "it is not necessarily the very lowest economic groups in the agricultural community that now contribute the most migrants to cities." [12] Zachariah also found that "migrants in Bombay City have much higher levels of educational attainment than the general population of the states from which they were drawn." [13] While noting that this migration deprives the villages of better educated persons, he mentions that "on the other hand, the talents and skills of these persons might well have been wasted in the rural areas," and that "migration may thus have helped in bringing skills to areas where they could most profitably be utilized and in contributing to a better utilization of the human resources of the country as a whole." [14] Zachariah also found that the "work participation ratios of migrants in Greater Bombay were higher than those of nonmigrants in the city. . . ." [15]

The problem posed by unemployed and underemployed squatters and slum dwellers and the concentration of employment opportunities in the tertiary sector may actually prove less serious if the time element is introduced. In contrasting the Western and the Asian situation, those who pose the problem are failing to consider the relative time dimensions involved. While slums are old features of Asian cities, for example, squatter colonies are relatively recent developments. In most cities, these colonies did not really become serious problems until after World War II. Considering the upwardly mobile nature of many migrants who live as squatters and the fact that it usually takes a couple of generations before a more stable socio-economic status is achieved by a migrant family, perhaps there is no reason to be too pessimistic.

Economic Integration

One of the observations often mentioned in the literature on slums and squatters is their lack of integration into the larger economic system. The high unemployment and underemployment rates found among these segments of the urban population are used as indices to this. Ray, in a study of Venezuelan *barrios,* for example, notes that "like a new barrio, which springs up without any apparent relationship to the economy of the surrounding city, the newly arrived migrant is also totally detached from the urban economy." [16] However, Ray recognizes that this may be peculiar to Venezuela. In a survey of Latin American squatters, Mangin comes to an opposite conclusion. He observes that "there is a tremendous proliferation of small enterprise" in the slums. In a study of a Lima *barriada,* Mangin observed banks, chain stores, movie houses, and small shops. Mangin also cites studies by Hoenack and Leeds which listed bars, restaurants, garages, repair shops, barber shops, school supply stores, bakeries, groceries, fruit stores, and newsstands in *favelas* of Rio de Janeiro.[17]

Making a distinction between "open-end" and "dead-end" slums, Frankenhoff concludes that the former have greater economic potentials when considered in relation to the urban system. Using a Rio *favela* as a case study, Frankenhoff states that "the very existence of a slum suggests that it serves some market." [18]

> "The normal slum worker is a lower echelon worker for private firms in the urban center, for the city government, and in legitimate travelling salesman activities. The labor product exported by the slum community serves a valid demand by the center community. More than this, the slum community is a significant market for goods and services produced by the urban center, including consumer durables such as sewing machines, refrigerators and television sets. At the same time, the slum community is undertaking a program of capital formation."

After concluding that "the economic potential of a slum community therefore involves key economic variables of investment, fiscal contribution, market for consumer goods and the supply of unskilled and semi-skilled labor," Frankenhoff raises the question: "Can a city or urban center grow properly without a slum component?" He answers this question by saying that the slum community has a "right to exist" although this does not imply that "the slum community as such is desirable in itself." Since slums already exist in almost all cities, he concludes that "no urban development policy is complete without a slum community integration policy." [19]

In regard to the integration of the slum with the city and the national economy, it is obvious that conditions vary from country to country. It may be proposed that relatively new slum and squatter settlements, especially when they are built in a hurry (such as the "invasions" found in Latin America), may originally be isolated from the total economic system. However, with time, links are established between the slum and squatter communities and the larger urban community. Since the development of these linkages is beneficial to the whole economy, one can only agree with Frankenhoff's proposal for an integration policy.

Capital Formation

One of the main problems of underdeveloped economies is capital formation. At relatively low income levels, a higher percentage of goods and services are consumed instead of being saved for investment. Since slum dwellers and squatters tend to belong to lower income groups, their capacity to save and contribute capital to the economic system may be questioned.

A study of slum-squatter community made by the author in Manila shows, however, that there are certain elements in slum life that contribute to saving abilities. The biggest saving for squatter and slum dwellers, of course, is on housing. In the community studied, 67.0 percent of the households owned their homes, 26.6 percent were renting and 6.4 percent had free housing because they were doubled up with relatives or staying with friends. Among renters, 82.8 percent paid Pesos 20 per month, which on the average, amounted to less than 5 percent of monthly income. For those who owned their homes, no additional housing costs aside from depreciation or improvements were involved because by definition, squatters pay no land rental.[20]

Life in the slums is also full of informal arrangements designed to stretch the family's peso. Cheap, ready to eat food is sold at the corner store. Transportation costs to work are minimal because of proximity of residence to work place. Many family heads even walk home for lunch, thus enabling the whole family to share the meager food available. Many vendors selling fish, vegetables, meat, and other things hawk their wares from door to door, making it possible for housewives to buy small quantities for each meal. The corner stores extend credit, carefully finding out each family's earning capacity and then setting debt limits on this basis. During periods of critical need, one can always count on relatives, friends, and neighbors for food or a small loan. Also, since one's neighbors are also poor, there is little tendency to keep up with them by conspicuous consumption and income dissipation.

An income and expenditure study conducted in the community showed that on the average, savings for a family of six starts to become possible at an income of Pesos 170 a month (minimum legal wage in the Philippines was Pesos 180). The study, which asked 62 households to note their daily expenses and income, also revealed that expenditures for necessities (food, clothing, fuel, transportation, water, medicine, and household needs) tend to stabilize at a certain level. Thus, with increased income, the percentage of funds devoted to necessities tends to decrease. The balance between income and expenses for necessities may be spent for semi-optionals (snacks, extra food ingredients, allowances, etc.) or optionals (liquor, cigarettes, comic books, ornaments, movies, toys, etc.). If these expenditures are postponed or foregone, however, savings occur. For families who had some savings, the study probed into what these savings would be spent for. The answers included: purchase of furniture and household appliances, repair and improvement of the house, children's education, food, clothing, and "business investment." The importance given to buying durable goods, to home improvement, and the education of children showed that squatters and slum dwellers, at least in this community, sought economic and social betterment.

Another important thing to consider in slum economics is the fact that, as in the rural areas, the whole family and not just the breadwinner is a productive entity. Aside from the father, the mother and children also earn. The mother may wash clothes, work as a cleaning woman, run a small store, peddle things or even seek regular employment (leaving the children with relatives). The children pick up extra money by selling newspapers, lottery tickets or cigarettes, shining shoes, or engaging in the watch-your-car racket downtown. The whole family may be engaged in handicrafts such as making doormats from discarded rubber tires, making lanterns, weaving baskets, etc. Many of the families in the community studied were engaged in scavenging—making the rounds of the city's garbage heaps and dumps salvaging tin cans, newspapers, bottles, wood, metal, and other scraps which they sell to junk dealers.

The important thing about the above observations is that there are various opportunities available to the squatter and slum dweller to stretch his income and accumulate savings—opportunities which would not be possible if he were forced to live outside the milieu of the slum and squatter area. To the extent that these opportunities are availed of by the slum dweller, they improve his chances for mobility. Thus, the more ambitious and aggressive slum dwellers need not look at the slum as a dead end. To them it may very well be a "zone of transition" where they only have to tarry a while on their way upward.

Economic Mobility

There is not too much concrete evidence on the upward mobility of squatters and slum dwellers, although the little that is available makes for some optimism. Zachariah, in comparing the occupational and industrial composition of migrant and nonmigrant groups in Bombay, found that the former tended to be employed in industries and occupations requiring less skill and education. There are indications, however, that the longer one stayed in the city, the greater is the tendency to have a better job. This mobility, of course, may be realized at the expense of older city dwellers; Zachariah observed that "if migration contributed to the rate of unemployment in the city, it was probably because of the displacement of nonmigrants by migrants." [21] While this may be harmful for individuals concerned, it may be beneficial for the whole economy for it fosters more rational and economical use of labor and skills.

Looking at the pattern of variation in occupational composition by duration of residence, Zachariah found this related to industrial distribution. He noted that: [22]

> "As duration of residence increased, there were declines in the proportion in service occupations and in unskilled labor. On the other hand, the proportion in crafts and white-collar occupations increase with duration of residence. Between duration of residence of less than I year and 15 years or over, the percentages employed in unskilled occupations and in service were nearly halved, while the percentages of craftsmen and those in clerical occupations were nearly doubled."

These survey findings moved Zachariah to conclude that ". . . the longer the exposure to city life, the greater the resemblance to city-born persons and the greater the dissimilarity with populations in states of origin." Zachariah found this observation to be true for the characteristics he examined—marital status, educational attainment, industrial detachment, and occupational composition. [23]

Evidence gathered by surveys and censuses, such as the ones cited by Zachariah, is greatly wanting, of course, because it does not show trends. At present, the author is conducting a longitudinal study of a slum area in Manila which had been surveyed in 1962, 1967, and will be surveyed again in 1972. While data from the more recent survey are still being analyzed, they do point to the improved economic and social status of the people who had moved out of the slum area. There is a possibility that, for people who are left behind, socio-economic conditions may not be as good. However, the fact that some people moved out of the slum area does not mean that conditions there have deteriorated further. More recent migrants have taken the place of those who left and the slum still plays its role of escalator up to the present.[24]

SOCIAL DEVELOPMENT

Social Stability

One of the "standard myths" about squatters in Latin America cited by Mangin is the belief that these areas are "breeding grounds for" or "festering sores of" radical political activity, particularly communism, because of resentment, ignorance, or a longing to be led.[25] Mangin explodes this myth, and, after making a survey of the studies of Latin American squatter areas, arrived at the conclusion that the value system or ideology of squatters seem to be very similar to the beliefs of the operator of a small business in nineteenth-century England or the United States: [26]

"Work hard, save your money, trust only family members (and then not too much), outwit the state, vote conservatively if possible but always in your own economic self-interest; educate your children for their future and as old age insurance for yourself."

With a value system such as this it is very difficult to conceive of the squatter or slum dweller as a radical or revolutionary. One of the strongest values mentioned by Mangin, which the author has also noted in Philippine slum communities, is the importance attached to the local community. Aspirations of squatters and slum dwellers (next to family improvement) are tied to the betterment of the local community. The most immediate concern, of course, is the very survival of this community—its protection against threats of eviction on the part of the government or private landowner. When this danger is not too immediate, most slum dwellers and squatters cooperate and participate in community activities such as *fiestas,* school functions, religious and social functions, etc. Some of them are even extremely concerned and sensitive to the "image" of their community, taking pains to prove that it is not tough, troubled, or dangerous.

It is readily apparent that the local-oriented interests mentioned above are closely related to "folkvalues" associated with community life in the rural areas. The prevalence of such values which provide comfort for rural-urban migrants add a significant amount of stability to social life in the slums. In a study of two slum communities in Manila conducted by the author, depth interviews showed a significant degree of satisfaction and enjoyment of life in the slums. When asked why respondents in the two communities chose to remain there in spite of repeated moves to evict them, many of them said that they liked the place because it was peaceful, their friends and relatives lived there, they had found social acceptance in the community, the place was safe for children, it had a good environment, it offered better living conditions, it was close to their place of work, and it was not crowded. These reasons were all contrary to "objective indices" applied by the police, social workers, and even researchers who knew the community. What the residents called a "safe place" was on the "critical spot" list of the Manila Police Department and was frequently in the newspapers for gang wars and killings. The community that was "safe" for children was flooded all year round, had no playgrounds (children played on the catwalks and plank bridges), and had criminal and prostitute residents. The places that were "not too crowded" had the highest population densities in the city.

In spite of these contradictions, however, the people had a reasonable explanation for their answers. The area was peaceful for most of them because troubles usually meant gang wars—and if you were not involved, you were safe. Children were safe because no automobiles intruded and relatives, friends, and neighbors could watch them all the time. The place was not crowded because not all the people were home all the time. Besides, there was great warmth and comfort in being with friends and neighbors who could help you and share everything with you.

With this value system emphasizing personal and family improvement, community solidarity, conformity with social norms and identification with the common good, it is very difficult to picture the slum and squatter areas as hotbeds of revolution. It is true that in Kalua Lumpur, Djakarta, and Manila, some people from the slums participated in riots and turbulent demonstrations. However, in proportion to other social sectors (students and labor union groups), considerably less people from the slum and squatter areas were involved in these affairs. It is also highly probable that their participation was triggered not by their coming from slums but by their association with other groups (labor unions, student groups, political parties, etc.).

The fear that squatters and slum dwellers will be involved in riots and revolution and other forms of destructive behavior is based on the theory that they flock to the city filled with hopes and aspirations only to end up in the slums. Thus, frustrated with their lives, unable to adjust to the fast pace of urban life, forced to live in filthy surroundings and lost in their strange urban environment, they become dehumanized and join the lawless elements. Most studies of slum life, however, do not seem to support this theory. Instead, they find that slum communities are usually primary group communities characterized by neighborly intimacy, a strong "we" feeling, and psycho-social security. Being mainly composed of recently migrated rural folk who bring with them their traditional institutions and means of social control, they resemble transplanted rural villages. Hence the feared Marxist tendencies to dehumanization, arising from industrialization and urbanism, are kept in check by social norms. Instead of anomic behavior and social isolation, most squatter and slum communities display coopera-tive activities and social cohesion.

Social and Political Participation

One of the persistent images of the slum dweller and squatter is social isolation, anomie, rootlessness, and lack of interest in the social and polit-ical life around him. This image closely supports the fear of disruption and revolution, because when large segments of an urban population do not share the common value system and are barred from participating in it be-cause of internal and external causes, the probability of a political explo-sion is high.

Contrary to common beliefs, however, most studies of life in these communities reveal a very high level of social and political participation. This shows itself in social activities reminiscent of rural life: religious festivities, communal work cooperation, kinship-related activities, various types of secular celebrations, etc. Squatters and slum dwellers, as com-munity development and social workers have consistently found out, tend to be joiners. In most instances, they set up organizations without any ex-ternal stimuli, and patterns of mutual help and cooperation are widely practiced even in activities that do not directly involve traditional or rural-based celebrations.

In a community studied by the author in Manila, it was found that this settlement with a population of 2,625 families was tightly knit to-gether by more than 20 organizations. Probe questions in the survey and casual conversations with local leaders revealed that the primary objective in setting up and/or joining organizations was protection and safety. Some of the older leaders remembered the time when the community was just starting and it had a reputation for trouble. Fights among the residents themselves and threats from members of nearby slum communities were strong reasons for the people to get together and act as a unit.

The first organizations in the place were the *rondas,* voluntary associa-tions set up by the men as fire brigades and citizen police forces. The or-ganizations required not only the participation of able bodied men but the contribution of money as well. Later, when the threat of eviction from the government became serious, the community leaders expanded the organiza-tions to include "lobbying" activities and political relationships with vari-ous officials and politicians in the city. The self-protective associations, therefore, have gradually been transformed into political machines. In the process, the stakes for every community member have been raised. Hence, each person considers it his duty to be interested and involved in com-munity affairs, knowing full well that his very existence in the place de-pends on his efforts in combination with all the others.

Integration with the Larger Society

With the tendency of the squatters and slum dwellers to rally around local issues that defend their homes and communities, there is a fear that they will contribute to the disorganization and disintegrative tendencies in underdeveloped countries which find it so difficult to achieve a common sense of nationhood. Deutsch regards urbanization as a key process in social mobilization, "the process in which major clusters of old social, economic and psychological commitments are eroded or broken and people become available for new patterns of socialization and behavior." [27] If community organization in slum and squatter areas persists in supporting these particularistic tendencies, then, the task of nation-building would be much more difficult.

The important question to ask in regard to slum and squatter communities, then, is the extent to which the folk traditions as a basis of community organization and behavior persist in the urban setting. Through time, what happens to the elements that make up this folk configuration (e.g. deference to high status authority and old people, reliance on indirect methods of social control as seen through gossip, ostracism, and appeal to family honor, strong family and kinship ties as determinant of position, role, and function in the community, etc.)? Do these elements persist or are they altered because of the transfer to the city?

Mangin, in his study of Peruvian *barriadas,* concedes that many rural characteristics persist in the city but that a qualitative change occurs in the rural-urban migration because of the difference in functions served by the urban slum community. He proposes that "the squatter associations are defense organizations based on labor union rather than rural community models and function as pressure groups on the government as well as screening committees for new residents." [28]

This observation may be a function of the relative age and the special circumstances of the slum or squatter community. In the longitudinal study the author is pursuing in a Manila slum, there are indications that Mangin's observations are valid. When this study was first initiated in the early 1960s, the slum area was actually composed of two communities—an old one which was a traditional farming village, and a new one composed of squatters evicted from other places in Manila. In five years' time, however, considerable integration had already occurred between the two distinct communities. The mutual need for protection, to control the undesirable elements within both communities, to provide a stronger bargaining base with city officials and political leaders, and the normal social inter-relationships made necessary by physical propinquity drew the two communities closer together. One of the main effects of these developments was a change in the leadership pattern in the area. Where, before, the "old man" in the farming community had been the undisputed leader in the whole area, he had, after a while, to acknowledge the leaders of the new community. Thus, in 1967, leadership in the integrated communities resided in a small clique which still retained the old man as symbolic leader but which now made decisions as a group. It is also significant that the clique now included younger leaders in both communities and was no longer dominated by the old man's extended family.

If there is one element that contributes to this change in the nature of community associations in slums, it is the increased politicalization of life in the urban area. The need to deal with political leaders is strong because the very survival of the community depends on the ability of the slum area to be on the good side of the power structure. High unemployment and underemployment also forces many slum dwellers to seek jobs (honest or patronage) from politicians. Since the politicians need votes and support, they accept this mutually beneficial exchange of favors. In this way closer links are established between the power superstructure of the larger polity and the community's power structure.

In the light of these findings, it may be considered ironic by some that the integration of squatters and slum dwellers in the larger society is actually hastened by that structure which is considered an evil in urban politics—the political machine. Even scanty evidence in South and Southeast Asian cities reveals, however, that many of the problems confronted by slum and squatter communities are being solved with the use of a political power structure. Many communities are finding out that schools, medical facilities, community centers, roads, toilets, water, electricity, fire-fighting equipment, and other things they need can be provided only by politicians who can influence the bureaucracy. Ambitious individuals in the slums are also becoming aware that to improve their lot, and those of their relatives, they have to join political alignments.

It is in their linkage to the national parties that the local political machines perform a crucial integrative function. In most developing countries, the national parties tend to be dominated by upper class elites who use them to maintain themselves in power. Traditional elite roles, buttressed by the hangover of a patronizing colonial style, tend to make the national parties aloof and unresponsive. In the urban areas the political machines, usually concentrated in the slums, first force these parties to take heed. Initially, elite response takes the form of "capturing" the machines by coopting their leaders into the power elite. As machine members and minor leaders become aware of the situation, however, they challenge this captive leadership and force the parties to pay attention to their demands. Eventually, because of the parties' need for votes, they have to listen to machine demands. As such, the machines, in their own peculiar way, tend to make political activity more democratic and distributive.

Hopefully, as economic and social conditions in the slums improve, there will be a shift in the role of the machines from selfishly particularistic pressure groups to more general welfare oriented lobbies. As the personal economic and social conditions of machine leaders improve, they may tend to become less dependent on patronage and doles. Perhaps it may take a generation of educated, upwardly mobile slum children to accomplish this. With the improvement of slum conditions, community interest may become broadened and enlightened; it may even influence the national parties to change as it becomes more integrated into the political system.

Openness to External Values

Changes in the structures of community life are usually accompanied by changes in the value system of the people. In the urban areas, in spite of the persistence of folk beliefs, the new environment cannot help but be felt by the squatter and slum dweller. In the first place, the person in the slums is much more exposed to messages from the external world than the person in rural areas. Radio and television ownership is common in most slum and squatter areas in South and Southeast Asia. Newspapers and magazines are also relatively more widely read. The closer interrelationships and higher physical propinquity that exist in the slums also result in more intensive use of mass communication agents. Thus, in Manila's slums, one can walk from one end of the community to another without missing a single word of a soap opera to which almost all the housewives listen. At night, crowds gather in front of stores or inside the living rooms of people with television sets (some of whom charge a small fee for the privilege of watching). Newspapers, magazines, comic books, and other reading materials are transferred from hand to hand until they are dog-eared and fall apart. All the messages from the various media, of course, are discussed, debated, and interpreted by the local communication system that is a reminder of rural origins.

In this tremendous information system, the slum and squatter areas, perched as they are between rural-based beliefs and norms of the people and the more sophisticated and urban patterns around them, provide opportunities for playing a most important role as mediator of social change. The new messages, with their sometimes disruptive and unsettling effects, are toned down by the survival of traditionally cohesive social patterns. Because the atmosphere of the area is made up of this mix between the old and the new, the messages of change are given time to take effect and they do not create too much anxiety. In this way, the slum and squatter areas, because of their transitional nature, become agents of transition themselves.

CONCLUSION

In this brief review of the meager literature on slum dwellers and squatters in South and Southeast Asia, there is a persistent air of optimism in sharp contrast to the general mood of many who have written about the conditions in these cities. Partly, this optimism is based on a distinction between urban processes in South and Southeast Asian cities and those that occurred in the West. Mainly, however, it is founded on actual studies and surveys that have tried to view the slum and squatter communities as they are and as they are related to the larger urban and national processes of which they form a part.

Briefly, the evidence suggests that slums and squatter communities contribute to economic production in many ways. They seem to be more closely integrated with the larger economic system than a structural and sectoral economic analysis tends to show. Living in the slums and squatter areas provides many opportunities for saving and for capital formation. Finally, there is evidence of economic social mobility in the slums, sketchy though it may be.

Socially, slums and squatter areas do not seem as disruptive and unstable as their physical appearance usually suggests. The value system of the slums is still largely rooted in a traditional rural origin and it tends to be conservative even as it emphasizes personal and community improvement. Survivals of the rural value system account for high social and political participation, although as slum and squatter communities grow older, organizational behavior akin to labor unions and other more structured organizations replace the traditional practices. The politicalization of the slum contributes to its greater integration into the larger polity and society. The fact that it exists within an urban system which is characterized by intensive communication interaction also makes the slum and squatter community an agent for social change.

These optimistic assertions about the nature and function of the slum and squatter community in South and Southeast Asia, of course, can only be validated by more empirical studies and surveys of a truly comparative nature. Compared to slums and squatter areas in Latin America, those in South and Southeast Asia have not been greatly studied. Particularly important are those studies that look into a slum or squatter community through time, documenting the actual changes occurring therein.

The most important factor to consider in regard to slum and squatter areas in South and Southeast Asia is time. As mentioned previously, many of these communities (particulary the squatter settlements) are of relatively recent origin. As such it is extremely difficult not only to determine

their nature but to predict what they will become in the future. Comparison
with slum and squatter communities in a different era and a different place
are at best approximate, at worst, misleading. These communities, therefore,
should be studied as they exist, and as they change, through time.

NOTES

1. The International Association for Metropolitan Research and Development
(INTERMET) is presently engaged in a comparative study of squatters and slum
dwellers in ten cities all over the world under the direction of the author. The cities
included in the study are Bandung, Belgrade, Caracas, Ibadan, Istanbul, Kuala
Lumpur, Lima, Manila, Seoul, and Tunis. This study is looking into similarities
and differences found in squatter and slum communities and probes into their role
in economic and social development.

2. See, *e.g.* the various classificatory schemes mentioned by Marshall Clinard
in *Slums and Community Development.* New York: Free Press, 1966, chapter 3.
Charles J. Stokes has popularized the distinction between "slums of hope" and
"slums of despair." Charles J. Stokes, "A Theory of Slums." *Land Economics,* Vol.
38 (August, 1962), pp. 187–197. Frankhenhoff also makes a distinction between
"open-end" and "dead-end" slums. Charles Frankenhoff, "The Economic Potential
of a Slum Community in a Developmental Economy." Unpublished paper presented
at the United Nations Interregional Seminar on Improvement of Slums and Uncontrolled Settlements, Medellin, Colombia, 15 Febraury–1 March, 1970.

3. William Mangin, "Latin American Squatter Settlements: A Problem and a
Solution." *Latin American Research Review,* Vol. II, No. 3 (Summer, 1967), pp.
74–78.

4. For a fuller explanation of this phrase, see *ibid.*

5. A.P. Van Huyck and K.C. Rosser, "An Environmental Approach to Low-
Income Housing." *International Development Review,* September 1966.

6. Aprodicio A. Laquian, *Slums are for People.* Manila: College of Public Administration, 1969.

7. Mangin, *op. cit.,* p. 74.

8. See, *e.g.* D.J. Dwyer, "Problems of Urbanization: the Example of Hong Kong."
Transactions and Papers, The Institute of British Geographers, Publication No. 44,
1968.

9. D.J. Dwyer, "The City in the Developing World and the Example of Southeast Asia." Supplement to the *Gazette,* Vol. XV, No. 6, 20 August 1968, p. 8.

10. Elizabeth Wittermans, "Urban Patterns in Indonesia: Variety and Change."
Unpublished paper, Institute of Advanced Projects, University of Hawaii, 1969.

11. K.C. Zachariah, "Bombay Migration Study: a Pilot Analysis of Migration
to an Asian Metropolis." *Demography,* Vol. 3, No. 2, 1966, pp. 378–393.

12. Hamzah Sendut, "Contemporary Urbanization in Malaysia." *Asian Survey,*
Vol. VI, No. 9, pp. 484–492.

13. Zachariah, *op. cit.,* p. 383.

14. *Ibid.*

15. *Ibid.*

16. Talton Ray, *The Politics of Barrios in Venezuela.* Berkeley: University of
California Press, 1969.

17. Mangin, *op. cit.,* pp. 76–77.

18. Frankenhoff, *op. cit.*

19. *Ibid.*
20. Laquian, *op. cit.,* pp. 177–196.
21. Zachariah, *op. cit.,* p. 383.
22. *Ibid.*
23. *Ibid.*
24. A.A. Laquian, "Isla de Kokomo: Politics Among Urban Slum Dwellers." *Philippine Journal of Public Administration,* Vol. 8, No. 2, April, 1964.
25. Mangin, *op. cit.,* p. 66.
26. *Ibid.,* pp. 84–85.
27. Karl Deutsch, "Social Mobilization and Political Development." *The American Political Science Review,* Vol. LV, No. 3, September 1961.
28. Mangin, *op. cit.,* p. 80.

CHAPTER 8

Ved Prakash

LAND POLICIES FOR URBAN DEVELOPMENT

Introduction

Total land area on the surface of the earth is fixed, and with rapid increase in population land-man ratios have been declining continuously in almost all the countries. In theory, the supply of land for urban uses is limited only by the total supply of land in any given country. For the most part, however, the cities grow by absorbing farm lands. These additions in any particular urban area are determined by the relationship between agricultural and urban land values. The urban use must create a land value just higher than the farm value to induce a shift from one use to another. Of course, urban land values may—and do—rise much faster than the farm values. Although land for urban uses is available so long as urban land values exceed agricultural land values, there are competing demands on it for rural uses. Furthermore, any parcel of land in an urban or a rural area is capable of many uses, and like every rational economic choice, the determination of the most appropriate land use is an extremely vital social decision.

In the face of the magnitude and rapid rates of urbanization in South and Southeast Asian countries in recent decades and future projections, the land use patterns in each one of these countries is expected to undergo pervasive changes during the rest of the century. Provision of basic infrastructure, housing, and other facilities and services will make colossal demands for urban land. It is increasingly being recognized that strategies for economic growth in any country must also give adequate attention to spatial aspects of economic development. In other words, there is an urgent need for systematic coordination between economic development, locational, and land use and development policies.

In spite of the very substantial proportion of urban dwellers living in extremely over-crowded conditions, many of the Asian cities contain large pockets of vacant or undeveloped land within their incorporated limits. A recent survey of existing land use patterns in 103 cities and towns of India showed that spot densities varied from a minimum of 2 to 3 to a maximum of 300 to 600 persons per square acre in smaller cities, whereas in Calcutta, Bombay, and other metropolitan areas the figures ranged from 15 to a few thousand.[1] The same study estimated that, on the average, 41.2 percent of the total area of the cities surveyed was vacant or undeveloped. Less than 1 percent of the land within the Bangkok metropolitan area possesses densities higher than 50 persons per acre. Within Bangkok municipal boundaries, about a third of the area is under agricultural use and another 12 percent is undeveloped.[2] The situation may not be very different in other Asian cities. Lack of adequate planning and haphazard development in these cities is resulting in inefficient uses of urban land.

All over the world urban land values have been rising rapidly, and at rates much higher than the general price level changes. In most of the Asian cities, however, increase in land values seems to have been phenomenal. A recent study of several selected Indian cities revealed that land values during the 15 year period 1950–65 increased in some cases by 18, 43, or 68 times, whereas the consumer price index during the same period did not increase by more than 45 percent.[3] As J.P. Sah points out: [4]

"In Singapore, the price of residential land within a five-mile radius of the city centre was M$1.5 per sq. ft. in 1962 and M$3.0 per sq. ft. in 1964—a cent percent increase in two years. In the new township of Patalling Jaya near Kuala Lumpur, land prices increased by 375 percent in four years. In Karachi, the cost of land in the Pakistan Employees Housing Society area increased from Rs 3 to Rs 100 per square yard between 1951 and 1965, an increase of over 30 times. Land prices in central Seoul and Taegu, both in the Republic of Korea, have increased by 3,000 and 1,000 percent respectively within the last 10 years.

The sky-rocketing of urban land values unrelated to any perceivable economic factors is largely explained by speculation in land. In the absence of adequate investment opportunities in the productive sectors, finance capital—earned and unearned—finds real estate a lucrative business. Handsome returns are obtained either through rentals, which are high because of the severe shortages of all types of accommodation, or through capital gains on transfer. Also, ownership of land and property is a fascination with many people in the developing countries. It is a status symbol, a means of social security and a hedge against inflation in the face of rapidly declining purchasing power of money. Exorbitant land values have begun to prove a serious obstacle to housing and urban as well as economic development projects."

Urban land policy instruments include land acquisition, development, disposal, and taxation policies as well as regulatory devices such as zoning, subdivision, official mapping, etc. Although regulation of urban land is more or less universally accepted, it should be recognized that regulatory devices may at best be successful in achieving only certain negative restrictions on the use and development of land. As such, these techniques, "introduced from more advanced countries, are highly ineffective for solving the problems faced by rapidly growing urban areas in developing countries." [5] On the other hand, land taxation, acquisition, compensation, development, and disposal policies may be so devised and coordinated with regulatory devices as to play a positive role in promoting planned uses of land and achieving other land policy objectives.

This essay is primarily concerned with analysis of objectives of land policies for urban development in South and Southeast Asia and the role that alternative land taxation, acquisition, compensation, development, and disposal policies may play in achieving those goals.

Objectives of Urban Land Policies

The national as well as state or regional governments in several countries recognize the importance of an appropriate land policy for urban and regional development. These measures are mostly *ad hoc* and sporadic, however, and do not seem to form part of a well-knit overall policy [6] nor possess well defined objectives.

In India the subject of urban land policy has received a great deal of attention in the recent years. Thus, Chapter XXXIII of *The Third Five Year Plan* on "Housing and Urban and Rural Planning" devoted a separate section to "Urban Planning and Land Policy." [7] In order to reduce the high cost of urbanization, certain measures have been suggested.[8] The major concern, however, is with rising land values in urban areas, and a number of specific steps to check the rise in land values are outlined in the *Plan*.[9] The recommendations of the Planning Commission contained in the Five Year Plans are in the nature of broad suggestions and general directives.

Owing to the importance attached to the question of an urban land policy in *The Third Five Year Plan,* the Town and Country Planning Organization, Ministry of Health, Government of India, has been devoting its attention to this problem during the past few years. This organization prepared a "Note on Urban Land Policy" which it submitted for consideration to the Second Conference of State Ministers for Town and Country Planning held at Trivandrum on October 13–15, 1961.[10] This conference decided that the "Note" should be sent to the state governments and the different ministries of the Government of India for comment. The Third Conference of the State Ministers for Town and Country Planning held at Calcutta during November, 1962, recommended the appointment of a committee to make a thorough study of the problem. The constitution of this committee was postponed, due to the national emergency as a consequence of the border dispute with China. The matter was discussed again at the Joint Session of the Central Council of Local Self-Government and the State Ministers for Town and Country Planning held at New Delhi in September, 1963. Following resolution No. 11 of the above joint conference, the Committee on Urban Land Policy was set up. This Committee finalized its report towards the close of the year 1964.[11]

The recommendations of the Planning Commission as listed in *The Third Five Year Plan* are primarily concerned with reducing the high costs associated with urban development. To achieve this, the following measures were suggested: [12]

1. Control of urban land values through public acquisition of land and appropriate fiscal policies;

2. Physical planning of the use of land and the preparation of master plans;

3. Defining tolerable minimum standards for housing and other services to be provided for towns according to their requirements, and also prescribing maximum standards to the extent necessary; and,

4. Strengthening of municipal administrations for undertaking new development responsibilities.

The major emphasis is on checking the rise in land values and among others the following specific steps have been suggested in *The Third Five Year Plan:* [13]

1. Issue of notifications for freezing land values with a view to early acquisition of land by public authorities.

2. Acquisition and development of land by public authorities in accordance with the interim general plans is essential for preventing speculation. The land should be acquired in bulk, although, depending upon local circumstances, the programme of acquisition would have to be suitably phased. Acquisition proceedings should be speedy and legal procedures should be simplified as far as possible. It is important that development of the acquired lands should be expedited. The essential services have to be provided by public authorities. Besides development undertaken directly by them, under appropriate regulations, cooperative and private agencies should also be utilized.

3. Allotment of land on a lease-hold basis. As a rule, lands acquired by public authorities should be given out only on a lease-hold basis so that, besides the recurring income secured on account of the ground rent, a fair share in the increase in the value of land continues to accrue to the community.

The Town and Country Planning Organization in its "Note on Urban Land Policy" suggested, and the Committee on Urban Land Policy adopted, the following four principal social objectives of urban land policy:[14]

1. To achieve an optimum social use of urban land.

2. To make land available in adequate quantity, at the right time and for reasonable prices to both the public authorities and the individuals.

3. To encourage cooperative community effort and bona fide individual builders in the field of land development, housing, and construction.

4. To prevent concentration of land ownership in a few private hands and safeguard specially the interest of the poor and under-privileged sections of urban society.

The foregoing objectives of a suitable policy for urban land as recommended by the Planning Commission and the Committee on Urban Land Policy are noteworthy. In brief, they are aimed at: (1) achieving planned uses of urban land; and (2) curbing speculation as well as holding down urban land prices. The first objective is to be gained through urban and regional planning methods. The second objective is to be fulfilled by increasing the supply of urban land at "reasonable prices." [15] This, in turn, is to be done by large-scale public acquisition of land in advance, by freezing urban land values, and by limiting the compensation liability for land that may be acquired.

Pakistan's Second and Third Five Year Plans do not specify land policy objectives as such, although, these plans recognize the link between physical and economic planning and emphasize their integration. By implication, land policies there are expected to be consistent with planning and location of economic and other activities. Planning of balanced industrial development and urban complexes, encouragement of land development for housing and related facilities, especially for slum dwellers and squatters, and discouragement of speculation in land are some of the major objectives of land policies in Pakistan.[16]

According to Thailand's development plans, a major objective seems to be the prevention of excessive growth in its primate city, Bangkok. Thus, "preparation of regional plans for balanced geographical distribution of development projects, development of medium towns as the regional foci, alleviating the housing situation and clearing slums are the other proposed objectives to which land policy is to be geared." [17]

Given the severe physical constraints in Hong Kong, land conservation and maximum utilization of available space are logically the major objectives of its land policy. These in turn imply very high density urban and housing development.

Information on land policies in other South and Southeast Asian countries is not available. However, it may be safe to assume that many of these countries have not given sufficient attention to these questions. India seems to be the only country where explicit land policy objectives have been formulated, a number of measures are suggested to achieve these objectives, and attempts are being made to integrate them with overall goals of urban and regional development. Even in Pakistan, Thailand, and Hong Kong, where land policies are implicit in their development plans, they do not include such basic aspects as controls on speculation and mounting urban land values, and these objectives have not been integrated in the form of an overall multi-measure policy-frame.[18]

The goals conceived by the Planning Commission and the Committee on Urban Land Policy are worth striving for. The means they suggest to achieve these objectives suffer from two major shortcomings. The role of the market as a disciplined, self-adjusting mechanism is more or less ignored, and the inter-relationships between rural and urban land taxation systems, compensation policy and selling and/or leasing of urban land have not been given serious consideration. The Committee on Urban Land Policy feels that the market value of land in urban areas is higher than its actual worth since market prices reflect unearned increments in land values as well as potential development value. Payment of compensation to owners whose land may be acquired is unfair to public authorities and to the community which has to bear the financial burden of such compensation costs through taxes and other charges. In order to limit the compensation liability for land acquisition two measures are suggested: ante-dating compensation or freezing land values and advance acquisition of land.[19]

It is logical that the Planning Commission, the Town and Country Planning Organization, and the Committee on Urban Land Policy show concern for high land prices in urban areas; but to limit the compensation liability as suggested by them raises certain crucial constitutional, administrative, and economic questions. The implications of these important issues have been clouded by ideological considerations. When solutions to economic problems are sought through non-economic methods they are unlikely to be acceptable to society as a whole.

Government in any country is based on the implied assent of its members that individual private interest shall be subservient to the public interest. The private interest is subordinate to the common welfare of the country as a whole. *Necessitas publica major est quam privata.* The power of "Eminent Domain" or public acquisition of land is inherent in every governmental system. Except in the totalitarian countries—where such power may be confiscatory—this right of the State to appropriate private property for public purpose must provide for just compensation to persons for the loss of their properties. Article 31(2) of the Constitution of India dealing with the question of compensation for public acquisition of land does not qualify the word "compensation" by the word "just." The Supreme Court of India in *State of West Bengal v. Bela Banerjee* upheld the word "compensation" as meaning a "full and fair money equivalent" of the property taken and "any law which denies this must be held to be void." [20] In another case the Supreme Court declared that

". . . the provision for compensation is merely a cloak for confiscatory legislation. It is not open to the Legislature to lay down any principles which may result in nonpayment of compensation or which may result in not paying any compensation whatsoever." [21]

Because of Court decisions unfavorable to the government, Article 31(2) was amended in 1955.[22] According to Durga Das Basu, "This change is the result of the reaction to the decision in the case of *Bela Banerjee* to the effect that under existing clause (2), the Court had jurisdiction to enquire in every case of appropriation of private property by the State whether the person appropriated has been ensured the true value of the property appropriated." [23] The amended clause (2) of Article 31 reads: [24]

"No property shall be compulsorily acquired or requisitioned save for a public purpose and save by authority of a law which provides for compensation for the property so acquired or requisitioned and either fixes the amount of the compensation or specifies the compensation is to be determined and given; and no such law shall be called in question in any court on the ground that the compensation provided by that law is not adequate."

Under the Fourth Amendment the question of "adequacy" of compensation has been taken away from the Courts; the final judgment is with the Legislature. Basu feels that "Though it is extremely risky to hazard any opinion as to what view the Supreme Court will take as to the interpretation of the word 'adequate,' it would be wrong to suppose that all questions relating to the exercise of the power of eminent domain have been taken away from the jurisdiction of the Courts." [25] He further points out that: [26]

"Of course, the amended cl. (2) says that the 'adequacy of compensation' shall not be justifiable, but, according to canons of legal construction, this provision cannot be used to control independent provisions outside Art. 31(2). Thus, a law providing for 'deprivation,' 'acquisition' or 'requisition' shall still be open to be challenged on the ground that it offends against Art. 14."

Even if ante-dating or freezing of land values for compensation purposes were constitutionally permissible, it would be grossly inequitable unless all land was nationalized. The persons whose lands were acquired would be more disadvantaged than those whose land would continue to remain in private ownership. If large tracts of land were notified for simultaneous acquisition, the consistent upward trend in urban land market values would cause an increasing inequity in compensation at successive stages of the acquisition process in particular urban areas. Obviously all notifications to private owners could not be made at once, but would be stretched out over thirty or forty years, and possession of and compensation for lands covered by a single notification would be spread over several years.

Determination of compensation for public acquisition of land, if related to market value at some previous date, would create other very real administrative difficulties. Most of the land acquisition cases referred to the Courts under Section 18 of the Land Acquisition Act, 1894, pertain to compensation. These cases contest that the amount of compensation awarded by the Collector was different from the market value of the land on the date of notification under Section 4. The acquisition process in the absence of an intervention by a Court of Law may take about two years.[27] When a Court intervenes the process is much longer. Because determination of compensation from a predated market value is inequitable, it naturally increases the incidence of Court cases and prolongs the acquisition process. As success of an urban land policy largely depends upon the speed at which the acquisition process proceeds, the lengthening of the process frustrates its objectives.

The crux of the compensation problem, however, is determination of the market value of particular parcels of land. The market value may be described as the price which a prudent and willing owner (who is not obligated to sell) might reasonably expect to obtain from a prudent and willing purchaser (who is under no necessity to buy) on the open market. The market value in this context is invariably the imputed value. It may generally be derived on the basis of recent and *bona fide* sales of land in particular localities. In order to simplify the determination of compensation liability or the market value, it may be best to change the system of real estate taxation from annual ratable to capital or market value, in addition to making a separate assessment for land and improvements. Such a change is also warranted from the point of view of rationalizing the present system of property taxation in India and will be discussed subsequently. If this is done, then it may be stipulated that municipal assessment records shall be taken as proof of the market value of land.

The main argument offered by the Committee on Urban Land Policy in favor of paying compensation at less than market value of land at the time of acquisition is that the prevailing market prices of land are too high, as they reflect unearned increments accumulated over the past several years, and that the unearned increment consists of potential development value, most of which is of speculative nature. But to appropriate the unearned increments through an arbitrary compensation policy seems to be unsound and irrational. A better method of doing this would be through land disposal and taxation policies.

Development and Disposal of Land for Urban Planning

An urban land policy may function both as a planning device for guiding future urban development and as a resource for financing such development. Public acquisition of land is a first step in that direction and is one of the means to achieve rational land uses according to a plan, but care must be taken that public ownership of land does not become an end in itself. Land is generally acquired for the definite purpose of developing and making it available for appropriate uses at some future date. Development and disposal of urban land are thus a vital concern of public policy.

Land development by a public authority has all the advantages of subdivision control exercised by local governmental units. The major shortcoming of subdivision control is that it does not provide an effective means of promoting land development for uses other than those determined by a private developer. If development is carried out by a public agency according to the goals and policies of a master plan, the above deficiency can be overcome and coordination between land development and improvement programs in urban areas is insured.

Yet, the private sector may continue to play a significant role, for even if land development is entrusted to a public agency, it may be necessary to invite bids from private developers and contractors to carry out the development according to standards and conditions prescribed by a development agency. Land development through contractors has two distinct advantages: (1) it will make use of the expertise and trained manpower already available within the system; and (2) it will be much less of a financial burden for public agencies. This second advantage follows from the fact that if land development and disposal are properly phased, then the payments to the contractors can be synchronized and coordinated with the receipts from land that may be subsequently disposed. The quantum of public investment needed for acquisition and development of a specific amount of land may thus be greatly reduced.

Public policy for development and disposal of land deals, for example, with the type of development to be carried out before disposal by public agencies, the decision to lease or sell the land, and the appropriate pricing policies for land disposal. These questions are considered here.

What type of development should be carried out before the land may be subdivided into plots and disposed of by the development agency? This question has to be largely determined by the master plan. Generally speaking, land development may consist of the following items:

1. Survey and preparation of sites
2. Roads and streets
3. Water supply
4. Sewerage and storm water drainage
5. Electricity distribution and street lighting
6. Landscaping

Items 3 (water supply) and 5 (electricity) may consist of different components in different areas as well as different localities within the same area. For example, electricity distribution and street lighting may be the responsibility entirely of the utility company providing electricity in particular areas. Similarly, the laying down of water pipes and sewer mains may be carried out by the municipal government rather than through the development agency.

Selling v. Leasing

Once the land is acquired, developed, and subdivided, the individual plots can either be sold on a free-hold basis or transferred on a lease-hold basis. Both practices are being followed in India and other countries. The Indian *Third Five Year Plan* is emphatic in stating that publicly owned land should be leased and not sold.[28]

Both systems—leasing and selling—have advantages as well as disadvantages. It is preferable to transfer the land on a lease-hold basis for two main reasons:

1. the lease-hold transfer of land allows the public authorities to exercise a greater degree of control over land uses;
2. the lease-hold basis provides the public authorities with more effective means to exercise control over land values as well as appropriate unearned increments in land values through pricing and taxation policies.

In some cases it may be possible to sell the land and at the same time exercise suitable controls by incorporating in the conditions of sale appropriate clauses to recapture the land if it is not built upon according to prescribed regulations and within stipulated periods. The selling of land has the distinct advantage of a speedy recoupment of capital costs. Furthermore, under the lease-hold system, there may be certain psychological barriers. An individual who owns land (free-hold transfer) may not hesitate to carry out the improvements, whereas he may be apprehensive of building upon land which is not legally owned by him.

In theory, therefore, it is likely that both systems—selling and leasing —may work equally satisfactorily in achieving the suggested twin purposes of an urban land policy. The success would, however, depend upon the design of pricing and taxation policies under each method. It is not possible to forecast the superiority of one method over the other due to lack of evidence and limited experience. It may perhaps be best to experiment with these two methods in their varying forms in different urban areas.

Theory aside, however, one overriding realistic consideration warrants that a major portion of the developed land must not be sold, but transferred on a lease-hold basis. At the present time, and conceivably for many years to come, due to limited resources and the very low level of incomes, urban planning and development in India has to be primarily concerned with providing infrastructure and basic public health facilities in urban areas. A realistic appraisal of the urbanization problem would require a considerable scaling down of the standards of services in the urban areas and housing may have to be assigned a low priority. Flexible interim land use planning is needed to bridge the gap between the present and the time in the future when standards can be improved, and leasing provides a great deal more flexibility than do outright land sales. In addition, leasing itself can be made more flexible by writing leases for different periods of time, depending upon the circumstances.

Examining this consideration more closely, we can recognize that, for a majority of the urban population, it may be feasible to provide each family with a plot of developed land—with hard-surface lanes, community water taps and latrines, drainage and street lighting—to encourage self-help housing, and to permit the building of huts and shacks on such land by the low-income families. To put it in another way, slums will remain an important element in urban housing in India. The emphasis on slum clearance and redevelopment may have to give place mainly to slum improvement.

The flexibility in planning approach aimed at providing interim standards and land uses involves preparation of a series of land use plans and public service programs over the period covered by a master plan, say thirty or forty years. For example: the land use plan for an urban area may earmark certain areas for a central shopping district and a civic center. Since, during the ensuing ten years or so, only a part of the shopping and civic center may be built, the interim land use plan may permit the unbuilt area to be leased for shorter periods to small vendors and stall holders (equivalent to hot dog or hamburger stands). Similarly, under a phased program for housing the low-income group of families, parts of the developed land may be allotted for temporary settlement and construction of huts.

If the developed land is sold outright, the flexibility necessary for interim planning as suggested above may become extremely difficult and cumbersome. Future programs may largely become renewal and redevelopment, and land acquisition all over again may be very expensive and time consuming. These reasons are added justifications for leasing as opposed to selling as a general ingredient of land disposal policy.

To take advantage of the flexibility inherent in leasing, the leases themselves should be varied to suit particular circumstances of individual urban areas. The following different categories of leases or any combination of them may be considered as possibilities:

1. *Long-term Lease for Permanent Buildings and Structures.* This type of lease may be for thirty years. The lessee may be given two options for renewal of the lease for another thirty years each, and land leased under this category may be for permanent buildings and structures in accordance with the long-term land use plan.

2. *Medium-term Lease for Semi-permanent Structures.* The period of such leases may be fifteen to twenty years. The lessee may be permitted to put structures on such land whose life may not be expected to be longer than the duration of the lease. In this case interim land uses or development at interim standards may be permitted.

3. *Short-term Lease.* The duration of this lease may be five to ten years. The lessee may be permitted to construct temporary structures on such land. Such areas may be specifically earmarked for construction of huts and vegetable and groceries stalls (permitting use of such materials as bamboo, mud-walls, etc.).

4. *Monthly Tenancy.* Plots of developed land may be allotted on a monthly rental basis. A part of the central and sector or neighborhood shopping areas may be allotted to small vendors who may, e.g. bring commodities and wares to such areas, sell them during the day and return home with empty baskets.

The classification of different types of leases as well as their duration must be kept flexible and may have to be modified to suit the needs and purposes of urban planning programs of individual areas. The administration of the lease-system would largely depend on its coordination with and updating of municipal bylaws.

Price Policy for Land Disposal

Payment for publicly owned land transferred to individuals and organizations, either free-hold sales or leaseholds, generally takes one of the following forms:

1. For outright sales, a lump-sum payment may be realized at the time of the sale of the land. In some cases payment may be arranged in annual or half-yearly installments, when interest charges on the sale prices are also added.

2. In case of leasehold transfers, two methods are used:

    (a) A downpayment—generally known as a premium—is required at the time of the transfer of land. In addition, a fixed percentage of the premium amount or a fixed sum may be charged annually.

    (b) No downpayment is necessary. A fixed sum is charged each year.

(The annual charge both under (a) and (b) is invariably known as ground rent. Sometimes, lease deeds contain a provision under which annual charges may be revised after every 10, 20, or 30 years.)

In addition to the lump-sum payment under method 1 and the downpayment and annual lease charges under method 2, the usual municipal property taxes are also levied.

In the case of outright sales, the amount of the lump-sum payment for different plots of land is predetermined. The prices so fixed may take into account the locational advantages and permissible uses for different plots. Land earmarked for commercial and industrial uses may be priced higher than the residential plots. Again, residential plots may have different prices for high, medium, and low-income families. The assignment of individual plots is decided by a public authority—sometimes allotted by drawing lots.

The downpayment and annual lease charges are determined more or less in the same manner as the prices for land sold on a free-hold basis. Recently, development agencies have sometimes auctioned leaseholds for certain categories of plots, generally for non-residential uses. Annual charges even in these cases are either a fixed percentage of the average price of the land or are a fixed amount per acre of per plot of land (the amount in these latter cases dependent upon the size of the plot).

The Committee on Urban Land Policy in India briefly discussed the pricing policy for land disposal and observed that: [29]

"Another significant aspect of land disposal policy relates to prices charged for various uses and from different income groups. From the information at our disposal we observe that some sort of discriminating price policy is already being practised by most of the public authorities engaged in land development and disposal. The general practice seems to be to allot land by draw of lots to persons in the low income groups including the slum evictees and by public auction in the case of others. It is also observed that slightly higher prices are charged by the public authorities for commercial and sometimes industrial uses. We are glad that a discriminating price policy is already being practised. We, nevertheless, feel that there is need to refine the price policy further. We wish to add that the overall objective of price policy should be to help the poorer sections of society and to encourage uses which are in the larger interests of the community like those for schools and playgrounds. At the same time, the financial soundness of the scheme as a whole should not be jeopardised."

The Committee on Urban Land Policy disposed of the question of pricing policy in just the one paragraph reproduced above. Evidently, this aspect of urban land policy has not received the serious consideration it deserves. The Committee's recommendations are rather vague and do not follow a realistic appraisal of the problem.

The guiding principle for an urban land policy must be that the public authority entrusted with land development and its disposal should not incur any loss, and should perhaps make a reasonable profit. These operations have to be run in a business-like manner, which presupposes careful consideration of the standards at which the land may be developed, the costs associated with such developments, and the prices to be charged at the time of transferring the developed land. The first two items can be estimated on the basis of engineering and cost data, whereas the pricing system cannot work if isolated from the market mechanism.

Assuming that public authorities developing urban land are somehow in a position to put price tags on individual plots of land by taking into consideration the locational and land use aspects of each parcel of land, as well as safeguarding and promoting the interest of the weaker sections of the economy, buyers of land will be willing to pay the prices fixed in this manner only so long as they are either equal to or less than the market value of each plot of land. To put it in another way, if the prices fixed administratively are higher than the market values, it would not be possible to dispose of the land.

It must be emphasized that the redistributional aspects of income and wealth are primarily the concern of national expenditure and taxation policies. The State governments to a lesser extent may supplement national policies. The problems relating to subsidized housing for individuals and families in the lower income groups, even though closely tied to the attainment of social objectives, must also be seen in economic and financial terms. The suggestion of the Committee on Urban Land Policy that the overall objectives of price policy for each urban community should be to help the weaker sections of society may cloud the economic calculations which must be made by federal and state governments in order to arrive at correct welfare policy decisions. The market value of any parcel of land is largely determined by the forces of demand and supply. The preparation and implementation of a master plan, especially a land use plan, influences and alters urban land values. When land policy—public acquisition and development—is also utilized as a positive tool for urban development, public policy measures become important determinants of urban land values. The land values thus determined indirectly contribute to other objectives as well. If determining the value of each plot of disposable land is entrusted to public authorities, and if these agencies become guided by subjective considerations as well, the fair administration of pricing and land disposal policies will be highly problematical. The decision-making process will lack objectivity. In addition, there is an inherent danger that such wide discretionary powers may sometimes be abused.

If it is accepted in principle that prices to be charged for publicly owned land should not be below market values at the time of disposal, then how should sale prices be determined? The valuation and pricing system may necessitate the adoption of the following measures:

1. Setting up of regional real estate valuation organizations and establishment of a number of field offices.[30]
2. All disposable land—to be sold or leased, and irrespective of the method of recovering the land prices or charges—must be publicly auctioned.

The valuation organization must be manned by suitably qualified and well trained personnel in order to achieve effectively the objectives of an urban land policy. The organization may estimate the approximate values of land from two sets of data: (1) the cost of acquisition and development of land in large subdivisions provides the total amount that must be realized at the time of transfer of the disposable land in that particular subdivision; and (2) the value of individual parcels of land would be determined through valuation techniques based on such factors as measurement, location, frontage, depth, etc. of each lot.

The receipts at the time of transfer of land through public auctions would depend upon whether the land is sold or leased, the price in the latter case also varying with the duration of the lease period. When land is sold outright, buyers are often willing to offer amounts higher than they would bid if the same land were being offered for lease for a fixed number of years.

The two measures—establishment of valuation organizations and transfer of public land through public auctions—if adopted simultaneously would also provide a self-checking mechanism. It might at times indicate the directions in which some components of an urban land policy would need reconsideration and modification. To illustrate: if standards of development in a particular area are high, or the timing of development immature, it may not be possible to dispose of the land at prices suggested by the valuation organization. If this happens, the standards of development may have to be modified or the phasing of the land development program may have to be altered.

Coordination Between Land Taxation and Urban Land Policy

The effectiveness of the measures for achieving objectives of land policy, discussed in the preceding pages, would largely depend upon the relationship of those measures to the system of taxation of urban and urbanizable land. The objectives of curbing speculation in land, appropriating unearned increments in land values, and curbing land values, may be achieved more effectively through appropriate tax policies, especially the system of real estate taxation.

Tax on realty in one form or another has been utilized for financing urban services in most parts of the world. Tax levies on property generally takes, with some exceptions, one of the following three forms:

1. A tax on the capital value of the property (land and improvements). This is the basic system in the United States and a majority of the Latin American countries.

2. A tax assessed on the basis of annual value, which in turn is generally related to the annual rental value of the property (land and improvements). This is basically the system of real estate taxation in the United Kingdom, India, Pakistan, Ceylon, and many of the countries in our region.

3. A tax on the unimproved value of land (land value is taxed but the improvements or buildings are exempt). This system is common in many local bodies in New Zealand, Australia, and some provinces in Canada.

The different systems of property taxation—i.e. when the basis is capital or annual value, and when both land and improvements are taxed either at the same or different rates or improvements are exempt—lead to varying effects and incidences. In an analysis of the effects of real estate taxation, it must be recognized that land and buildings are two different things. "They are different factors of production—the economic concept of 'land' applying to the one, the concept of 'capital' to the other." [31]

Seligman clearly points out the differences between land and improvement from the viewpoint of analyzing the incidence of real estate taxation: [32]

"The urban real estate tax is either a pure land tax—for example laid on vacant lots—or a tax on both the land and the buildings. The latter is called in America the real estate tax, and on the continent the house tax; but both of these designations are, from the point of view of economics, incorrect. The continental term is wrong because the house tax really includes a tax on the site as well as a tax on the structure. The American term is inexact, because it confuses such entirely different taxes as the ground tax and the building tax, which are governed by different laws of incidence."

According to Seligman, if a tax is shifted, it cannot be capitalized. On the other hand, if a tax is capitalized, it cannot be shifted. He concluded that the tax on improvements is normally shifted, whereas the tax on land is capitalized. In other words, the tax on land leads to a reduction in the market price of land. It is on these grounds that substantially higher taxes on land are advocated to reduce speculation in land as well as encourage building activity and promote more efficient and economic uses of land.

Since the effect and incidence of taxation on land and improvements is significantly different, and in order to make the system more flexible for achieving rational land uses and other urban development goals, it is necessary to separate the assessments for land and improvements for real estate tax purposes. Different standards of measurement apply for valuation of land and improvements. In fact, "the techniques of appraising land and improvements differ so much that it is necessary for assessors to value each of these major divisions of real property separately." [33] The appraisal of land is generally made on the basis of sales analysis whereas the improvements are valued by the replacement-cost method.[34]

When the property tax base includes both land and improvements, the difference between capital value and annual value is a fundamental one.

However, in theory there is no significant difference between the sale or rental value bases of taxation when land alone is taxed and the improvements are exempt. The salable value of a piece of land is arrived at by capitalizing its future expected rents. However, there is a difference between the taxation of rent actually received and the taxation of rental value. In the former case, if a piece of land is kept out of use no tax is levied irrespective of the rent such land would yield if used:[35]

"In the other case the tax on unused land from which no rent was received by the owner would be as high as if the land were yielding its entire potential rent. Hence, in this latter case there would be a definite penalty against the holding of valuable land out of use."

Looked at in another way, any given piece of land at a particular point in time can be assumed to have specific ideal rental value based on its location in relation to other developing land and sites of economic activity, on existing levels of income and ways of doing business, and on existing transportation facilities and public improvements. In the event that a particular piece of land is developed below the highest permissible use, its actual rental value may be less than its ideal rental value. In this sense the ideal rental value is the true determination of capital or salable value.

In a majority of the South and Southeast Asian countries, where annual value is the basis of property taxation, unimproved parcels of land within the urban areas bear an extremely low burden because actual rent in such cases is very low or may even be zero. Under such a system it is very profitable for the land owners to keep the land out of use for speculative purposes. The position is substantially the same when a piece of land is developed at levels lower than what is warranted by land values. It is reasonable to believe that when land values are rising, this system encourages speculation and leads to less efficient and uneconomic uses of land. When the tax is based on actual rents, it is less productive of revenues and at the same time it necessitates higher costs for municipal services.

When economic aspects of annual value and capital value as the bases of property taxation are considered, both methods offer certain advantages and limitations over each other. On balance, however, property taxation based on capital value appears to be preferable for the developing countries. As U.K. Hicks points out: [36]

"The great advantage of an annual value which is closely related to rent is that it is easy to explain to the taxpayer what he has to pay is just a percentage of his rent. Overseas, many rating officers today hanker after annual value on this account, . . . but from the economic point of view it has three rather serious disadvantages in developing countries. The first of these (which is indeed common to annual value and capital value, but

not to unimproved value) is that every addition to the property, in principle at least, immediately gives rise to additional rate liability; the tax
thus tends to be disincentive both of new building and of the improvement of existing houses. Secondly, in countries where rent control exists
or is likely to be introduced in an emergency, rents will be frozen in
relation to other prices, so that the real value of the revenue fails to
keep pace with government needs. In the U.K. half-hearted attempts
were made to get round this difficulty by basing liability on what the
rent would have been if there had been no rent control; they were not
successful. The developing countries have been if anything more ardent
believers in rent control than the U.K., so if they insist on using annual
value they must expect this sort of trouble.

   "A more serious drawback to annual value, especially relevant to
developing countries, is that the rent in question and hence the valuation,
is closely tied to the existing building (or in valuation terminology, existing user). When a town is developing rapidly people will want to buy
property not so much for the use of the existing building, as for what
they could put there instead. For this reason the true value of the property
may far exceed the capitalized value of rent of the present hereditament,
and the true value will go on rising as the town develops, although the
rent may be fixed. On the annual value base the town authorities will thus
be foregoing the advantage they might be reaping from the rise in property
values."

The shift from rental to capital value as the basis of assessment would
make the real estate taxation more equitable and in all likelihood make
the system more productive. Besides, the proposed change would, to an
extent, tax the unearned increments in land values and at the same time
discourage speculation and promote more efficient use of land.

   Since property values generally rise much faster than the increase in
prices of other goods and services—and this is especially so in rapidly
growing urban centers in the developing countries, an efficient system of
property taxation requires that assessments must be kept up-to-date. It
may be desirable that state and/or regional governments set up real estate
valuation organizations and also establish a number of field offices. Since
these valuation organizations must employ suitably trained personnel,
they may also be entrusted with valuation of real estate for local purposes.
This would avoid duplication of administration at state and local levels,
and at the same time insulate local assessments from the political processes.
The centralization of assessment was recommended in India by the Local
Finance Enquiry Committee and the Taxation Enquiry Commission.

   The present system of real estate, whatever its form, is not entirely
satisfactory for urban development in our region. In addition to separa-
tion of assessments for land and improvements, adoption of capital value
as the basis of property taxation, and assessment procedures as discussed
earlier, some further modifications are necessary. Two such measures are:
(1) taxation of land at rates substantially higher than improvements, and

(2) a special ad valorem tax at the time of sale and/or transfer of properties. These changes would make the system of real estate taxation more equitable and more productive; they would discourage speculation in land, encourage building activity, and promote more efficient and economic uses of land; and they would tax the unearned increment in land values and in all likelihood result in a reduction of the market price of land.

Conclusion

In the face of rapid urban growth, a long-range land policy is indispensable for urban and regional planning and development in each of the developing countries. Urban land policies include government legislation and public policies concerned with land use planning, land acquisition, development and disposal, taxation, zoning, subdivision control, and other devices for regulation of land uses. Some countries in South and Southeast Asia have been concerned with these problems and in recent years have formulated land policies for urban development. However, these measures are mostly *ad hoc* and sporadic, and do not seem to form part of a well-knit overall policy. There is an urgent need to coordinate land use planning, public control over land, and land taxation.

NOTES

1. See J.P. Sah *et al., Land Use Patterns of India's Cities and Towns, A Draft Report*. New Delhi: Government of India, Town and Country Planning Organization, 1966.
2. See Litchfield, Whiting, Bowne and Associates, *Greater Bangkok Plan, 1990.*
3. See K.A. Ramasubramaniyam, "Steep Rise in Values of Urban Land." *Yojana,* January 26, 1966.
4. J.P. Sah, "Land Policies for Urban and Regional Development in the Countries of the ECAFE Region." Paper prepared for the Seminar on Planning for Urban and Regional Development, Including Metropolitan Areas, New Towns and Land Policies, held at Nagoya, Japan, October 10–20, 1966, sponsored by the United Nations and the government of Japan.
5. International Union of Local Authorities, *Urbanization in Developing Countries.* The Hague: Martinus Nijhoff, 1968, p. 53.
6. See J.P. Sah, "Land Policies for Urban and Regional Development in the Countries of the ECAFE Region," *op. cit.,* p. 14.
7. Government of India, Planning Commission, *The Third Five Year Plan.* New Delhi: Government of India Press, 1961, pp. 679–699.
8. *Ibid.,* p. 690.
9. *Ibid.,* p. 691.
10. Government of India, Ministry of Health, *Proceedings of the Second Conference of State Ministers for Town and Country Planning.* New Delhi: 1961, pp. 73–77; and Appendix III, "Note on Urban Land Policy," pp. 195–239. See also pp. 77–84 and Appendix IV, "An Act for Acquisition of Land for Town Planning Purposes," pp. 240–264.

11. Government of India, Ministry of Health, *Report of the Committee on Urban Land Policy*. New Delhi: 1965.

12. *The Third Five Year Plan, op. cit.*, p. 690.

13. *Ibid.*, p. 691.

14. *Report of the Committee on Urban Land Policy, op. cit.*, pp. 14–15.

15. *Ibid.*, p. 14.

16. J.P. Sah, "Land Policies for Urban and Regional Development in the Countries of the ECAFE Region," *op. cit.*, p. 16.

17. *Ibid.*, p. 17.

18. *Ibid.*, p. 17.

19. The Town and Country Planning Organization in its *Draft Model Act for Acquisition of Land for Town Planning Purposes* recommends that amount of compensation shall be the equivalent of the market value of land as of January 1, 1951, plus 25 percent of the difference between market value of land at the date of publication of the notification for acquisition and the market value on January 1, 1951.

20. *State of West Bengal v. Bela Banerjee* (1954), Supreme Court Appeals, 41(45). For a detailed analysis of constitutional issues and court decisions see B.N. Mukerjee and David L. Wilcox, "Development Planning and the Power of Acquisition of Property," a paper prepared for the Seminar on Law and Urbanization Held at Allahabad, December 26, 1967, sponsored by the Indian Law Institute, New Delhi.

21. *State of Bihar v. Kameshwar, All India Reports* (1952), Supreme Court Appeals, 252.

22. *The Constitution (Fourth Amendment) Act. 1955.*

23. Durga Das Basu, *Commentary on Constitution of India,* 3rd Ed. Calcutta: S.C. Sarkar, 1955, Vol. 1, p. 833.

24. *The Constitution (Fourth Amendment) Act. 1955.*

25. Durga Das Basu, *op. cit.*, p. 835.

26. *Ibid.*, p. 835. Article 14, concerned with the Right to Equality, provides that "The State shall not deny to any person equality before the law or the equal protection of the laws within the territory of India."

27. The Law Commission of India, as quoted in "Note on Urban Land Policy," *op. cit.*, p. 215.

28. *The Third Five Year Plan, op. cit.*, p. 691.

29. *Report of the Committee on Urban Land Policy, op. cit.*, pp. 79–80.

30. Establishing a valuation organization is justified from the viewpoint of rationalizing the system of municipal taxation of properties and of coordinating this with the urban land policy. The desirability of an independent agency or an independent assessing officer has long been accepted the world over. So long as both assessor and assessment procedures are related to the political process, incalculable mischief may be the natural outcome.

31. Mary Rawson, *Property Taxation and Urban Development*. Washington, D.C.: Urban Land Institute, 1961, p. 8.

32. Edwin Seligman, *The Shifting and Incidence of Taxation*, 4th ed. New York: Columbia University Press, 1921, p. 277.

33. International City Managers' Association, *Municipal Finance Administration*. Chicago: International City Managers' Association, 1962, p. 83.

34. *Ibid.*, p. 83.

35. Harry Gunnison Brown, "Taxing Rental Versus Taxing Salable Value of Land." *Journal of Political Economy,* Vol. 36 (February 1928), p. 164.

36. U.K. Hicks, *Development From Below*. Oxford: Clarendon Press, 1961, p. 356.

CHAPTER 9

Tarlok Singh

URBAN DEVELOPMENT POLICY AND THE ROLE OF GOVERNMENTS AND PUBLIC AUTHORITIES

Analysis and Action

Two processes are under way in all aspects of national planning. On the one hand, as its scope expands, concerns and activities which at first fell beyond its purview now move closer to the center. On the other, it contains a greater degree of complexity in the forces at work and of inadequacy in analysis, means, and resources. The growth of population, increasing discontents, and disparities—the products of change, impart to every major economic and social problem the character of an enormous drift. In looking at any important problem comparatively, across a group of different but not wholly dissimilar countries, we free ourselves from some part of the detail and observe more clearly both its uniqueness in each different setting and its common features which may perhaps admit of a more general approach. However, a wide gulf separates analysis from action. By its nature, analysis draws attention to causes and principles, action to the limitations in each separate situation and to the institutional devices by which they may be overcome. The study of urbanization as embodying certain processes in social and economic change belongs to analysis; formulation of urban development policy, as a way of influencing the future, perhaps even of controlling it, falls within the sphere of action.

In the present phase of national development in South and Southeast Asia, the urban process is expressed through the growth of numbers and the concentration at key points of an increasing variety of economic, social, and administrative activities. Except for a limited number of people, ur-

Author's Note: The writer wishes to express his gratitude to Professor Gerald Breese and Mr. Albert Mayer for having read this paper in draft and favored him with their suggestions.

banization is not yet a pattern of culture which has been chosen and assimilated. The populations of cities and towns present levels of living, income, and opportunity which are often too diverse and too full of contrasts to possess any notion of inner unity, whether social and economic, or civic and cultural. We see in each country cities and towns growing at varying rates and under different impulses, but so far without a design as to where these processes lead and why. Cities and towns find themselves overtaken by pressures they are neither equipped to receive nor able to halt. In these conditions, the objectives of urban development policy have to be, firstly, to propose long- and short-term goals consistent with the potentials available and, secondly, to assist the transition to planned and ordered growth. These are essential aims if each community is to evolve a culture and economy in keeping with its size and location and its possibilities as a center for growth and development within the wider national and regional economy of which it forms a part.

Urban Situations

In the more advanced countries, urban policy can be considered almost entirely in the urban context, but even there this approach is proving to be insufficient. Overwhelmingly industrial, directly or indirectly, the urban way of life governs the entire population. Countries in South and Southeast Asia—and indeed all those which are counted as underdeveloped—are overwhelmingly agricultural and rural. Even four or five decades from now, the rural sector will still form the basic segment in their life and economy. In these countries, therefore, urban problems have to be considered, not in isolation, as if their main characteristics had already been determined, but both for what they are and for their relation to the hinterland of each city and town. The nature of the hinterland is one of the principal factors distinguishing towns and cities of different sizes, complexions, and economic and social origins. The influences at work between a city and its hinterland, including its system of communications and supply areas, are among its "external" relations. From these and from the "internal" conditions within each municipality we can derive much of the substance of urban development policy. The weight of "internal" influences in shaping urban policy is greater the larger the size of a city, the longer the course it has run, and the more varied the sources from which its population is drawn.

For purposes of analysis, four main development situations may be distinguished, namely, (a) metropolitan cities, along with their satellite towns, (b) middle-sized cities, (c) industrial towns, and (d) towns serving resource-development regions and other rural hinterlands. From the nature of the underlying urban problems, metropolitan and large cities fall into a class of their own while there are also several common features among middle-sized cities, industrial towns, resource towns, and rural nuclei. In these latter nuclei, the primary cause of growth can be traced frequently to some external stimulus in the recent past—e.g. a significant economic development or a piece of public investment. On the other hand, because of the size of their populations and the range of activities they encompass, metropolitan and large cities once they come of age have a marked capacity to generate growth and expansion from within. Even if the external stimuli were smaller, these cities would continue to grow rapidly. Urban development policy is required, therefore, not only to guide and influence the future pattern of growth, but also to control and restrict the present rate of growth. In the other urban situations, it is sufficient if the broad directions of future growth are clearly established.

A second distinction arises from the fact that costs of development in metropolitan cities are exceptionally high. Their growing gap between requirements and availabilities in all essential services cannot be reduced without large capital expenditures over a considerable period. In middle-sized cities and other towns, there are certainly substantial arrears in the services needed, but with advance planning and phased action, the tasks may remain within manageable bounds.

Thirdly, by their very nature, metropolitan and large cities represent social situations of great complexity. Even if they had the minimum amenities (from which they now stand a great distance away), they would still be centers of strife and instability. Compared to them, there is greater opportunity for smaller cities and towns, based on the growth of industry and natural resources, to mould their populations into fairly homogenous urban communities.

Finally, because of the dynamics of development in the past, metropolitan and large cities have considerable private economic activity and do not depend greatly for their future growth on the economic role which public authorities may assume. However, the tasks of providing them with the essential public services are of dimensions so large that they can only be fulfilled as a national responsibility. The social and civic problems presented by middle-sized cities, industrial and resource towns, and rural nuclei can be dealt with at the state, regional, and local level without invoking specific national responsibility. While drawing this distinction, it is necessary to stress that both sets of development situations demand efficient systems of urban administration based on increasing participation and response at the level of the community.

Interaction with National Planning

The interaction between urban problems and national planning comes about in somewhat different ways in the two situations described above. The pervasive elements are of course increases in population and the growth and diversification of the national and regional economies. Focussing their main attention on the size and pattern of investment and on possibilities of growth, national plans are generally weak in visualizing and providing for the complementary investments and inputs needed at each level and in integrating the spatial aspects of development. Invariably, they fail to spell out the impact and direction of action in different urban and regional situations.

The pressures of economic development upon urban centers which have already assumed metropolitan proportions come from several directions. The volume of transport, both in goods and passengers, increases rapidly, and commuting over long distances becomes a daily necessity. Supply zones lengthen, and there is considerable increase in wholesale and retail trade. Industries already established diversify, find themselves choked, and must either specialize further or move out. Small ancillary industries grow and link up more closely with the larger industries and are no less difficult to shift. Metropolitan and large cities like Calcutta, Bombay, Madras, Delhi, Karachi, Bangkok, and Singapore, attract both economic enterprise and social, political, administrative, and other activity. There are pressing demands for the expansion of housing, water supply, electricity, transportation, industrial areas, marketing and educational facilities.

Middle-sized cities react differently to economic development, depending on the nature of their primary cause of growth and the local and regional influences which bear on them. Frequently, they also serve as the headquarters of state and regional administrations. Three groups of middle-sized cities can be distinguished, namely, (a) those which are essentially new cities, for instance, new capitals, (b) cities with roots back in history, now coming under the influence of new developments, and (c) cities which owe their importance to their role in industry, trade, or transportation. Chandigarh and Islamabad may be cited in the first group; Bhubaneshwar, Bhopal, and Jaipur in India, Lahore, Hyderabad (Sind), and Dacca in Pakistan, or Kuala Lumpur in Malaysia in the second group; and Kanpur, Ahmedabad, Jamshedpur, and Asansol in India, and Chittagong, Khulna or Lyallpar in Pakistan in the third group. In middle-sized cities, the industrial component generally tends to increase. To a large extent this factor will determine which of the cities expands and how fast, and which of them may remain in a state of relative stagnation or slow growth.

New industrial townships and towns which grow as a result of the development of natural resources are essentially a post-World War II phenomenon. A country like India has an increasing number of such towns, for instance, new steel towns like Durgapur, Rourkela, and Bhilai, expansions of older centers like Bhadravati, Visakhapatnam, and Ranchi, and towns like Nangal, Sambalpur, and Ganganagar which gain from being at the center of important resource regions. A feature common to both groups of towns is that once the initial investment has done its work, their economic base is found to be too narrow. Their prosperity comes to depend too heavily on a limited range of industrial, commercial, or agricultural activities; there is need to give them both a broader economic foundation and to develop them purposefully as focal points for the regions around them.

Towns which lie at the center of rural and agricultural areas are numerous everywhere, but there has been little systematic effort to link their growth and expansion with plans for the development of agriculture and other economic and social activities in the surrounding countryside. Many of these towns continue to function like overgrown villages without a positive role assigned to them in the scheme of development.

The summary conclusion to be drawn from this discussion is that so far, in countries in South and Southeast Asia, the concepts of national planning have not become broad enough to comprehend the scale and intensity of problems which cities and towns now present. Although some progress has been made, anything like a clearly worked out view of the shape of the rural and urban economies, and of the rural and urban social systems to be evolved through planning and the strategies associated with it, has yet to emerge. Economic development plans commonly give a global treatment to projections of population growth. Numbers are seen in relation to the economy as a whole; beyond this there is much less precision, and plans fail to indicate the order of the population for which each regional economy and each urban economy should provide within specified periods. More inclusive concepts in planning and a firmer view of numbers are indispensable conditions for giving an operational meaning to several of the priorities associated with urban planning. Without these, policies bearing on location of industrial and other activities and on inputs by way of public organization, community effort, and investment complementary to the main productive investments remain vague. The central problem in setting goals for urban development is one of defining the range and the instruments of national planning itself. Along with limitations of resources which, whatever the magnitudes, will always persist, this is perhaps the main explanation for the failure, as in India in recent years, to give effect to the proposal for city development plans as an organic element in state and national planning. For the same reason, the concept of urban community development, which presents more difficult questions than in the setting of rural communities, has not proceeded much beyond the phase of preliminary experience.

Major Problems of Policy

As suggested earlier, the growth of cities and towns, as a general movement related to economic and social development in the poorer countries, is essentially a post-war development. Its acceptance as a vital area of economic and social policy has come more slowly than recognition of the general problem of economic backwardness of which it forms part. In themselves, even where they arise from lags in rural areas more than from any other factor, trends towards rapid urban growth represent a hastening of the pace of change. Rural lags alone could not have produced them. In viewing these trends as part of a larger change affecting both urban and rural areas and, progressively, increasing segments of the entire national economy, we may find some useful clues to urban development policy and its implications for planning.

The factors which are now most decisive in urban growth relate primarily to industrialization and economic development, the growth of education, dissatisfaction with existing conditions, and the search for new opportunities. These factors are qualitatively different from those which greatly influenced the growth of cities and towns in the past. In the West, urban and industrial development occurred when populations were small and levels of income, education, and well-being were already much higher than those now prevailing in the under-developed countries. Pressure of population, low levels of income and savings, and the day-to-day struggle for a bare livelihood which goes with them, make it especially difficult to deal with new problems. Neither past experience in countries in South and Southeast Asia, nor even the work of the past decade, which itself has been a period of preparation, provides sufficient guidance for future policy. Therefore, in the main, urban development policy has to evolve from a careful analysis of the kind of problems which have emerged in the post-war years and from an effort to relate them to one another, both at the level of the city or the town, and within the total national structure.

In the more advanced countries, with the spread of industrial and economic activities, improved transportation, and the growing preference of middle-income groups to shift to suburbs and areas further away from central cities and business centers, it appears there are certain inherent limits to the growth of large cities. There is no evidence of a similar tendency in metropolitan and other large cities in South and Southeast Asia. Poverty, slums, poor housing, and lack of amenities are not yet a bar to more and more persons eager to seek their fortunes in the cities. Neither industries nor individuals are moving away from the larger urban centers. While this is the general situation, each city has its own distinguishing features as to the types of migrants it attracts, the areas from which they come, how they choose to huddle together and gain their pittance and how, in turn, the neighboring rural areas may be affected by these and other developments. We see in an overall way that the burdens which come upon metropolitan and large cities are to be traced to regions situated far apart from them and from one another. Even the growth of satellite towns adds to these burdens and can only be part of a solution. Within each city we observe some parts growing, others stagnating and neglected—with little in either case by way of anticipatory or corrective action. New towns, which could make a fruitful contribution if they were not limited to just a few enterprises (however large), are protected only superficially from developments in their immediate neighborhood.

Anyone looking at the urban scene is struck by the intimate connection, on the one hand, between economic, social, and fiscal problems at the level of the city or the town and, on the other, by the extent to which the future of each municipality is determined by overall economic, development, and location decisions which remain largely outside its influence. The crux of an adequate urban development policy lies, therefore, in identifying each major problem and the nature of its relationship to other problems, and in establishing systems for joint planning and for specifying responsibilities and obligations for continuing action on the part of the civic institutions of a city and public authorities functioning at higher regional, state, and national levels.

The demand for extended public services and the economic and social dislocations which each city faces arise from causes not wholly within its control. Its civic institutions are seldom equipped to see the situation in all its aspects and to forecast future developments, much less to provide for them. Working together, with an understanding of the issues at stake— with all their limitations—public authorities at different levels can at least hope to make a more nearly adequate approach to their common concerns and go some distance towards overcoming the obstacles due to drifts in policy, lack of community, and want of resources. While the largest area of responsibility must lie within the city itself, public authorities at higher levels and public opinion generally can do much to strengthen the constructive role of local leadership and institutions. If this has not happened thus far in any South and Southeast Asian country, the failure does not belong to the cities alone; it runs through policy and planning, and holds equally for the higher levels of government and administration.

Social Goals

Surveys of urban conditions—of which there have been a large number in recent years, particularly in India—depict conditions which give cause for greater disquiet than they have in fact evoked. The picture is increasingly one of inequalities and contrasts which hurt, of a society in which callousness and exploitation abound, and in which evils such as beggary, prostitution, and crime can thrive. The larger the size of a city, the more intense its social problems. Economic and social factors act and inter-act upon one another. Poverty and lack of education versus skill are two sides of the same coin and are equally widespread. If urban development policy is to go beyond the current symptoms it must grapple with the underlying conditions of economic and social life.

It needs to be stressed that for the bulk of the urban population living conditions have been deteriorating steadily. All cities, and some doubtless more than others, are full of slums. In the largest cities, where numbers sleep on pavements, there is now an insensitiveness born of familiarity. The additions to low-income and middle-income housing each year are meagre, while the need increases. Urban housing and the improvement and clearance of slums are among those crucial areas of policy in which, without a national program and commitment, state, regional, and local authorities can make little headway. This is equally true of water supply and drainage and not much less true of education, health, sanitation, and other public amenities.

The malaise of the cities has also certain other aspects. A large city is composed of numerous social groups, or "sub-societies," as it were, which have little communication with one another. Whatever cohesion and community of purpose exist are largely concentrated within these sub-societies. The numbers of people with identification and pride in the tra-ditions of even an old city are small. For many decades, in countries which came under foreign rule, prestige was attached to the "civil lines," where the higher bureaucracy—and those nearest to them—lived. Moreover, in the nature of things, much the larger proportion of a city's population everywhere is a comparatively recent addition. The incoming populations bring with them their links of caste, language, and region, sometimes also of occupation, and frequently these are powerful factors in securing em-ployment or the means of subsistence. In comparison with the older divisions, new groupings based on occupation or neighborhood or on the membership of trade unions or other secular associations are weak; in course of time, under favorable conditions they might become stronger but, without marked improvement in urban life, they can remain without real influence. Under existing circumstances, instead of being a means to greater equality and to assimilation within a wider and more varied cul-tural setting, even for those firmly within the established urban occupation and residence patterns, urbanization carries forward most of the earlier values, attitudes, and obligations to families and groups. Therefore, even more than the provision of essential amenities, the role of urban develop-ment policy is fundamental in guiding change in social and economic life.

To design the future development of a city is a segment of the larger task of reconstructing urban society, of reshaping it from its very founda-tions. In every city economic factors sharpen social disparities and, in their turn, the latter determine the next phase in development. For life in the city to be satisfying to the bulk of the population, the principal criterion of social policy must be to enable persons of small and modest means to live and work in dignity and in an environment which will add to the quality of life. Therefore, the unifying concepts of policy in the cities have to be derived, not so much from the diverse and contradictory elements which operated in the past, as from the new goals and ideals for which urban communities, viewed as *communities,* may be helped to strive in the future. This is the central social premise of urban development policy.

The principal instruments for realizing the social as well as other goals in urban development include, (a) action at the level of the community, notably through an expanding role for voluntary social work and urban community development, (b) action through public authorities at different levels and more specially through civic bodies, as in city development plans, and (c) action at the regional level, the region varying according to the purpose in view. Each line of advance calls for an adequate framework of national policy, for institutions which express a national ethos, and for hard decisions bearing on the harnessing and allocation of scarce resources. For the same reasons for which economic and social conditions are closely connected with one another, the means through which the social and economic aims of urban development policy are to be attained have much in common.

Economic Objectives

Whatever the dominant causal factors in the expansion of a city, once under way, growth itself stimulates further growth. Expansion of facilities for distribution, marketing, processing, manufacture, transportation, and the hiring of skilled workers, and the development of enterprises with common interests provide economies which support expansion. More than at any time in the past, pressures for economic development strengthen those factors which pull new economic and other activities in the direction of cities and towns. The so-called "push" factors are in a sense only a response to the capacity of cities and towns to serve as growth points with an intrinsic vitality of their own.

It is only too apparent that in the ordinary course cities and towns are unable to meet the mounting demands which press upon them. City transportation systems are congested, inefficient, and uneconomic, and need to be developed and reorganized with all the modes of transport integrated as a common service. New industrial sites are essential as a measure for "de-concentration." The serious lags on all sides are compulsions in favor of extension, decentralization, and dispersal of economic, administrative, and other activities. The high proportions of the community recorded in most urban surveys as being "poor" and "destitute," the prevailing levels of unemployment, the large numbers in tertiary and other activities shown as "self-employed" earners—who are in fact only partially employed—and the small ground occupied in the urban economy by the corporate or the organized sector and by productive and expanding activities, are unmistakeable indications that the economic base of cities and

234 TARLOK SINGH

towns is too weak to support existing populations, much less the numbers who will pour into them in the future.

In these conditions, an adequate urban policy cannot emerge until the role of cities and towns as a positive ingredient in national economic development has been visualized clearly. There are two questions to be answered. How can the gains which the concentration of population at given points, and the development of economic and social overhead associated with them, be turned into a dynamic and constructive force for progress? How can the possible adverse consequences of this process be kept to the minimum? There are perhaps four principal directions along which appropriate economic premises for an urban development policy will need to be established:

(1) In metropolitan cities, and in the larger cities generally, it is essential to determine broadly the population for which a city is to be prepared over a defined period. This implies both a long-range view and restrictions upon growth of new activities except in accordance with clearly established plans of land use and location;

(2) Depending on its size and situation, a city or town has always to be considered as the center of a "planning region." Basically, this will be the area with which the economy of the city or the town is most intimately connected. Its hinterland, however, does not alone by any means delimit all the relevant relationships. For the services which the "central" institutions of a city render, and for purposes such as transportation, power, or distribution, there may well be different planning areas. However flexible the notion of "region" in relation to different objectives, it should be adhered to firmly in detailed planning until circumstances justify a fresh and comprehensive review, for the "politics" of urban administration only too often present exigencies which run counter to policy;

(3) For cities and regions to maintain consistent schemes of expansion, it is essential to secure a broad design of location for major economic, industrial, and other activities in terms of the wider national and state economy. Such a blueprint may be modified as economic development proceeds, but must apply equally whether an activity is undertaken by a public authority or by a private agency; and

(4) Within the economic framework outlined above, city development plans, to which reference has been made earlier, can provide the precise instrument for viewing the present and future requirements of a city and its environment in an integrated manner and for enabling public authorities at different levels to cooperate in planning and execution.

The Fiscal Base

Consideration of the social conditions and economic requirements of urban areas, especially of the larger cities, brings into sharp relief the continuing shortage of resources which impedes the best intentions. There is little doubt that, with better management and administration, existing resources could yield more and could be employed to greater purpose. The problem of administrative and technical efficiency and integrity in civic institutions is not to be minimized. Yet, it remains true of all countries in South and Southeast Asia that the tax resources available to civic bodies are far too meagre, both absolutely and in relation to needs. In countries with unitary systems of government, the problem of local resources can be met to an extent through direct allocations and grants. However, in a federal structure, the fiscal system must be viewed as a whole, and each level in the structure of government and administration must share in resources which will increase in some fair proportion in relation to national economic growth. In their turn, civic institutions owe it to themselves and to the cities they serve to make the maximum use of taxes on property and of changes in the values of land and of other immovable property. How the processes of valuation are arranged is a matter of detail, but sufficient experience exists to indicate the improvements needed in the operation of each city. The rising costs of local services are closely associated with national economic policies. Therefore, devices such as provision for a *local* element in the taxation of personal and corporate income are founded on equity and on a larger view of administrative effectiveness than has been recognized in the past at the higher levels of government.

The gaps between requirements and availability in all essential services are growing to unmanageable proportions in most cities. With their tax systems already inadequate, the capital markets too are not organized so that any substantial savings could be channelled to civic institutions. Therefore, for the greater part, capital expenditures have to be found through loans and grants from government at the national and state levels. These latter also have not done enough to mobilize domestic savings for development. The deeper cause of fiscal weaknesses at the city level, however, is that the provision of essential amenities for urban populations is not felt to be a high enough priority in the total scheme of development. The higher levels of administration under-estimate the nature of civic needs and their bearing on wider economic and social objectives. To treat the fiscal problems of urban areas apart from these objectives is an error of policy which has yet to be corrected. In stressing this, we should not, of course, overlook the acknowledged obligations which, on their part, civic institutions have invariably failed to observe.

Most civic bodies do not charge adequately for the services which they provide, so that they are scarcely able to service the loans they obtain and to repay them. The higher levels of government have a duty to insist upon high fiscal and administrative standards on the part of civic institutions. Inefficiency destroys autonomy—and has to be prevented sooner rather than later. The management aspects of civic enterprises, such as municipal transportation, markets, and other services, frequently leave much to be desired. As a necessary complement of its social and economic goals, urban development policy should also provide for independent and continuous evaluation and exposure of administrative, technical, and financial performance on the part of civic bodies, especially in the larger cities. Existing laws make it possible for state authorities to supercede civic bodies and to replace them by appointed administrators when their financial and administrative management deteriorates seriously. However, the use of such powers as a last resort overlooks important prior steps which would have still greater value, namely, early warning, mobilizing public opinion against waste and mismanagement, and proposals for precise remedial action. Evaluation, education and training, and the enforcement standards in fiscal and other matters are a duty cast upon the higher levels of government equally with the obligation to enlarge and strengthen local autonomy and local responsibility.

Policy for Land

The prices at which lands situated within a city and its environs sell and the manner in which they are utilized are an essential touchstone of effective urban policy and planning. As it happens, speculation in land, encroachments, unsightly slums, ribbon development, prices far beyond the reach of ordinary people, and grossly wasteful uses of land are examples of failures encountered almost everywhere, with little successful effort to control them. The main reasons for failure in this critical area of urban policy as well as the main remedies are known well enough. Vested interests are often able to hold up effective action for long periods. Land-use plans tend to be too restricted in scope. The pace at which they are drawn up and incorporated into agreed master plans is too slow, and much mischief occurs while these processes are under way. Areas under different authorities which need to work together—and with equal vigor— in drawing up and implementing common plans remain apart, thus providing opportunities for windfall gains.

Failures in the use of land occur, with respect to developed and undeveloped lands, both within and outside city limits. Many half measures have been taken, invariably leaving some serious loopholes. This has happened in attempts to declare lands liable to public acquisition, to freeze land values, to realize part of the increment in values, to extend leasehold systems and to restrict freehold in the disposal of land, and to pursue complementary housing and other development policies. Many serious students of urban land problems have, therefore, come to the view that land within the master plan covering a city and its immediate environs, even when more than one civic authority may be concerned, should be acquired and turned into social property. Various means of acquisition are possible, including the issue of bonds redeemable over a period. Once acquired, other action must follow swiftly. The lands should be used for substantial development of housing and other facilities though, at a later stage, smaller additions at short and regular intervals may prove adequate. Control over the use of property and public acquisition of lands have to go hand in hand with active development policies for housing, markets, and industrial areas, with the provision of adequate housing finance, improved building techniques, and better administrative procedures. To pursue firmly a land policy committed to mass welfare—against the pressures of powerful local interests—is a task beyond the capacity of political parties alone. Political parties have their own fears and are unable to see far enough unless deeper community and public sanctions are also forged and brought into play, not on occasion, but continuously as part of the values of urban society itself.

Planning and Implementation

Successive plans of economic development have not yet made any significant impact on urban problems. They have, however, helped to show many problems in clearer relation to one another and to propose strategies more broad-based than would have been otherwise possible. In relation to urban development, national plans have four distinct functions. These are:

(1) to provide for allocations of resources for sectors which bear directly on the welfare of cities and towns, such as health services, water supply and sewage disposal, improvement and clearance of slums, education, transportation systems, etc.;

(2) to offer an institutional framework in which city development plans may be formulated and executed in organic and continuing relationship with state and national plans;

(3) to identify the regional implications of planning, especially in relation to large cities and resource development regions; and

(4) to indicate the steps required for organizing the planning process and the measures to be taken for strengthening the agencies which are charged with the tasks of planning and implementation.

These functions are only just beginning to be clearly visualized. In common with social services generally, the resources allocated to the main urban needs are relatively small. The bulk of the resources provided through national plans go to a few big cities, without obligations being sufficiently placed on them to mobilize their own resources and to charge adequately for the services which they provide. At the same time, middle-sized cities are given little support, and the smaller towns are left in a state of neglect. Where resources are devoted to the construction of new towns in association with some major industrial or other enterprises, there is little effort to ensure broad-based civic and industrial development, and much of the potential developmental value of new towns fails to be realized.

If the concept of city development plans has not yet taken root, only part of the explanation lies in resource constraints and in the customary short-term view in national plans and policies on the fiscal problems of cities and towns. In India, city development plans were viewed as a counterpart of rural area plans and, together with them, they were expected to lead on to more composite rural-*cum*-urban development at the local and regional level in different parts of the country. The features emphasized in city development plans were land acquisition and development; urban water supply, sanitation and drainage; housing, particularly industrial and low-income housing; slum clearance and improvement; medical and health facilities; primary and secondary education; and the development of rural areas falling within city limits. By its nature, a city development plan must entail a great deal of joint planning and execution between agencies at the city level and at higher levels of administration. A condition for such joint action is the recognition that within a city the primary responsibility for planning and coordinating the use of resources drawn from diverse sources rests on the city's own institutions. Another condition is that these institutions must possess or be helped to acquire the necessary skills and experience. Both conditions have been wanting, and too little has yet been done to make good the present deficiencies.

The regional approach in planning, now better understood in its broad principles, has still to advance beyond the phase of exploratory studies. It is more easily adopted where a region lies wholly within the territory of a single political administrative unit. When more than one such unit is concerned, successful regional planning calls for national initiatives supported by allocations of resources in finance and personnel. In this more difficult, yet more vital, area of regional planning and development, loose arrangements attempted in the past have not proved durable or effective. A permanent body with a continuing nucleus staff, for instance, a regional planning board or commission, drawing authority from the constituent units and support from the national government, is essential if there is to be a common development design and the regional approach is not to be discarded when difficult decisions are required.

One important reason for the small attention given in action both to city development plans and to the regional approach is the very weakness of local government viewed as an institution in national life. Traditionally, the concept of responsibility on the part of the people for their own affairs remained confined in countries in South and Southeast Asia, unlike the West, to small rural communities, seldom beyond the village. The governors of cities and towns were appointed from above. On the Indian subcontinent, when local self-government was put forward as official policy, it was felt to be a poor substitute for real transfer of power at the higher levels of the structure. In the post-war period, countries in South and Southeast Asia have been too greatly preoccupied with unsolved problems of political stability and economic development to have found the energy for building up stable processes and institutions for articulating the needs of local urban and rural communities and for enabling them to take a dynamic and autonomous part in national life. In a few countries the beginnings have been made, but far greater understanding and public leadership must be invoked before local development and local effort can become movements with vitality and capacity to achieve results. Perhaps the political, social, and economic goals of each nation have to emerge more sharply and with a greater sense of urgency before local planners and institutions will attract the human and other resources needed to give them their real share in the scheme of development.

Weak local institutions also lie at the bottom of some of the problems of planning and implementation which have been met with not only in middle-sized cities and in the smaller urban centers, but also in large and metropolitan cities. In these latter, there has been a certain tendency to set up new and parallel organizations for allied objects rather than to broaden and strengthen the scope and capability of the main municipal institutions themselves and to enable them to create the specialized organs which would be required to serve new needs. Thus, in the Indian context, improvement trusts, development authorities, and planning organizations, with whatever specific contribution they might have made, have also had the effect of fragmenting and limiting the intrinsic responsibility of the principal municipal or civic institutions. In taking away from them some part of the challenge and the opportunity inherent in new tasks of development, and in their desire for quick results, authorities at higher levels have even caused civic institutions at the base to weaken and stagnate. Moreover, when functions are loosely and unwittingly separated and occasions for differences and confusions in policy are frequent, attempts to secure coordination between parallel institutions inevitably strengthen control from above, diminish effective local initiative and responsibility, and dilute the sanctions of popular opinion and criticism.

From what has been described above, it will not be unfair to conclude that post-war developments in South and Southeast Asia have not yet led to sufficient creative exploration of the forms and possibilities of local government and local planning. Under one impulse or another, a few steps have been taken, but little has been done to strengthen the foundations. Civic institutions remain weak in organization and starved of resources, and are still largely outside the orbit of comprehensive economic and social planning.

The issues which any meaningful urban development evoke are fundamental. They involve a view of the future economic and social order, the social structure and economic basis—equally of cities and towns and of rural areas, and of the mutual relationship between them. Planning and administration, and the institutions charged with these functions, must succeed in expressing this larger concept in the form of practical and continuing action. Hitherto, urban development policies have been conceived in the framework of past patterns of economic and social development and of administrative behavior and organization, rather than of the underlying problems and the future goals. Therefore, it has become all the more important to define the goals of urban development in terms of the welfare of the people as a whole, to create for them the sanctions of public concern and response, to integrate them fully into the processes of economic and social planning, and to build up the necessary skills, organization and resources.

CHAPTER 10

Leo Jakobson and Ved Prakash
A SELECTED BIBLIOGRAPHY

GENERAL

ABRAMS, CHARLES, *Man's Struggle for Shelter in an Urbanizing World.* Cambridge, Mass.: M.I.T. Press, 1964.

——, "Regional Planning Legislation in Underdeveloped Areas." *Land Economics,* 35 (May, 1959), pp. 83–103.

ADAMS, FREDERICK J., "Techniques of Regional Planning as Applied to Metropolitan Areas." Seminar on Regional Planning, Tokyo, 28 July to 8 August, 1958.

ADELMAN, IRMA and CYNTHIA TAFT MORRIS, *Society, Politics and Economic Development.* Baltimore: Johns Hopkins Press, 1967.

ALCOCK, ALFRED E.S., "The Application of Regional Planning Techniques to Rural Development Programmes." Seminar on Regional Planning, Tokyo, 28 July to 8 August, 1958.

ALONSO, WILLIAM, *Urban and Regional Imbalances in Economic Development.* Harvard Regional Development Program, 1967. Unpublished Manuscript.

APPLEYARD, R.T., "Immigration Policies and Economic Development in South Eastern and Eastern Asia." Paper for the International Union for the Scientific Study of Population, Sydney Conference, Australia, 21 to 25 August, 1967, pp. 798–804.

"Aspects of Urbanization in ECAFE Countries." *Economic Bulletin for Asia and the Far East,* 4 (May, 1953), pp. 1–15.

BECKMANN, MARTIN J., "City Hierarchies and the Distribution of City Size." *Economic Development and Cultural Change,* 6 (April, 1958), pp. 243–248.

BERRY, BRIAN J.L., "Cities as Systems Within Systems of Cities," in John Friedmann and William Alonso (eds.), *Regional Development and Planning.* Cambridge, Mass.: M.I.T. Press, 1964, pp. 116–137.

——, "City Size Distribution Development." *Economic Development and Cultural Change,* 9 (July, 1961), pp. 573–587.

——, *Geography of Market Centers and Retail Distribution.* Englewood Cliffs, N.J.: Prentice-Hall, 1967.

Authors' Note: Our thanks are due to Janet Kline and Thomas Eisemon who assisted in the compilation of this bibliography.

——, "An Inductive Approach to the Regionalization of Economic Development," in N. Ginsburg (ed.), *Essays on Geography and Economic Development*. Chicago: University of Chicago, Department of Geography, Research Paper 62, 1960, pp. 78–107.

——, "Interdependency of Spatial Structure and Spatial Behavior: A General Field Theory Formulation." *Papers of the Regional Science Association*, 21 (1968), pp. 205–227.

——, "Some Relations of Urbanization and Basic Patterns of Economic Development," in Forrest R. Pitts (ed.), *Urban Systems and Economic Development*. Eugene, Ore.: University of Oregon, School of Business Administration, 1962, pp. 1–15.

BERRY, BRIAN J.L., and WILLIAM L. GARRISON, "Alternate Explanations of Urban Rank-Size Relationships." *Annals of the Association of American Geographers*, 48 (March, 1958), pp. 83–91.

BERRY, BRIAN J.L. and ALLEN PRED, *Central Place Studies: A Bibliography of Theory and Applications*. Philadelphia: Regional Science Research Institute, Bibliography Series, No. 1, 1961.

BONIFAT, 'YEVA L.I., "Some Characteristics of the Urban Network in the Countries of South and South East Asia." *Soviet Geography* (New York), 7 (January, 1966), pp. 60–70.

BOSE, ASHISH, "Problems of Urbanization in Countries of the ECAFE Region." Paper prepared for ECAFE Expert Working Group on Internal Migration and Urbanization, Bangkok, 1967.

BREESE, GERALD (ed.), *Urbanization in Newly Developing Countries*. Englewood Cliffs, N.J.: Prentice-Hall, 1966.

BROWNING, CLYDE E., "Primate Cities and Related Concepts," in Forrest R. Pitts (ed.), *Urban Systems and Economic Development*. Eugene, Ore.: University of Oregon, School of Business Administration, 1962, pp. 16–27.

BROWNING, H.L., "Recent Trends in Latin American Urbanization." *The Annals of the American Association of Political and Social Science*, 316 (March, 1958), pp. 111–120.

CLARK, COLIN, *Population Growth and Land Use*. New York: St. Martin's Press, 1967.

CLINARD, MARSHALL B., *Urban Community Development and Slums*. New York: Free Press, 1965.

"Conclusions and Recommendations of the United Nations Seminar on Regional Planning." Seminar on Regional Planning, Tokyo, 28 July to 8 August, 1958.

DAVIS, KINGSLEY, "The Origins and Growth of Urbanization in the World." *American Journal of Sociology*, 60 (March, 1955), pp. 429–437.

DAVIS, KINGSLEY and HILDA HERTZ GOLDEN, "Urbanization and the Development of Preindustrial Areas." *Economic Development and Cultural Change*, 3 (October, 1955), pp. 6–26.

DOTSON, A.T., "Summary of Discussion." Seminar on Metropolitan Planning in Asia, June 7–14, 1964.

ENSMINGER, DOUGLAS, "Urbanism—A Growing World Problem," in William R. Polk (ed.), *Developmental Revolution*. Washington, D.C.: The Middle East Institute, 1963, pp. 24–35.

ESTALL, R.C. and BUCHANAN OGILVIE, *Industrial Activity and Economic Geography*. London, 1961.

"Freedom from Hunger Campaign." *National Development Efforts: Basic Study Number 13*. New York: United Nations, 1962.

FRIEDMAN, JOHN, "Locational Aspects of Economic Development." *Land Economics*, 32 (September, 1956), pp. 262–282.

——, *Regional Development Policy: A Case Study in Venezuela*. Cambridge, Mass.: M.I.T. Press, 1966.

——, "Regional Economic Policy for Developing Areas." Regional Science Association, *Papers*, 11 (1963), pp. 41–61.

——, "Regional Planning and Nation Building: An Agenda for International Research." *Economic Development and Cultural Change*, 16 (October, 1967), pp. 119–129.

——, "The Urban-Regional Frame for National Development." *International Development Review*, 8 (September, 1966), pp. 9–15.

——, "Towards a National Urbanization Policy: Problems, Decisions and Consequences." A paper presented at UN Interregional Seminar on Financing of Housing and Urban Development, Copenhagen, May 25–June 10, 1970.

FRYER, D.W., "The Million City in Southeast Asia." *Geographical Review*, 43 (October, 1953), pp. 474–494.

GERMANI, GINO, "Migration and Acculturation," in *Handbook for Social Research in Urban Areas*. Paris: UNESCO, 1964.

GINSBURG, LESLIE B., "Current Trends Influencing Regional Planning." Seminar on Regional Planning, Tokyo, 28 July to 8 August, 1958.

GINSBURG, NORTON S., *Atlas of Economic Development*. Chicago: University of Chicago Press, 1961.

——, "Cities Without Nations?" *Pacific Viewpoint*, 9 (September, 1968), pp. 208–210.

——, "The Great City in Southeast Asia." *American Journal of Sociology*, 60 (March, 1955), pp. 455–462.

——, "The Regional Concept and Planning Regions." Seminar on Regional Planning, Tokyo, 28 July to 8 August, 1958.

——, "Urban Geography and 'Non Western Areas,' " in Philip M. Hauser and Leo F. Schnore (eds.), *The Study of Urbanization*. New York: John Wiley, 1965, pp. 311–346.

HAUSER, PHILIP M., "Implications of Population Trends for Regional and Urban Planning in Asia and the Far East." Seminar on Regional Planning, Tokyo, 28 July to 8 August, 1958.

——, "Summary Report of the General Rapporteur," in Philip M. Hauser (ed.), *Urbanization in Asia and the Far East*. Calcutta: UNESCO, 1957, pp. 3–32.

——, Urbanization: An Overview," in Philip M. Hauser and Leo F. Schnore, (eds.), *The Study of Urbanization*. New York: John Wiley, 1965, pp. 1–47.

—— (ed.), *Urbanization in Asia and the Far East: Proceedings of the Joint UN/UNESCO Seminar (In Cooperation with the International Labour Office) on Urbanization in the ECAFE Region, Bangkok, 8–18, August, 1956*. Calcutta: UNESCO Research Centre on the Social Implications of Industrialization in Southern Asia, 1957.

HIRSCHMAN, A.O., *The Strategy of Economic Development*. New Haven: Yale University Press, 1958.

HOOVER, EDGAR, M., "The Concept of a System of Cities: A Comment on Rutledge Vining's Paper." *Economic Development and Cultural Change*, 3 (January, 1955), pp. 196–198.

HOSELITZ, BERT F., "Generative and Parasitic Cities." *Economic Development and Cultural Change*, 3 (April, 1955), pp. 278–294.

——, "Urbanization: International Comparisons," in Roy Turner (ed.), *India's Urban Future: Selected Studies from an International Conference on Urbanization in India held at the University of California in 1960*. Berkeley: University of California Press, 1962, pp. 157–181.

——, "Urbanization and Economic Growth." Unpublished paper presented to the Conference on the City as a Centre of Change in Asia, University of Hong Kong, June, 1969.

——, "Urbanization and Economic Growth in Asia." *Economic Development and Cultural Change*, 6 (October, 1957), pp. 42–54.

"Industrial Development in Asia and the Far East." *Industrial Development News*, 1 (1965).

Institute of Public Administration, *The Urban Challenge to Government: An International Comparison of Thirteen Cities*. New York: Frederick A. Praeger, 1969.

International Union for the Scientific Study of Population, *Variables for Comparative Fertility Studies: A Working Paper*. Liege: 1967.

International Union of Local Authorities, "The Government of Metropolitan Areas." Seminar on Metropolitan Planning in Asia, Tokyo, June 7 to 14, 1964.

——, *Urbanization in Developing Countries*. The Hague: Martinus Nijhoff, 1968.

International Urban Research, *The World's Metropolitan Areas*. Berkeley and Los Angeles: University of California Press, 1959.

ISOMURA, EIICHI, "Concluding Report." Seminar on Metropolitan Planning in Asia, June 7 to 14, 1964.

JEFFERSON, MARK, "The Anthropography of Some Great Cities." *Bulletin of the American Geographical Society*, 41 (No. 9, 1909), pp. 537–566.

——, "Distribution of the World's City Folk." *Geographical Review*, 21 (July, 1931), pp. 446–465.

——, "The Law of the Primate City." *Geographical Review*, 29 (April, 1939), pp. 226–232.

KAUNITZ, RITA D., "National and Regional Development Policies and Planning in Relation to Urbanization." Inter-Regional Seminar on Development Policies and Planning in Relation to Urbanization, University of Pittsburgh, Pittsburgh, 24 October to 7 November, 1966, pp. 1–112.

KEYFITZ, NATHAN, "Political-Economic Aspects of Urbanization in South and Southeast Asia," in Philip M. Hauser and Leo F. Schnore (eds.), *The Study of Urbanization*. New York: John Wiley, 1965, pp. 265–309.

KOENIGSBERGER, OTTO H., "Regional Planning in Asia." Seminar on Regional Planning, Tokyo, 28 July to August, 1958.

KONDRACKI, G. and J.P. SAH, "Population Trends and Urban and Regional

Planning in the ECAFE Region." Paper prepared for ECAFE Expert Working Group on Internal Migration and Urbanization, Bangkok, 1967.

KURODA, TOSHIO, "Demographic Aspects of Urbanization in Japan: The New Dimension of Internal Migration and Urbanization." Paper for International Union for the Scientific Study of Population Conference, London, September, 1969.

KURODA, TOSHIO, *Rural Urban Migration and Social Mobility in Asia.* Tokyo: Institute of Population Problems, English Pamphlet Series Number 66, 1968.

KUZNETS, SIMON, "Consumption, Industrialization and Urbanization," in Bert F. Hoselitz and Wilbert E. Moore (eds.), *Industrialization and Society.* New York: UNESCO-Monton, 1963.

LACQUIAN, A.A., "The Asian City and the Political Process." Unpublished paper submitted to the Conference on the City as a Centre of Change in Asia, University of Hong Kong, June, 1969.

——, *The City in Nation-Building.* Manila: 1966.

LEWIS, ARTHUR, "Unemployment in Developing Countries." *The World Today,* 23 January, 1967), pp. 13–22.

LINSKY, ARNOLD S., "Some Generalizations Concerning Primate Cities." *Annals of the Association of American Geographers,* 55 (September, 1965), pp. 506–513.

LOTKA, ALFRED J., "The Law of Urban Concentration." *Science,* 94 (August 15, 1941), p. 164.

MABOGUNJE, AKIN L., "Urbanization in Nigeria—A Constraint on Economic Development." *Economic Development and Cultural Change,* 13 (July, 1965), pp. 413–438.

McGEE, T.G., "An Aspect of the Urbanization of Southeast Asia: The Process of Cityward Migration." *Proceedings of the Fourth New Zealand Geographical Conference.* Dunedin: 1965, pp. 207–218.

——, *The South East Asian City.* New York: Frederick A. Praeger, 1967.

MARBLE, DUANE F., "Some Cultural and Social Aspects of Transport Impact in the Underdeveloped Areas," in Forrest R. Pitts (ed.), *Urban Systems and Economic Development.* Eugene, Ore.: University of Oregon, School of Business Administration, 1962, pp. 29–43.

MARTIN, GEFFREY J., "The Law of the Primate Cities 'Re-Examined.'" *Journal of Geography,* 60 (April, 1961), pp. 165–172.

MEHTA, SURENDER K., "Some Demographic and Economic Correlates of Primary Cities: A Case for Re-Evaluation." *Demography,* 1 (No. 1, 1964), pp. 136–147.

MEYERSON, MARTIN, "Planning for Movement in Developing Countries." Seminar on Regional Planning, Tokyo, 28 July to 8 August, 1958.

MYRDAL, GUNNAR, *Asian drama: An Inquiry into the Poverty of Nations.* New York: Pantheon, 1968.

MOORE, FREDERICK T., "A Note on City Size Distribution." *Economic Development and Cultural Change,* 7 (July, 1959), pp. 465–466.

MURPHEY, R., "New Capitals of Asia." *Economic Development and Cultural Change,* 6 (January, 1957), pp. 216–243.

——, "Urbanization in Asia" in Gerald Breese (ed.), *The City in Newly*

Developing Countries: Readings on Urbanism and Urbanization. London: Prentice-Hall, 1969, pp. 58–75.

The New Urban Debate. A Conference Report, Pacific Conference on Urban Growth, Honolulu, Hawaii, May 1–2, 1967. Washington, D.C.: Agency for International Development, 1968.

NIGAM, M.N., "Collection of Data about Rural-Urban Migration in Developing Countries." Paper for 21st International Geographical Congress, held at New Delhi, November to December, 1968, in S.P. Das Gupta and T.T. Lakshmanan (eds.), *Abstract of Papers.* Calcutta: National Committee for Geography, 1968, p. 231.

REDFIELD, ROBERT and MILTON B. SINGER, "The Cultural Role of Cities." *Economic Development and Cultural Change,* 3 (October, 1954), pp. 53–73.

REINER, T.A., "Sub National and National Planning: Decision Criteria." Regional Science Association, *Papers,* 14 (1965), pp. 107–136.

RESOURCES FOR THE FUTURE, *Design for a Worldwide Study of Regional Development.* Baltimore: Johns Hopkins Press, 1966.

RIVKIN, MALCOM D., "Urbanization and National Development," in *Urbanization: Development Policies and Planning.* New York: United Nations, 1968.

——, "Urbanization and National Development: Some Approaches to the Dilemma." Inter-Regional Seminar on Development Policies and Planning in Relation to Urbanization, University of Pittsburgh, Pittsburgh, 24 October to 7 November, 1966, pp. 1–72.

ROBOCK, STEFAN H., "Strategies for Regional Economic Development." Regional Science Association, *Papers,* 16 (1966), pp. 129–141.

RODWIN, LLOYD, "Metropolitan Policy for Developing Areas," in Walter Isard and John H. Cumberland, *Regional Economic Planning.* Paris: OEED, 1961, pp. 221–232.

——, *Nations and Cities: A Comparison of Strategies for Urban Growth.* Boston: Houghton Mifflin, 1970.

SCHNORE, LEO F., "The Statistical Measurement of Urbanization and Economic Development." *Land Economics,* 37 (August, 1961), pp. 221–245.

SENDUT, HAMZAH, "City Size Distribution in Southeast Asia." *Journal of Asian Studies,* 4 (August, 1966), pp. 268–280.

SJOBERG, GIDEON, "The Preindustrial City." *American Journal of Sociology,* 60 (March, 1955), pp. 438–445.

——, *The Preindustrial City: Past and Present.* New York: Free Press, 1960.

STANFORD RESEARCH INSTITUTE, *Costs of Urban Infrastructure as Related to City Size in Developing Countries.* Menlo Park, Calif.: 1968.

STOLPER, WOLFGANG, "Spatial Order and the Economic Growth of Cities: A Comment on Eric Lampard's Paper." *Economic Development and Cultural Change,* 3 (January, 1955), pp. 137–146.

TINBERGEN J., "Links Between National Planning and Town and Country Planning," in International Union of Local Authorities, *Urbanization in Developing Countries; Report of a Symposium, Noordwijk, Netherlands, December, 1967.* The Hague: Nijhoff, 1968.

TURNER, JOHN F.C., "Uncontrolled Urban Settlement: Problems and Policies."

Inter-Regional Seminar on Development Policies and Planning in Relation to Urbanization, University of Pittsburgh, Pittsburgh, 24 October to 7 November, 1966, pp. 1–112.

UNESCO, *Report by the Director-General on the Joint UN/UNESCO Seminar on Urbanization in the ECAFE Region.* Paris: UNESCO, 1956.

UNESCO, Research Center on Social and Economic Development in Southern Asia, *Urban-Rural Differences in Southern Asia.* New Delhi: Allied Publishers, 1964.

UNESCO, Research Centre on Social Implications of Industrialiation in Southern Asia, *Report on Social and Cultural Factors Affecting Productivity of Industrial Labour.* Calcutta: UNESCO, 1958.

——, *The Social Implications of Industrialization and Urbanization: Five Studies of Urban Populations of Recent Rural Origins in Cities of Southern Asia.* Calcutta: UNESCO, 1956.

UNITED NATIONS, *Basic Statistics for Formulating and Implementing Plans of Economic and Social Development in Countries of Asia and the Far East.* New York: United Nations, 1965.

——, *Decentralization for National and Local Development.* New York United Nations, 1962.

——, *Economic Survey of Asia and the Far East.* New York: United Nations, 1968.

——, *Public Administration Problems of New and Rapidly Growing Towns in Southern Asia: Report.* New York: United Nations Technical Assistance Programme, 1962.

——, *Report on the World Social Situation Including Studies of Urbanization in Underdeveloped Areas.* New York: United Nations, 1957.

UNITED NATIONS, Bureau of Social Affairs, "Urban-Rural Population Distribution and Settlement Patterns in Asia: Their Relation to Development Policies." Inter-Regional Seminar on Development Policies and Planning in Relation to Urbanization, University of Pittsburgh, Pittsburgh, 24 October to 7 November, 1966, pp. 1–36.

UNITED NATIONS, Bureau of Social Affairs, Population Division, "World Urbanization Trends, 1920–1960 (An Interim Report on Work in Progress)." Inter-Regional Seminar on Development Policies and Planning in Relation to Urbanization, University of Pittsburgh, Pittsburgh, 24 October to 7 November, 1966.

—— (In Collaboration with Sidney Goldstein), "Urbanization and Economic and Social Change (An Exploratory Investigation, with Special Reference to Developing Countries)." Inter-Regional Seminar on Development Policies and Planning in Relation to Urbanization, University of Pittsburgh, 24 October to 7 November, 1966.

UNITED NATIONS, Centre for Housing, Building and Planning, "Economic Aspects of Urbanization." Inter-Regional Seminar on Development Policies and Planning in Relation to Urbanization, University of Pittsburgh, Pittsburgh, 24 October to 7 November, 1966.

UNITED NATIONS, Department of Economic and Social Affairs, *Growth of the World's Urban and Rural Population 1920–2000.* New York: United Nations, 1968.

UNITED NATIONS, ECAFE Secretariat, "The Demographic Situation and Prospective Population Trends in Asia and the Far East," in Gerald Breese (ed.), *The City in Newly Developing Countries: Readings on Urbanism and Urbanization*. London: Prentice-Hall, 1969, pp. 76–80.

——, "Report of the Expert Working Group and Problems of Internal Migration and Urbanization." Paper prepared for ECAFE Expert Working Groups on Internal Migration and Urbanization, Bangkok, 1967.

——, "Statistics on Internal Migration." Paper prepared for ECAFE Expert Working Group on Internal Migration and Urbanization, Bangkok, 1967.

UNITED NATIONS, Population Division, "Urbanization and Economic and Social Change." Paper prepared for the ECAFE Expert Working Group on Internal Migration and Urbanization, Bangkok, 1967.

"Urbanization: Development Policies and Planning." *International Social Development Review*, Number 1 (1968).

VAN HUYCK, A.P. and K.C. ROSSER, "An Environmental Approach to Low Income Housing." *International Development Review*, 8 (September, 1966), pp. 15–18.

WALSH, ANNMARIE HAUCK, *The Urban Challenge to Government: An International Comparison of Thirteen Cities*. New York: Frederick A. Praeger, 1969.

WERTHEIM, W.F., *East-West Parallels*. The Hague: W. Van Hoeve, 1964.

WHEATLEY, PAUL, "What the Greatness of a City Is Said to Be." *Pacific Viewpoint*, 4 (September, 1963), pp. 163–188.

WILLIAMSON, G.G., "Regional Inequality and the Process of National Development: A Description of the Patterns." *Economic Development and Cultural Change*, 13 (July, 1965), pp. 3–84.

YLVISAKER, PAUL, "Administrative Considerations in Regional Planning." Seminar on Regional Planning, Tokyo, 28 July to 8 August, 1958.

ZACHARIAH, K.C., "Estimation of Return Migration from Place of Birth and Duration of Residence Data." Paper presented to the International Union for the Scientific Study of Population, Sydney Conference, 1967.

BURMA

CHA CUN OO, "Rangoon Sample Survey." *Burma*, 1 (July, 1951).

MEHTA, SURINDER K., "Comparative Analysis of the Industrial Structure of the Urban Labour Force of Burma and the United States." *Economic Development and Cultural Change*, 9 (January, 1961), pp. 164–179.

ORR, KENNETH G., "Tenasserim: Ancient City in Lower Burma." *Burma*, 2 (Ocbtoer, 1951), pp. 25–35.

SANDRUM, R.M., *Urbanization: The Burmese Experience*. Rangoon: Rangoon University, Economic Research Centre, Economic Paper Number 16, 1957.

SPATE, O.H.K. and L. TRUEBLOOD, "Rangoon: A Study in Urban Geography." *Geographical Review*, 32 (January, 1942), pp. 56–73.

CAMBODIA

EBIHARA, MAY, "Village Relations with Town and City in Cambodia." Paper presented at a meeting of the Association of Asian Studies, 1962.

——, "Interactions of the Village and Its Milieu in Cambodia." Paper presented at a meeting of the American Anthropological Association, 1961.

CEYLON

BOSE, ASHISH, "Internal Migration in India, Pakistan and Ceylon." Paper for the World Population Conference, Belgrade, 30 August to 10 September, 1965.

CEYLON, Department of Census and Statistics, *Demographic Study of the City of Colombo.* Colombo: Ceylon Government Press, Monograph Number 2, 1954.

CEYLON, Inter-Department Committee on the Reclamation and Utilization of the Swamps in and Around City of Colombo, *Report.* Colombo: 1960.

GOVINDARAJ, MANOHARAN, "The Town of Kegalle: A Brief Sketch of Its Geography." *Ceylon Geographer,* 13 (January–December, 1959), pp. 1–14.

PANDIT THARATNA, B.L., "Trends of Urbanization in Ceylon, 1901–1953." *Ceylon Journal of Historical and Social Studies,* 7 (July–December, 1964), pp. 203–217.

SARKAR, N.K., *The Demography of Ceylon.* Colombo: Ceylon Government Press, 1957.

VAIYAPURI, PILLAI, "Life in an Ancient Tamil City: Kaveripatanam." *Ceylon Historical Journal,* 2 (July–October, 1952), pp. 5–7.

VAMATHEVAN, S., *Internal Migration in Ceylon, 1946–53.* Colombo: Ceylon Department of Census and Statistics, Monograph Number 13, 1961.

INDIA

ABHAY KUMAR, "Urban Employment in Bihar." *Economic Papers,* (1957), pp. 49–50.

ACHARYA, H., "Urban Rural Relation." Paper submitted to the 39th Conference of the Indian Economic Association, Cuttack, December, 1956.

——, "Urbanizing Role of a One-Lakh City." *Sociological Bulletin,* 5 (September, 1956), pp. 89–101.

ADAMS, HOWARD and F. GREELY, *Report on the Revised Plan for the Town and Region of Gandhidham.* New Delhi: 1952.

AGARWAL, S.C., *Industrial Housing in India.* New Delhi: 1952.

——, *Recent Developments in Housing.* New Delhi: 1958.

AGARWALA, S.N., "A Method for Estimating Decade Internal Migration in Cities from Indian Census Data," *Indian Economic Review,* 4 (February, 1958), pp. 59–76.

——, "Socio-Economic and Demographic Characteristics of the Rural Migrants and the Non-Migrants." *Journal of the Institute of Economic Research* 3 (July, 1968), pp. 1–15.

AGNIHOTRI, V.D., *Housing Conditions of Factory Workers in Kanpur.* Lucknow: Fine Press, 1954.

——, "Poverty Among Factory Workers in Kanpur." *Labour Bulletin, Uttar Pradesh,* 14 (October, 1954), pp. 13–19.

——, "Standard of Living of the Factory Workers in Kanpur." *Labour Bulletin,* 14 (September, 1954), pp. 12–22.

——, "Unemployment in Uttar Pradesh." *Labour Bulletin,* 15 (March, 1955), pp. 15–18.

AHMAD, ENAYAT, "Distribution of Cities in India and Pakistan." *Geographical Outlook,* 2 (1956).

——, "Origin and Evolution of the Town of Uttar Pradesh." *Geographical Outlook,* 1 (January, 1956), pp. 38–58.

——, "Town Study with Special Reference to India." *Geographer* (Aligarh), 5 (1952), pp. 1–26.

AHMAD, MOBIN, *Patna—Redevelopment of Central Area.* New Delhi: 1961.

AHMED, LATEEF N., "A Comparative Inquiry in Rural Urban Dichotomy vs. Rural Urban Continuum in the United States and India: Its Administrative Challenges." *Indian Journal of Social Research,* 7 (August, 1966), pp. 101–109.

AHMED, QAZI, *Indian Cities: Characteristics and Correlates.* Chicago: University of Chicago Press, 1965.

ALAM, SHAH MANZOOR, *Hyderabad–Secunderabad (Twin Cities).* New Delhi: Allied Publishers, 1965.

——, "Masulipattam, a Metropolitan Port in the Seventeenth Century, A.D." *Indian Geographical Journal,* 34 (July–December, 1959), pp. 33–42.

——, "An Outline Plan to Conduct Urban Survey and to Map Urban Features." *Geographer* (Aligarh), 11 (1959), pp. 50–57.

ALAM, SHAH MANZOOR, K.N. GOPI, and W.A. KHAN, "Planning a Metropolitan Region: A Case Study (Hyderabad)." Paper for 21st International Geographical Congress, held at New Delhi, November–December, 1968. Unpublished.

ALEXANDER, P.C., *Industrial Estates in India.* Bombay: Asia Publishing House, 1963.

ALI, ASHRAF, *The City Government of Calcutta: A Study of Inertia.* New York: Institute of Public Administration, 1966.

ALI, S.H., *Banaras: The Holy City Kashi.* Banaras: 1955.

ALI, S.M., "The Towns of the Indian Desert." Proceedings of International Geography Seminar, India, 1956.

ALTBACH, PHILIP G., *Student Politics in Bombay.* Bombay: Asia Publishing House, 1968.

AMES, MICHAEL M., "Modernization and Social Structure: Family, Caste, and Class in Jamshedpur." *Economic and Political Weekly,* 4 (July, 1969), pp. 1217–1224.

ANAND, MULK RAJ, "Jaipur." *Illustrated Weekly of India,* 82 (8 October, 1961).

ANANTAPRADMANABHAN, N., "Functional Classification of Urban Centres in Madras State." Proceedings of Summer School in Geography, Simla, 1962.

ANITA, F.P., "The Final Test." *Seminar,* 79 (March, 1966), pp. 14–19.

——, "India's Urban Population—the City Dweller: A New Deal." *Population Review,* 7 (January, 1963), pp. 17–32.

ARUNACHALAM, B., "Bombay City: Stages of Development." *Bombay Geographical Magazine,* 3 (December, 1955), pp. 34–39.

ASSAM, Department of Economics and Statistics, *Report on the Sample Survey of Consumer Expenditure of the Urban Middle Class Households of Assam.* Assam: Assam Department of Economics and Statistics, 1961.

AYYAR, N.P., "Towns of the Upper Narmada Valley, Madhya Pradesh." Paper for 21st International Geographical Congress, held at New Delhi, November–December, 1968, in S.P. Das Gupta and T.R. Lakshmanan (eds.),

Abstract of Papers. Calcutta: National Committee for Geography, 1968.

AZIZ, A., "Impact of Urbanization on Rural Land." *Geographer* (Aligarh), 15 (November, 1968), pp. 50–56.

——, "Physical Setting of Delhi." *Geographer* (Aligarh), 8, 9, (1956, 1957), pp. 37–44.

——, "A Study of Indian Towns." *Geographer* (Aligarh), 7 (Summer, 1955), pp. 9–18.

——, "Delhi Metropolitan Area." Paper for 21st International Geographical Congress, held at New Delhi November–December, 1968, in S.P. Das Gupta and T.R. Lakshmanan (eds.), *Abstract of Papers.* Calcutta: National Committee for Geography, 1968, p. 255.

BADGAIYAN, S.D., "Social Consequences of Industrialization—A Typological Study." Paper prepared for the Seminar on Sociology of Economic Development at Institute on Economic Growth, Delhi, 25 November to 20 December, 1968. Mimeo.

BAHADUR, P., "Urban Unemployment: An Estimate Based on Employment Exchange Statistics," in V.K.R.V. Rao (ed.), *Employment and Unemployment.* New Delhi: Allied Publishers, 1968, pp. 57–79.

BAGAI, S.B., "Trends of Urbanization and Rural-Urban Migration in India, 1901–51." Paper submitted to the 39th Conference of the Indian Economic Association, Cuttack, December, 1956.

BAGAL, S.B., "Trends of Urbanization and Rural-Urban Migration in India, 1901–51." Paper read at the 39th Annual Conference of Indian Economic Association, Bombay, 1958, pp. 15–29.

BAGCHI, K. and USHA SEN, "Giridih: Its Growth and Land Use." *Geographical Review of India,* 25 (December, 1963), pp. 243–250.

BAJWA, GURBACHAN SINGH, "Political Democracy and Administrative Efficiency in City Government, in Indian Institute of Public Administration." *Improving City Government: Proceedings of a Seminar 13–14 September, 1958.* New Delhi: Indian Institute of Public Administration, 1959, pp. 7–8.

BAKRE, V.K., "Replanning and Redevelopment of Cities and Towns: Statement of the Problem." *Journal of the Institute of Town Planners, India* (January–April, 1959), pp. 53–57.

BALA, SUNDARAM, D., "Some Special Aspects of Citizen Participation," in Indian Institute of Public Administration, *Improving City Government: Proceedings of a Seminar 13–14 September, 1958.* New Delhi: Indian Institute of Public Administration, 1959, pp. 163–165.

BALAKRISHNA, R., *Economic Survey of Madras.* Faridabad: Government of India Press, 1961.

BALASUNDARAM, D., *History of Slums in Madras City and Scheme for Improvement.* Madras: 1957.

BALI, RAJESHWAR, *Faridabad: An Industrial Township (a Study of Some Social Implications of Industrialization.* Bombay: University of Bombay, Department of Sociology, 1967. Ph.D. Thesis.

——, "The Social Consequences of Industrialization—The Family in Faridabad." Paper prepared for the Seminar on Sociology of Economic Growth, Delhi, 25 November to 20 December, 1968. Mimeo.

BANDOPADHYA, S., "Low Cost Housing—Some Attempts at Implementation." *Indian Builder* (December, 1957), pp. 79–84.

BANERJEE, NIRMALA, "The Role of Fiscal Planning in Metropolitan Development: A Partisan View." Paper submitted at the 16th Annual Town and Country Planning Seminar, 1967.

——, "What Course for Urbanization in India?" *Economic and Political Weekly,* 4 (July, 1969), pp. 1173–1176.

BANERJEE, S.K., "Ring Towns—An Answer to the Fringe Area Growth of the Metropolitan City of Delhi." Paper for 7th Annual Town and Country Planning Seminar, New Delhi, October, 1968. Mimeo.

BANGALORE, City Improvement Trust Board, *Plans of Improvement and Expansion.* Bangalore: Bangalore City Improvement Trust Board, 1951.

BARVE, S.G., *An Outline of a Master Plan for Greater Poona.* Poona: Municipal Corporation, 1952.

——, "Towns of Tomorrow." *Seminar,* 79 (March, 1966), pp. 20–24.

——, "Urbanization in Maharashtra State: Problems and a Plan of Action," in Roy Turner (ed.), *India's Urban Future.* Bombay: Oxford University Press, 1962, pp. 347–360.

BASU, SIBDAS, *Verdict on Calcutta.* Calcutta: 1950.

BAUER, CATHERINE, "The Pattern of Urban and Economic Development: Social Implications." *Annals of the American Academy of Political and Social Science,* 305 (May, 1956), pp. 60–69.

BERRY, BRIAN J.L., "Policy Implications of an Urban Location Model for the Kanpur Region." in *Regional Perspective of Industrial and Urban Growth.* Bombay: Macmillan, 1969, pp. 203–219.

BERRY, BRIAN J.L. and *V.L.S.* PRAKASA RAO, "Urban-Rural Duality in the Regional Structure of Andhra Pradesh: A Challenge to Regional Planning and Development." Paper for 21st International Geographical Congress, held at New Delhi, November–December, 1968. Unpublished.

BERTOLI, JACK VICAJEE, "Towards City Planning and Architecture for New Needs." *Design,* 2 (February, 1959), pp. 12–18, 41.

BETEILLE, ANDRE, "Our Villages and Cities: A Contrast Between Tradition and Modernity." *Yojana,* 9 (January 26, 1965), pp. 27–28.

BHAGAT, D.G., "Regional Concept for Highway Planning—The Web and Lattice Pattern." *Journal of the Institute of Town Planners, India* (January–April), pp. 36–42.

BHARGAVA, GOPAL, "Central Village in the National Capital Region Guidelines for Their Development." Paper for 7th Annual Town and Country Planning Seminar, New Delhi, October, 1968. Mimeo.

BHASIN, M.G., "Akuluj, a Declassified Town in Maharashtra." Paper for 21st International Geographical Congress, held at New Delhi, November–December, 1968, in S.P. Das Gupta and T.R. Lakshmanan (ed.), *Abstract of Papers.* Calcutta: National Committee for Geography, 1968, p. 256.

——, "Capital Cities in Indian History." Paper for 21st International Geographical Congress, held at New Delhi, November–December, 1968, in S.P. Das Gupta and T.R. Lakshmanan (ed.), *Abstract of Papers.* Calcutta: National Committee for Geography, 1968, pp. 291–292.

——, "Slums and Slum Rehabilitation in India." *A.I.C.C. Economic Review,* 9 (15 July, 1957), pp. 16–18.

BHATIA, SHYAM SUNDER, "Historical Geography of Delhi." *Indian Geographer,* 1 (August, 1956), pp. 17–43.

——, "A Reconsideration of the Concept of the Primate City." *Indian Geographical Journal,* 37 (January–March, 1962), pp. 23–30.

BHATT, G.S., "The Chamar of Lucknow." *Eastern Anthropologist,* 8 (September–November, 1954), pp. 27–41.

BHATT, HARSHA J. (ed.), *Ports of India: Reference Manual.* Bombay: 1959.

BHATTACHARY, MOHIT, "Government in Metropolitan Calcutta." *Indian Journal of Public Administration,* 11 (October–December, 1965), pp. 702–720.

BHATTACHARYA, ANIMA, "Declining Trends in Residential Towns of West Bengal, 1871–1961." Paper presented in Seminar on Trends of Socio-Economic Change in India, 1871–1961, 11 to 23 September, 1967, Simla, Indian Institute of Advanced Study.

——, "Migration Pattern of West Bengal." Paper for 21st International Geographical Congress, held at New Delhi, November–December, 1968, in S.P. Das Gupta and R.T. Lakshmanan (eds.), *Abstract of Papers.* Calcutta: National Committee for Geography, 1968, p. 227.

BHATTACHARYA, ARDHENDU, "Industries in Poona Metropolitan Region: A Preliminary Study." *Economic and Political Weekly,* 4 (July, 1969), pp. 1191–1196.

BHATTACHARYA, A.N. and LODHA, R., *Evolution of Chittorgarh.* Udaipur: University of Udaipur, Research Studies, 2, 1964.

BHATTACHARYA, BHABANI, "Nagpur." *Illustrated Weekly of India,* 82 (15 October, 1961).

BHATTACHARYA, BIMALENDU, "Functional Classification of Towns in West Bengal." Paper for 21st International Geographical Congress, held at New Delhi, November–December, 1968, in S.P. Das Gupta and T.R. Lakshmanan (eds.), *Abstract of Papers.* Calcutta: National Committee for Geography, 1968, p. 256.

——, "Locational Analysis of Siliguri and Kalimpong Towns in West Bengal." Paper for 21st International Geographical Congress held at New Delhi, November–December, 1968, in S.P. Das Gupta and T.R. Lakshmanan (eds.), *Abstract of Papers.* Calcutta: National Committee for Geography, 1968, p. 256.

BHATTACHARYA, M., M.M. SINGH, and FRANK J. TYSEN, *Government in Metropolitan Calcutta: A Manual.* New York: Institute of Public Administration, 1966.

BHATTACHARYA, MOHIT, *Rural Self Government in Metropolitan Calcutta.* New York: Institute of Public Administration, 1966.

BHATTACHARYA, N.D., "Geographical Zones of Murshidabad." *Indian Geographical Journal,* 36 (October–December, 1961), pp. 132–139.

BHATTACHARYA, N.D., "Murshidabad: A Study in Urban Geography." *National Geographical Journal of India,* 5 (March, 1959), pp. 38–51.

BHATTACHARYA, R. and USHA SEN, "Rajgir and Its Surroundings." *Geographical Review of India,* 28 (June, 1966), pp. 41–46.

BHATTACHARYYA, S.K., *Passenger Transport Problem in Calcutta: A Study of Rush-Hour Traffic.* Calcutta: Bookland, 1958.

BOGUE, DONALD J. and K.C. ZACHARIAH, "Urbanization and Migration in

India," in Roy Turner (ed.), *India's Urban Future: Selected Studies from an International Conference on Urbanization in India, held at the University of California in 1960.* Berkeley: University of California Press, 1962, pp. 27–56.

BOMBAY, Bureau of Economics and Statistics, *Report on the Survey into the Economic Conditions of Middle Class Families in Bombay City.* Bombay: 1952.

BOMBAY, Directorate of Publicity, *Bombay 1955: An Authentic Reference Annual.* Bombay: 1955.

BOMBAY, Study Group for Greater Bombay, *Report.* Bombay: 1959.

BOPEGAMAGE, A., *Delhi: A Study in Urban Sociology.* Bombay: University of Bombay, 1957.

——, "Demographic Approach to the Study of Urban Ecology." *Sociological Bulletin,* 9 (March, 1960), pp. 82–95.

——, "A Methodological Problem in Indian Urban Sociological Research." *Sociology and Social Research,* 50 (January, 1966), pp. 236–240.

——, "Neighborhood Relations in Indian Cities—Delhi." *Sociological Bulletin,* 6 (March, 1957), pp. 34–42.

——, "Village Within a Metropolitan Aura." *Sociological Bulletin,* 5 (September, 1956), pp. 102–110.

BOSE, ASHISH, "An Appraisal of Data on Internal Migration and Urbanization in India." Paper prepared for the ECAFE Expert Working Group on Internal Migration and Urbanization, Bangkok, 1967.

——, *Delhi's Urban Sprawl, Land Development and Land Prices, 1947–67.* Delhi: Institute of Economic Growth, 1967. Unpublished.

——, "Delimitation of the Kanpur Regional and Urban Growth," in P.B. Desai (ed.), *Regional Perspective of Industrial and Urban Growth: The Case of Kanpur—Papers and Proceeding of the Kanpur International Seminar, 28 January–4 February, 1967.* Bombay: Macmillan, 1968, pp. 132–137.

——, "The Estimation of Working Force and Unemployed Population from Census Data," in 1st Indian Conference on Research in National Income, Calcutta, July, 1958. Mimeo.

——, "Housing the Rich in Delhi." *Economic and Political Weekly,* 2 (3 June, 1967), pp. 997–999.

——, "India's Basic Industrial Region—The Durgapur-Ranchi-Rourkela Development Region," *A.I.C.C. Economic Review,* 17 (February, 1966), pp. 150–156.

——, "Internal Migration in India, Pakistan and Ceylon." Paper for the World Population Conference, Belgrade, 30 August to 10 September 1965.

——, "Land Policy Thinking too Tentative." *Yojana,* 10 (18 September, 1966), pp. 19–21.

——, "Land Speculation in Urban Delhi." Lecture delivered at the National Building Organization, New Delhi, 8 and 9 July, 1968. Mimeo.

——, "Migration Streams in India." Paper presented to International Union for the Scientific Study of Population, Sydney Conference, 1967.

——, "A Note on the definition of 'Town' in the Indian Census: 1901–1961." *Indian Economic and Social History Review,* (January–March, 1964), pp. 84–94.

——, "The Pace of Urbanization in India," in *Papers Read at the 39th Annual Conference of the Indian Economic Association.* Bombay: 1958, pp. 30–42.

—— (ed.), *Patterns of Population Change in India, 1951–61.* New Delhi: Allied Publishers, 1967.

——, "Patterns of Urban Growth in India, 1951–61," in Ashish Bose (ed.), *Patterns of Population Change in India, 1951–61.* New Delhi: Allied Publishers, 1967, pp. 48–86.

——, "Population Growth and the Industrialization-Urbanization Process in India." *Man in India,* 41 (October–December, 1961), pp. 255–275.

——, "The Population Puzzle in India." *Economic Development and Cultural Change,* 7 (April, 1959), pp. 230–48.

——, *The Process of Urbanization in India, 1901–1951.* Delhi: University of Delhi, 1959. Unpublished Ph.D. Thesis.

——, "The Role of Small Towns in the Urbanization Process of India and Pakistan." Paper for International Union for the Scientific Study of Population Conference, London, September, 1969.

——, "Six Decades of Urbanization in India, 1901–1961." *Indian Economic and Social History Review,* 2 (January, 1965), pp. 23–41.

——, "Some Economic Aspects of Urban Redevelopment." *Journal of the Institute of Town Planners, India* (January–April, 1959), pp. 2–5.

——, *Studies in India's Urbanization.* Delhi: Institute of Economic Growth, 1966. Mimeo.

——, "The Study of City Population Growth on the Basis of Census Statistics." *Journal of the Institute of Town Planners, India* (January–April, 1959), pp. 40–44.

——, "Urban Characteristics of Towns in India—A Statistical Study." *Indian Journal of Public Administration,* 14 (July–September, 1968), pp. 457–465.

——, "Urban Employment Pattern: Materials, Methods, and Problems." *Seminar on Urbanization in India.* Berkeley: University of California, 1960. Mimeo.

——, "Urban Planning and Policy in India." *A.I.C.C. Economic Review,* 13 (22 September, 1961), pp. 19–22.

——, "Urbanization and Our Youth." *Illustrated Weekly of India,* 88 (15 October, 1967), pp. 26–27.

——, "Urbanization in the Face of Rapid Population Growth and Surplus Labour—The Case of India." *Indian Population Bulletin* (No. 3, 1966).

——, "Why Do People Migrate to Cities?" *Yojana,* 9 (January, 1965), pp. 25, 26, 37.

——, *Urbanization in India: An Inventory of Source Materials.* Bombay: Academic Books Limited, 1970.

BOSE, DEB KUMAR, *Regional Disparities in India—An Analysis Through National Sample Survey Data.* Calcutta: Indian Statistical Institute, 1967. Mimeo.

——, "Urbanization, Industrialization and Planning for Regional Development." *Economic and Political Weekly,* 4 (July, 1969), pp. 1169–1172.

BOSE, NIRMAL KUMAR, *Calcutta—A Social Survey.* Bombay: Lalwani, 1967.

——, "Calcutta: Premature Metropolis." *Scientific American,* 213 (September, 1965), pp. 91–102.

——, *Culture and Society in India.* Bombay: Asia Publishing House, 1967.

——, "Effect of Urbanization on Work and Leisure." *Man in India,* 37 (January–March, 1957), pp. 1–9.

——, "Social and Cultural life of Calcutta." *Geographical Review of India,* 20 (December, 1958), pp. 1–46.

——, "Social and Cultural Life of Calcutta: A Social Survey of Calcutta," in *Culture and Society in India.* Bombay: Asia Publishing House, 1967, pp. 309–357, 373–402.

——, "Some Problems of Urbanization." *Khadi Gramodyog,* 7 (September, 1961), pp. 20–35.

——, "Note on Tension Study in Calcutta Industrial Area." *Indian Journal of Psychology,* 27 (Parts I–IV, 1952), pp. 153–157.

BRAHME, SULABBA and PRAKASH GOLE, *Deluge in Poona—Aftermath and Rehabilitation.* Bombay: Asia Publishing House, 1968.

BREDO, WILLIAM, "Industrial Decentralization in India," in Roy Turner (ed.), *India's Urban Future.* Bombay: Oxford University Press, 1962, pp. 240–260.

BRIJ, MOHAR, "Municipal Government in the Capital." *Civic Affairs* (May, 1963), pp. 12–14.

BRUSH, JOHN E., "Application of a Model to the Analysis of Population Distribution in Indian Cities." Paper for 21st International Geographical Congress, held at New Delhi, November–December, 1968, in S.P. Das Gupta and T.R. Lakshmanan (eds.), *Abstract of Papers.* Calcutta: National Committee for Geography, 1968, p. 257.

——, "The Morphology of Indian Cities, in Roy Turner (ed.), *India's Urban Future.* Bombay: Oxford University Press, 1962, pp. 57–70.

——, "Some Dimensions of Urban Population Pressure in India." Paper presented for a Symposium, International Geographic Union's Commission on the Cartography and Geography of World Population, Pennsylvania State University, September, 1967.

——, "Spatial Pattern of Population in Indian Cities." *Geographic Review,* 58 (July, 1968), pp. 362–391.

BULSARA, JAL F., *Problems of Rapid Urbanization in India.* Bombay: Popular Prakashan, 1964.

——, "Urban Community Development." *Indian Journal of Public Administration,* 14 (July–September, 1968), pp. 746–752.

——, "Urban Community Development." *Indian Journal of Social Work,* 18 (March, 1957), pp. 245–259.

BURMAN, B.K. RAY, "Observations on the Effect on Work and Leisure." *Man in India,* 37 (July–September, 1957), pp. 217–229.

BUSCHMANN, K.H., "Some Remarks on Urban Geography." *Indian Geographical Journal,* 32 (July–December, 1957), pp. 101–105.

"Calcutta's Development Plans." *Civic Affairs* (October, 1963), pp. 9–14.

CALCUTTA METROPOLITAN PLANNING ORGANIZATION, *Basic Development Plan —Calcutta Metropolitan District, 1966–1986.* Culcutta: Calcutta Metropolitan Planning Organization, 1966.

——, *Bustee Improvement Programme.* Calcutta: Calcutta Metropolitan Planning Organization, 1967.

——, "The Problem of Housing." *Quarterly Journal of the Local Self Government Institute* (July, 1963), pp. 121–135.

——, *Regional Planning for West Bengal*. Calcutta: Calcutta Metropolitan Planning Organization, 1965.

"Census: A Symposium on What the Great Counting of 1961 Will Teach Us." *Seminar*, 18 (February, 1961). Complete issue.

CHADDA, V.N., "Planning in the States: A Development Plan for Lucknow." *Journal of the Institute of Town Planners, India* (July, 1955), p. 3.

CHAKRABARTI, P.M., *Banaras and Sarnath*. Banaras: 1957.

CHAKRABARTTY, SYAMAL, *Housing Conditions in Calcutta*. Calcutta: Bookland, 1958.

CHAKRABORTY, SATYESH C., "Population Density Gradients of Cities." Paper for 21st International Geographical Congress, held at New Delhi, November–December, 1968, in S.P. Das Gupta and T.R. Lakshmanan (eds.), *Abstract of Papers*. Calcutta: National Committee for Geography, 1968, p. 258.

CHAKRAVARTI, P.C. and V.L.S. PRAKASA RAO, "Planning and the City of Calcutta." *Modern Review*, 82 (August, 1947), pp. 129–133.

CHAKRAVARTI, SUBHASH, "Calcutta's Water Supply." *Civic Affairs* (July, 1963), pp. 23–24.

CHAKRAVARTI, S.K. et al., "Share of Urban and Rural Sectors in the Domestic Product in India in 1952–53," in V.K.R.V. Rao, et al. (eds.), *National Income and Allied Topics*. Bombay: Asia Publishing House, 1960, pp. 151–158.

CHAKRAVORTI, B., "A Broad View of Social Class Mortality Indices for Calcutta City." Paper presented to a Seminar on Problems Relating to Demography in West Bengal, 27 to 28 September, 1967, State Statistical Bureau, 1967. Mimeo.

CHAND, S.K., "Urban Development Trends in India." Paper submitted to the 39th Conference of the Indian Economic Association, Cuttack, December, 1956.

"Chandigarh: A New Planned City." *Marg*, 15 (December, 1961). Complete issue.

"Chandigarh: Punjab's New Capital." *Asian Review*, 50 (January, 1954), pp. 56–57.

CHANDRASEKARAN, C., "Survey of the Status of Demography in India," in Philip M. Hauser and Otis Dudley Duncan (eds.), *The Study of Population: An Inventory and Appraisal*. Chicago: University of Chicago Press, 1959, pp. 249–258.

CHANDRASEKHAR, A., "Regional Variation in Development as Reflected by Certain Population Characteristics in Andhra Pradesh, India." Paper for 21st International Geographical Congress, held at New Delhi, November–December, 1968, in S.P. Das Gupta and T.R. Lakshmanan (eds.), *Abstract of Papers*. Calcutta: National Committee for Geography, 1968, p. 218.

——, "Some Aspects of the Urbanization of Population in India." Paper for International Union for the Scientific Study of Population Conference, London, September, 1969.

CHANDRASEKHAR, S., "Growth of Population in Madras City, 1639–1961." *Population Review*, 8 (January, 1964), pp. 3–45.

——, "How India Is Tackling Her Population Problem." *Foreign Affairs*, 47 (October, 1968), pp. 138–150.

——, "India's Population: Fact, Problem and Policy," in S. Chandrasekhar (ed.), *Asia's Population Problems*. London: Allen & Unwin, 1967, pp. 72–99.

——, *Infant Mortality in India, 1901–1955: A Matter of Life and Death*. London: Allen & Unwin, 1959.

——, "Infant Mortality in Madras City." *Population Review*, 9 (January, July, 1965), pp. 100–105.

——, "Population Trends and Housing Needs in India." *Population Review*, 1 (January, 1957), pp. 12–18.

CHANDRASEKHARA, C.S., "The Central Vista: Effective Use for Recreation." Paper for 7th Annual Town and Country Seminar, New Delhi, October, 1968. Mimeo.

——, "Delhi Master Plan-Proposal for a New Spatial Pattern." Paper for 7th Annual Town and Country Seminar, New Delhi, October, 1968. Mimeo.

——, "Land Use and Zoning." *Indian Journal of Public Administration*, 14 (July–September, 1968), pp. 652–669.

——, "Regulations for Central Areas in Replanning and Redevelopment of Cities and Towns." *Journal of the Institute of Town Planners, India* (January–April, 1959), pp. 89–92.

——, "Role of City Planning in City Government," in Indian Institute of Public Administration, *Improving City Government: Proceedings of a Seminar 13–14 September, 1958*. New Delhi: Indian Institute of Public Administration, 1959, pp. 78–117.

——, "Workshop on Problems of Planning Administration of the National Capital." Paper for 7th Annual Town and Country Planning Seminar, New Delhi, October, 1968. Mimeo.

CHANDRASEKHARA, C.S. and SUNDARAM, K.V., "Urban Morphology and Internal Structure of Indian Towns—Some Case Studies." Paper presented at the Annual Seminar of the Institute of Town Planners, India, Bhopal, 1962.

CHANDRASEKHARIAH, N.G., "Urbanization in Mysore State." *Khadi Gramodyog* (August, 1967), pp. 780–784.

"Changing Patterns in Regional Personal Income Distribution 1955–56 to 1965–66: A Revised Analysis of the Structure of Urban and Rural Inequalities of Income." *Quarterly Economic Report of the Indian Institute of Public Opinion*, 8 (January, 1962), pp. 19–34.

CHATTERJEE, AMIYA BHUSAN, "Demographic Features of Howrah City." *Geographical Review of India*, 20 (December, 1958), pp. 150–169.

——, "Ecological Structure of Calcutta's Twins." *National Geographical Journal of India*, 11 (June, 1965), pp. 59–62.

——, "Hinterland of a Symbiotic City (Howrah): A Case Study." *Geographical Review of India*, 27 (June, 1965), pp. 55–71.

——, "Industrial Landscape of Howrah." *Geographical Review of India*, 26 (December, 1963), pp. 211–242.

——, "Morphology of Howrah," in M.R. Chaudhure (ed.), *Essays in Geography (S.P. Chatterjee Volume)*. Calcutta: Geographical Society of India, 1965, pp. 45–76.

CHATTERJEE, CHITTARANJAN, "Calcutta, 1863–1963." *Civic Affairs* (August, 1963), pp. 26–34.

CHATTERJEE, LATA, "Urban Hinterlands in West Bengal." Paper for 21st International Geographical Congress, held at New Delhi, November–December, 1968, in S.P. Das Gupta and T.R. Lakshmanan (eds.), *Abstract of Papers*. Calcutta: National Committee for Geography, 1968, p. 258.

CHATTERJEE, MARGARET, "The Town/Village Dichotomy in Delhi." *Man in India,* 48 (July–September, 1968), pp. 193–200.

CHATTERJEE, M.K., "Traditional and Leontief Capital Output Ratios in the Calcutta Industrial Region." *Artha Vijnana,* 6 (September, 1964), pp. 180–186.

CHATTERJEE, S., "Problems of Slum." *Quarterly Journal of the Local Self Government Institute* (July, 1962), pp. 1–12.

CHATTERJI, B., "Urban Community Development." *Civic Affairs* (January, 1963), pp. 22–27.

CHATTERJI, M., "Municipal Costs and Revenues in the Calcutta Industrial Region." *Quarterly Journal of the Local Self Government Institute* (January–March, 1966), pp. 251–262.

CHATTOPADHYAY, GOURANGA and ANIL A. SENGUPTA, "Growth of Disciplined Labour Force: A Case Study of Social Impediments." *Economic and Political Weekly,* 4 (July, 1969), pp. 1209–1216.

CHATTOPADHYAYA, K.P., "City Surveys—Methods and Analysis of Data." Paper presented in seminar on Impact of Urbanization, Industrialization and Social Change in India, 3 to 5 August, 1961, Ranchi, Ranchi University, 1961.

CHATTOPADHYAYA, K.P. and G.S. RAR, *Municipal Labour in Calcutta*. Calcutta: 1947.

CHATURVEDI, S.C., *Rural and Urban Wages in Uttar Pradesh*. Lucknow: 1956.

CHAUDHURI, MANORANJAN, "Recent Trends in Industrialization and Urban Growth in West Bengal." Paper for 21st International Geographical Congress, held at New Delhi, November–December, 1968, in S.P. Das Gupta and T.R. Lakshmanan (eds.), *Abstract of Papers*. Calcutta: National Committee for Geography, 1968, p. 258.

CHAUDHURI, M.R., "Durgapur, Future Ruhr of India," in M.R. Chaudhuri (ed.), *Essays in Geography (S.P. Chatterjee Volume)*. Calcutta: Geographical Society of India, 1965, pp. 165–180.

CHAUDHURI, SACHIN, "Centralization and the Alternate Forms of Decentralization: A Key Issue," in Roy Turner (ed.), *India's Urban Future*. Bombay: Oxford University Press, 1962, pp. 213–239.

CHAUHAN, D.S., "Caste and Occupation in Agra City." *Economic Weekly,* 12 (July 23, 1960), pp. 1147–1149.

——, "Cultural Assimilation of Immigrants in Cities: A Study in Social Interaction." *Journal of Social Sciences,* 2 (July, 1959), pp. 9–35.

——, "Social Costs of Migration." *Journal of Social Science,* 1 (July, 1958), pp. 29–51.

——, *Trends of Urbanization in Agra*. Agra: 1966.

CHAUHAN, I.S., "Social Groups and Categories in a Small Town." *Man in India,* 47 (January–March, 1967), pp. 15–34.

CHAWIA, INDER NATH, "Urbanization of the Punjab Plains." *Indian Geographer,* 3 (December, 1958), pp. 29–48.

CHELLASWAMI, T., "Population Trends and Labour Force in India, 1951–66." *Population Review,* 2 (July, 1958), pp. 42–48.

CHERRY, G.E., "Social Planning for Urban Communities." *Journal of the Town Planning Institute* (June, 1965), pp. 252–257.

CHERUKUPALLE, NIRMALA DEVI, *Demographic Correlates of Urban Size in India.* Cambridge, Mass.: Harvard University, 1967. Ph.D. Dissertation.

CHILDYAL, U.C., "Urbanization and Rural Government." *Indian Journal of Public Administration,* 14 (July–September, 1968), pp. 514–532.

CHITNIS, SUMA, "Urban Concentration of Higher Education." *Economic and Political Weekly,* 4 (July, 1969), pp. 1235–1238.

CHOPRA, R.N., "Municipal Personnel Administration." *Indian Journal of Public Administration,* 14 (July–September, 1968), pp. 538–550.

CHOWDHURY, J.G., "New Industrial Township: Nangal." *Urban and Rural Planning Thought,* 1 (January, 1958), pp. 12–27.

CHOWDHRY, U., "Chandigarh—The Work of Pierre Jeanneret." *Design,* 1 (1957), pp. 3–8.

"Civic Bombay: Its Lapses." *Economic Weekly,* 14 (25 August, 1962), pp. 1363–1364.

CLINARD, MARSHALL B. and B. CHATTERJEE, "Urban Community Development in India: The Delhi Pilot Project," in Roy Turner (ed.), *India's Urban Future.* Bombay: Oxford University Press, 1962 pp. 71–93.

CRANE, DAVID, "Chandigarh Reconsidered." *American Institute of Architects Journal,* 34 (May 1960), pp. 32–39.

CRANE, ROBERT I., "Urbanism in India." *American Journal of Sociology,* 60 (March, 1955), pp. 463–470.

DAMLE, Y.B., *College Youth in Poona—A Study of Elite in the Making.* Washington, D.C.: National Institute of Mental Health, 1966. Mimeo.

——, "The Hindu Family in Its Urban Setting." *Indian Journal of Social Work,* 25 (April, 1964), pp. 89–93.

DANGAYACH, K.B., "Unemployment in Urban Areas." *Indian Journal of Commerce,* 7 (December, 1954), pp. 29–38.

DANI, L.V., *Middle Class Social Survey of Ahmedabad City.* Ahmedabad: Gujerat Vepari Mahamandal, 1955.

DATTA, ABHIJIT, "Centre-State Relations and the National Capital Region." Paper for 7th Annual Town and Country Planning Seminar, New Delhi, October, 1968. Mimeo.

——, *Metropolitan Calcutta: Inter-Governmental Grants.* New York: Institute of Public Administration, 1968.

——, "Financing Municipal Services." *Indian Journal of Public Administration,* 14 (July–September, 1968), pp. 551–567.

——, "Financing Urban Development." *Economic and Political Weekly,* 4 (July, 1969), pp. 1177–1184.

DATTA, ABHIJIT and MOHIT BHATTACHARYA, *Centre-State Relations in Urban Development.* New Delhi: Indian Institute of Public Administration, 1966. Mimeo.

DATTA, ABHIJIT and DAVID C. RANNEY, *Municipal Finances in the Calcutta*

Metropolitan District: A Preliminary Survey. New York: Institute of Public Administration, 1966.

DATTA, J.M., "Urbanization in Bengal." *Geographical Review of India,* 18 (December, 1956), pp. 19–23.

DATTAR, B.N., "Housing Problems in Cities." *Yojana,* 9 (January 26, 1965), pp. 21–24.

DAVIS, KINGSLEY, *Population of India and Pakistan.* Princeton: Princeton University Press, 1951.

——, "Urbanization in India: Past and Future," in Roy Turner (ed.), *India's Urban Future.* Bombay: Oxford University Press, 1962, pp. 3–26.

DAYAL, P., "Population Growth and Rural-Urban Migration in India." *National Geographical Journal of India,* 5 (December, 1959), pp. 179–185.

DEB, AMAL KUMAR, "Orientation of Industries in Howrah." *Geographical Review of India,* 19 (December, 1957), pp. 59–66.

DEKNADAYALU VAIDU, U.P., "Is Bangalore a Self-Contained City." *Journal of the Institute of Town Planners, India* (January–April, 1961), pp. 78–79.

DELHI, Directorate of Industries and Labour, *Report on an Industrial Survey Conducted in 1950–51 in Delhi State.* Delhi: 1952.

DELHI, Municipal Corporation, Department of Urban Community Development, *Evaluation Study of the Formation and Working of the Vikas Mandals (Agency for Urban Community Development).* Delhi: 1962.

DELHI, Town and Country Planning Organisation, "Land Use Patterns of India's Cities and Towns." *Indian Journal of Public Administration* (July–September, 1968), pp. 606–651.

"Delhi, Agra, Sikri." *Marg,* 20 (September, 1967).

DELHI DEVELOPMENT AUTHORITY, *Draft Master Plan for Delhi.* Delhi: 1960.

——, *Master Plan for Delhi.* Delhi: 1962.

——, "Some Salient Features of the Draft Master Plan for Delhi." *Journal of the Institute of Town Planners, India* (January–April, 1961), pp. 104–110.

DELHI PRADESH, BHARAT SEVAK SAMAJ, *Slums of Old Delhi.* Delhi: Atma Ram & Sons, 1958.

DESAI, A.R., "Urbanization and Social Stratification." *Sociological Bulletin,* 9 (September, 1960), pp. 7–14.

DESAI, B.G., *The Emerging Youth.* Bombay: Popular Prakashan, 1967.

DESAI, C.C., "Urban Land Policies in India," in United Nations, *Urban Land Problems and Policies.* New York: United Nations, 1953, pp. 76–82.

DESAI, I.P., "Small Towns: Facts and Problems." *Economic Weekly,* 16 (April 13, 1964), pp. 725–730.

DESAI, P.B., *Development of West Bengal: Some Considerations for Regional Planning.* Delhi: Institute of Economic Growth, 1967. Mimeo.

——, "A Draft Master Plan for Delhi." *Shakti,* 5 (February, 1968), pp. 9–18.

——, "Economy of Indian Cities." *Indian Journal of Public Administration,* 14 (July–September, 1968), pp. 449–456.

—— (ed.), *Regional Perspective of Industrial and Urban Growth: the Case of Kanpur—Papers and Proceedings of the Kanpur International Seminar, 28 January–4 February, 1967.* Bombay: Macmillan, 1968.

——, *Size and Sex Ratio Composition of Population of India, 1901–1961.* Bombay: Asia Publishing House, 1968.

DESAI, P.B. and ASHISH BOSE, *Planning and Development of Satellite and*

New Towns in Developing Countries: Economic Considerations. Delhi: Institute of Economic Growth, 1963. Mimeo.

——, "Satellite and New Towns." *Shakti,* 2 (February, 1965), pp. 11–20.

DESAI, P.B. and M.N. PAL, "Economic Activity in Kanpur Region: Present Pattern and Perspective Dimension," in P.B. Desai (ed.), *Regional Perspective of Industrial and Urban Growth: The Case of Kanpur—Papers and Proceedings of the Kanpur International Seminar, 28 January–4 February, 1967.* Bombay: Macmillan, 1968, pp. 6–33.

DESAI, S.V. and B.B. GODAMBE, "Physical Aspects of Replanning Bombay City." *Journal of the Institute of Town Planners, India* (January–April, 1959), pp. 66–69.

DE SARKAR, SYAMAL KUMAR, "Calcutta: A Problem Metropolis." *Journal of the Institute of Town Planners, India* (January–April, 1961), pp. 63–73.

DESHMUKH, M.B., "Delhi: A Study of Floating Migration," in *The Social Implications of Industrialization and Urbanization.* Calcutta: UNESCO Research Centre on the Social Implications of Industrialization in Southern Asia, 1956, pp. 143–226.

DESHPANDE, C.D., "Cities and Towns of Bombay Province, Aspects of Urban Geography." *Indian Geographical Journal,* 16 (July–September, 1941), pp. 268–286.

——, "A Regional Plan and Its Physical Setting—The Poona Metropolitan Region—A Case Study." Paper for 21st International Geographical Congress, held at New Delhi, November–December, 1968. Unpublished.

——, "Visnagar." *Geographical Outlook,* 1 (No. 2, 1956).

DEV RAJ, "Kanpur Plan and Its Implementation." *Journal of the Institute of Town Planners, India* (January–April, 1961), pp. 110–115.

DEVANANDAN, P.D. and M.M. THOMAS (eds.), *Community Development in India's Industrial Urban Areas.* Bangalore: Committee for Literature on Social Concerns, 1958.

DEVASAGAYAM, A., "A Survey of Dwellings and Living Conditions of Industrial Employees in Madras City." *Indian Journal of Social Work,* 26 (April, 1965), pp. 69–86.

DEVKI, NANDAN, "Urban Development Trends in India." Paper submitted to the 39th Annual Conference of the Indian Economic Association, Cuttack, December, 1956.

DEY, S.K., *Nilokheri.* Bombay: Asia Publishing House, 1962.

——, "Urban Community Development." *Civic Affairs* (June, 1962), pp. 60–61, 64.

DHABRIYA, S.S., "Evolution of Cultural Landscape in Udaipur City." *National Geographical Journal of India,* 10 (June, 1964), pp. 92–99.

——, "Evolution of Six Principal Cities of Rajasthan." Paper for 21st International Geographical Congress, held at New Delhi, November–December, 1968, in S.P. Das Gupta and T.R. Lakshmanan (eds.), *Abstract of Papers.* Calcutta: National Committee for Geography, 1968, p. 261.

——, "Manpower in Six Principal Cities of Rajasthan." Paper for 21st International Geographical Congress, held at New Delhi, November–December, 1968, in S.P. Das Gupta and T.R. Lakshmanan (eds.), *Abstract of Papers.* Calcutta: National Committee for Geography, 1968, pp. 262–263.

————, "Udaipur: A Study in Urban Planning." *Annals of the Social Sciences* (*Jodhpur*), 1 (No. 1, 1965).

DHAMIJA, R.P., "City at Capital's Door Step: Ghaziabad." *Yojana*, 1 (July, 1955), pp. 6–9.

DHANKAR, G. H. and D.G. DESHPANDE, "Effect of Urbanization of Balutedar System of a Village." *Khadi Gramodyog*, 11 (March, 1965), pp. 443–446.

DHAR, P.N. and H.F. LYDALL, *Role of Small Enterprises in Indian Economic Development*. Bombay: Asia Publishing House, 1961.

DHAR, P.N., *Small Scale Industries in Delhi: A Study in Investment, Output and Employment Aspects*. Bombay: Asia Publishing House, 1958.

DHAR, R., *An Input-Output Table for West Bengal and the Calcutta Metropolitan Districts; First and Second Interim Reports*. New York: Institute of Public Administration.

DHAR, R., WALLACE E. REED, and P.N. SINHA, *The Economic Hinterland of Calcutta*. New York: Institute of Public Administration.

DHEKNEY, B.R., *Hubli City: A Study in Urban Economic Life*. Dharwar: Karnatak University, 1959.

DI BONA, JOSEPH, "Elite and Mass in Indian Higher Education," in *Turmoil and Transition: Higher Education and Student Politics in India*. Bombay: Lalvani Publishing House, 1968, pp. 131–168.

DIXIT, K.R., "City of Sholapur: A Brief Study of Some Aspects of Its Urban Geography." *Journal of the Maharaja Sayajirao University of Baroda*, 8 (March, 1959), pp. 51–56.

DOIG, DESMOND, *Calcutta—An Artist's Impression*. Calcutta: The Statesman, 1968.

DOSHI, HARISH C., "Industrialization and Neighbourhood Communities in a Western Indian City—Challenge and Response." *Sociological Bulletin*, 17 (March, 1968), pp. 19–34.

DREW, JANE, "Chandigarh and Planning Development in India." *Asian Review*, 51 (April, 1955), pp. 110–123.

D'SOUZA, V.L., "Socio-Economic Factors in Urban Planning." *Urban and Rural Planning Thought*, 2 (July, 1959), pp. 109–113.

D'SOUZA, VICTOR S., *Social Structure of a Planned City—Chandigarh*. Bombay: Orient Longman's Limited, 1968.

DUBEY, BASUDEO, "Functional Classification of Towns in Narmada River Basin, Central India." Paper for 21st International Geographical Congress, held at New Delhi, November–December, 1968, in S.P. Das Gupta and T.R. Lakshmanan (eds.), *Abstract of Papers*. Calcutta: National Committee for Geography, 1968, p. 263.

DUTT, A.K., "An Analysis of Commutation to the Metropolis of Calcutta." *National Geographical Journal of India*, 10 (September–December, 1964), pp. 194–206.

————, "A Critique of Town Plans of Jamshedpur." *National Geographical Journal of India*, 5 (December, 1959), pp. 205–211.

————, "Evolution of Jamshedpur City: A Historical Approach to Urban Study." *Indian Geographical Journal*, 41 (January–June, 1966), pp. 19–28.

————, "The Neighborhood Unit Plan and Its Impact on Jamshedpur City." *Journal of Social Research*, 5 (March, 1962), pp. 109–113.

DUTT, A.K. and S.C. CHAKHABORTY, "Reality of Calcutta Conurbation." *National Geographical Journal of India,* 9 (September–December, 1963), pp. 161–174.

DUTT, K.L., "Urban Bones of India." *National Geographical Journal of India,* 13 (June, 1967).

DUTT, MAYA, "Growth of Jamshedpur in Bihar." Paper for 21st International Geographical Congress, held at New Delhi, November–December, 1968, in S.P. Das Gupta and T.R. Lakshmanan (eds.), *Abstract of Papers.* Calcutta: National Committee for Geography, 1968, p. 263.

DUTTA, A.K., "Umland of Jamshedpur." *Geographical Review of India,* 26 (June, 1963), pp. 84–98.

DWIVEDI, R.L. "Allahabad: A Study in Industrial Development." *National Geographer* (May, 1961), pp. 53–60.

——, "Delimiting the Umland of Allahabad." *Indian Geographical Journal,* 39 (July–December, 1964), pp. 123–139.

——, "Demographic Features of Allahabad City." *Geographical Review of India,* 27 (December, 1965), pp. 163–188.

——, "Replanning an Existing City—Allahabad." *National Geographer* (Allahabad), 5 (1962).

EAMES, E., "Some Aspects of Urban Migration from a Village in North Central India." *Eastern Anthropologist,* 8 (September–November, 1954), pp. 13–26.

——, "Urbanization and Rural-Urban Migration." *Population Review,* 9 (January–July, 1965), pp. 38–47.

ECHEVERRIA, EDWARD, "How Much Do Indian Cities Need." *Urban and Rural Planning Thought,* 1 (October, 1958), pp. 253–263.

——, "Need for New Density Patterns for Indian Cities." *Design,* 2 (July, 1958), pp. 58–61, 90.

EKAMBARAM, S.K. et al., "Growth of Population in Mysore City: A Critical Examination." *Asian Economic Review,* 3 (May, 1961), pp. 236–242.

ELLEFSEN, RICHARD A., "City-Hinterland Relationships in India (with Special Reference to the Hinterlands of Bombay, Delhi, Madras, Hyderabad, and Baroda)," in Roy Turner (ed.), *India's Urban Future.* Bombay: Oxford University Press, 1962, pp. 94–116.

——, "Land Use in Delhi State." *Indian Geographer,* 5 (December, 1960), pp. 59–68.

EVENSON, NORMA, *Chandigarh.* Berkeley and Los Angeles: University of California Press, 1966. Ph.D. Thesis.

FERNANDEZ, B.G., *Analysis of the Land Use Provision of the Greater Bombay Plan.* Berkeley: University of California, 1952. M.A. Thesis.

"First Thoughts on Census." *Yojana,* 5 (16 April, 1961), pp. 3–7, 9–14.

FOLKE, STEEN, "Central Place Systems and Spatial Interaction in Nilgiris and Coorg (South India)." Paper for 21st International Geographical Congress, held at New Delhi, November–December, 1968, in Niels Kingo Jacobsen and Ruth Helkiaer Jensen (eds.), *Collected Papers—Denmark.* Denmark: Kobenhaavns Universitets Geografiske Institut, 1968, pp. 55–69.

——, "Evolution of Plantations. Immigration and Population Growth in Nilgiris and Coorg (South India)." Paper for 21st International Geographical Congress, held at New Delhi, November–December, 1968, in Niels

Kingo Jacobsen and Ruth Helkiaer Jensen (eds.), *Collected Papers—Denmark.* Denmark: Kobenhaavns Universitets Geografiske Institut, 1968, pp. 16–54.

"Forthcoming India Census of 1961." *Indian Population Bulletin,* 1 (April, 1960), pp. 222–231.

FRY, E. MAXWELL, "Chandigarh: A New Town for India." *Town and Country Planning Institute,* 21 (May, 1953), pp. 217–221.

——, "Chandigarh: The Capital of East Punjab." *Royal Institute of British Architects Journal,* 62 (January, 1955), pp. 87–94.

——, "Chandigarh Architecture." *Urban and Rural Planning Thought,* 2 (October, 1959), pp. 117–126.

FRY, E. MAXWELL and J.B. DREW, "Chandigarh and Planning Development in India." *Asian Review,* 51 (April, 1955), pp. 110–123.

GADGIL, D.R., "Housing and Slums in Poona." *Economic Weekly,* 11 (April 4, 1959), pp. 496–488.

——, *The Industrial Evolution of India in Recent Times.* Calcutta: Oxford University Press, 1959.

——, *Poona: A Socio-Economic Survey, Parts I and II.* Poona: Gokhale Institute of Politics and Economics, 1945, 1952, respectively.

——, *Sholapur City: Socio-Economic Studies.* Poona: Gokhale Institute of Politics and Economics, 1965.

GADGIL, GANGADHAR, "Poona." *Illustrated Weekly of India,* 83 (April 1, 1962).

GAGNON, GILES, "Chandigarh," *Royal Architecture Institute of Canada Journal.* 34 (June, 1957), pp. 193–200.

GANANATHAN, V.S., "Distribution of Urban Settlements in India," in proceedings of the 17th International Geographical Congress, Washington, 1952, pp. 742–745.

——, "Phursinghi." *Indian Geographical Journal,* 36 (April–June, 1961).

——, "Some Aspects of Rural-Urban Relationships in India." *Indian Geographer,* 4 (December, 1959), pp. 29–36.

GANDHI, N.K., "New Towns Construction in India." *Quarterly Journal of the Local Self-Government Institute,* 18 (October, 1957), pp. 435–445.

GANGULI, B.N., "Classification of Indian Cities, Town Groups and Towns," in M.R. Chaudhuri (ed.), *Essays in Geography (S.P. Chatterjee Volume).* Calcutta: Geographical Society of India, 1965, pp. 82–93.

——, "Need for a Rational Policy on Urbanization." *Economic Weekly,* 13 (July 29, 1961), p. 1215.

——, "Rent Regulation and the Housing Problem in Delhi." *A.I.C.C. Economic Review,* 12 (January 16, 1960), pp. 91–93.

GARG, B.B., "Use of Electronic Digital Computer in Planning of Housing Developments." Paper for 7th Annual Town and Country Planning Seminar, New Delhi, October, 1968.

GAUDINO, ROBERT L., *The Indian University.* Bombay: Popular Prakashan, 1965.

GEORGE, M.V., "A Decade of Unprecedented Population Growth in India." *Asian Survey,* 2 (February, 1962), pp. 31–33.

——, *Internal Migration in Assam and Bengal, 1960–61.* Australia: Australian National University, 1965. Unpublished Ph.D. Thesis.

GHOSE, BENOY, "Colonial Beginnings of Calcutta: Urbanization Without Industrialization." *Economic Weekly,* 12 (July–December, 1960), p. 1260.

——, "The Economic Character of the Urban Middle Class in 19th Century Bengal," in B.N. Ganguli (ed.), *Readings in Indian Economic History.* Bombay: Asia Publishing House, 1964, pp. 137–147.

——, "Town Improvement in Old Calcutta: Its Impact on Property Owners." *Economic Weekly,* 10 (July, 1958), pp. 873–876.

GHOSH, AMBICA, *Calcutta: The Primate City.* Delhi: Registrar General, 1966. Census Monograph No. 2.

——, "Immigration from Rural Areas into Calcutta Metropolitan Region: Analysis and Projections," in *Papers Contributed by Indian Authors to the World Population Conference, Belgrade, 1965.* New Delhi: Register General, 1965, pp. 153–161.

——, "A Note on Internal Migration and Urbanization (with Particular Reference to the Calcutta Metropolitan Area)." Paper prepared for ECAFE Expert Working Group on Internal Migration and Urbanization, Bangkok, 1967.

——, "A Study of Demographic Trends in West Bengal During 1901–1950." *Population Studies,* 9 (March, 1956), pp. 217–236.

GOSH, BIJIT, "Cities of Our Making." *Indian Journal of Public Administration,* 14 (July–September, 1968), pp. 446–473.

——, "City Builders Need a Philosophy." *Yojana,* 9 (January 26, 1965), pp. 43–45.

——, "Delhi 1981 A.D.: Promises for a Better City." *Shakti,* 5 (February, 1968), pp. 19–26.

——, *Indian Cities: Selected Papers on Contemporary Town Planning.* New Delhi: 1966.

GHOSH, D., "Census Economic Classification." *Economic Weekly,* 10 (October 18, 1958), pp. 1335–1339.

GHOSH, P.K., "Some Aspects of Unemployment in Urban Areas." *Indian Journal of Commerce,* 7 (December, 1954), pp. 39–46.

GHOSH, SHANKAR, "Calcutta: The Shame of Our Cities." *Civic Affairs,* (December, 1966), pp. 5–7.

GHOSH, SUDHIR, "The Challenge of Calcutta." *Civic Affairs,* (January, 1962), pp. 72–75, 77.

GHURYE, G.S., *Cities and Civilization.* Bombay: Popular Prakashan, 1962.

——, "Cities of India." *Sociological Bulletin,* 2 (May, 1953), pp. 47–71.

——, "Prolegomena to Town and Country Planning." *Sociological Bulletin,* 9 (September, 1960), pp. 73–91.

GIAN, PRAKASH, "Organizing City Government." *Indian Journal of Public Administration,* 14 (July–September, 1968), pp. 498–513.

GIST, NOEL P., "Ecological Structure of an Asian City: An East-West Comparison." *Population Review,* 2 (January, 1958), pp. 17–25.

——, "Ecology of Bangalore, India: An East-West Comparison." *Modern Review,* 101 (May, 1957), pp. 357–364.

——, "Selective Migration in South India." *Sociological Bulletin,* 4 (September, 1955), pp. 147–160.

——, "Selective Migration in Urban South India," in *Proceedings of the World Population Conference, 1954.* New York: United Nations, 1955, pp. 811–822.

GOOPTU, A., "Bankura." *Geographical Review of India,* 13 (September, 1951), pp. 40–42.

GOPALAN, B.S., *Kolar Gold Fields: A Study of Mining Towns.* New Delhi: 1961.

"Gorakhpur—A Study in Historical Geography." *Geographical Thought* (Gorakhpur), 3 (No. 1, 1967).

"Gorakhpur Master Plan." *Civic Affairs* (September, 1963), pp. 6–8.

GORE, M.S., "Language in Metropolitan Life." *Economic and Political Weekly,* 4 (July, 1969), pp. 1225–1234.

——, "Some Problems of Urban Growth." *Social Welfare,* 9 (November, 1962), pp. 6–7.

——, *Urbanization and Family Change.* Bombay: Popular Prakashan, 1968. Ph.D. Thesis.

GOSAL, GURDEV SINGH, "Redistribution of Population in Punjab during 1951–61," in Ashish Bose (ed.), *Patterns of Population in India, 1951–61.* New Delhi: Allied Publishers, 1967, pp. 107–129.

GOSWAMI, SATYA BRATA, "Regional Setting of Dalmianagar." *Geographical Outlook,* 1 (1956), pp. 28–37.

Greater Bombay Municipal Corporation, *Report on the Development Plan for Greater Bombay, 1964.* Bombay: Bombay Greater Municipal Corporation, 1964.

"Greater Calcutta in Making." *Civic Affairs* (May, 1962), pp. 7–8.

"Growth of Industrial Towns in India." *Eastern Economist,* 29 (January 11, 1957), pp. 52–54.

GUHA, A., "Evolution of Serampur, a Study in Urban Geography." *The Earth* (1964).

GUHA, B., "Morphology of Burdwan." *The Earth* (1964).

GUHA, M. and A.B. CHATTERJEE, "Serampore: Study in Urban Geography." *Geographical Review of India,* 16 (December, 1954), pp. 38–41.

GUHA, MEERA, "The Definition of an Indian Urban Neighbourhood (Calcutta)." *Man in India,* 46 (January–March, 1966), pp. 59–65.

——, "The Development of the Urban Functions of Calcutta." *Journal of Social Research,* 5 (March, 1962), pp. 91–96.

——, "Growth of Calcutta's Business District." *Economic Weekly,* 17 (November 13, 1965), pp. 1695–1698.

——, "Morphology of Calcutta." *Geographical Review of India,* 15 (September, 1953), pp. 20–28.

——, "Spatial Aspects of Ecological Organization in Calcutta." Paper for 21st International Geographical Congress, held at New Delhi, November–December, 1968, in S.P. Das Gupta and T.R. Lakshmanan (eds.), *Abstract of Papers.* Calcutta: National Committee for Geography, 1968, p. 265.

——, "Transport In and Around Calcutta." *Geographical Review of India,* 17 (June, 1955), pp. 1–8.

——, "Urban Regions of West Bengal: a Few Examples." *Geographical Review of India,* 19 (September, 1957), pp. 31–44.

GUHA, S., "Socio-Economic Impact of Urbanization." *A.I.C.C. Economic Review,* 10 (June 1, 1958), pp. 16–18.

GUHA, UMA, *A Short Sample-Survey of the Socio-Economic Conditions of*

Saheb Bagan Bustee, Rajabazar, Calcutta. Calcutta: Government of India, Department of Anthropology, 1963.

GUJRAL, I.K., "Civic Administration." *Seminar,* (March, 1966).

GUPTA, BELA DATT, "Urbanization in Bengal and Rural-Urban Interaction Pattern." Paper presented in Seminar on Trends of Socio-Economic Change in India, 1871–1961, 11 to 23 September, 1967. Simla: Indian Institute of Advanced Studies, 1967.

GUPTA, J. DATTE, "Immigration into Calcutta." Paper for 21st International Geographical Congress, held at New Delhi, November–December, 1968; in S.P. Das Gupta and T.R. Lakshmanan (eds.), *Abstract of Papers.* Calcutta: National Committee for Geography, 1968, p. 229.

GUPTA, J.P., "Kanpur: The City of Slums." *Social Welfare,* 10 (December, 1963), pp. 28–30.

——, "Social X-Ray of the City of Lucknow." *Social Welfare,* 10 (May, 1963), pp. 6–7.

GUPTA, MURARILAL, "A Note on Demography of Jabalpur City." *Journal of Geography* (University of Jabalpur), (November, 1959), pp. 10–16.

GUPTA, O.P., "Migration in Delhi." Paper for the Annual Town and Country Planning Seminar, New Delhi, October, 1968. Mimeo.

GUPTA, P. SEN, "Some Characteristics of Internal Migration in India." Paper prepared for the ECAFE Expert Working Group on Internal Migration and Urbanization, Bangkok, 1967.

GUPTA, SATISH CHANDER, "New Towns in India—A Socio-Economic Experiment." *Journal of the Institute of Town Planners, India* (March, 1964), pp. 4–5.

GUPTA, S.K., "Redevelopment of Central Area." *Journal of the Institute of Town Planners, India* (January–April, 1959), pp. 69–75.

——, "Second Thoughts on Problems of Muncipal Relationships," in Indian Institute of Public Administration, *Improving City Government: Proceedings of a Seminar 13–14 September, 1958.* New Delhi: Indian Institute of Public Administration, 1959, pp. 23–27.

GUPTOO, A., "Nainital, U.P." *Geographical Review of India,* 16 (September, 1954), pp. 12–18.

GURNER, W., "Town Planning in India with Discussion." *Asiatic Review,* 43 (July, 1947), pp. 209–219.

HANUMANTHARAYAPPA, P., "The Use of Household Data in the Analysis of Housing Situation in India." *Asian Economic Review,* 8 (November, 1965), pp. 88–102.

HAR SWARUP, "Growth and Functioning of Municipalities in Rohilkhand Division, U.P." *Quarterly Journal of the Local Self Government Institute,* (October, 1962), pp. 129–171; (January, 1963), pp. 246–288; (July, 1963), pp. 41–81; (October, 1963), pp. 191–223.

——, "Survey of the Municipalities in Rohilkhand Division, U.P." *Quarterly Journal of the Local Self Government Institute* (October, 1961), pp. 191–196.

HARRIS, BRITTON, "Urban Centralization and Planned Development," in Roy Turner (ed.), *India's Urban Future.* Bombay: Oxford University Press, 1962, pp. 261–276.

——, "Urban Problems in the Third Plan: Some of Their Implications." *Economic Weekly,* 12 (June, 1960), pp. 875–878.

——, "Urbanization Policy in India," in Gerald A.P. Carrothers (ed.), *Papers and Proceedings of the Regional Science Association,* 5 (1959), pp. 181–203.

HARRIS, M., *Problems that Urbanization Faces.* Ahmedabad: Harold Laski Institute of Political Science, 1960.

HART, HENRY C., "Bombay Politics: Pluralism and Popularization; A Symposium on Urban Politics in a Plural Society." *Journal of Asian Studies,* 20 (May, 1961), pp. 267–274.

——, "Urban Politics in Bombay: The Meaning of Community." *Economic Weekly,* 12 (January–June, 1960), pp. 983–988.

HAZLEHURST, LEIGHTON W., "The Middle-Range City in India." *Asian Survey,* 8 (July, 1968), pp. 539–552.

HIRT, HOWARD F., *Aligarh, Uttar Pradesh: A Geography of Urban Growth.* Syracuse: Syracuse University, 1955. Unpublished Ph.D. Dissertation.

——, "Spatial Aspects of the Housing Problem in Aligarh, U.P., India." *Population Review,* 2 (January, 1960).

——, "Study of Urban Geography of Aligarh." *Geographer* (Aligarh), 5 (December, 1952), pp. 24–27.

HONRAO, M.S., "Urban Features of Bijapur." *Journal of the Karnatak University* (June, 1960), pp. 64–74.

HONRAO, M.S. and V.R. PRABHU, "Karwar, the Port Town." *National Geographical Review of India,* 3 (Nos. 3, 4, 1967).

HOSELITZ, BERT F., "The Cities of India and Their Problems." *Annals of Association of American Geographers,* 49 (June, 1959), pp. 223–231.

——, "Indian Cities: The Surveys of Calcutta, Kanpur and Jamshedpur." *Economic Weekly,* 13 (July, 1961), pp. 1071–1078.

——, "The Role of Urbanization in Economic Development: Some International Comparisons," in Roy Turner (ed.), *India's Urban Future.* Bombay: Oxford University Press, 1962, pp. 155–181.

——, "A Survey of the Literature on Urbanization in India," in Roy Turner (ed.), *India's Urban Future.* Bombay: Oxford University Press, 1962, pp. 155–181.

——, "Urban-Rural Contrast as a Factor in Socio-Cultural Change." *Economic Weekly,* 12 (January, 1960), pp. 145–162.

——, "Urbanization and Town Planning in India." *Confluence,* 7 (Summer, 1968), pp. 115–127.

——, "Urbanization in India." *Kyklos,* 3 (June, 1960), pp. 361–372.

"Housing Problem in Bombay City." *Economic Weekly,* 11 (July–December, 1959), pp. 1289–1290.

HUSSEY, CHRISTOPHER, *The Life of Sir Edwin Lutyens.* New York: Charles Scribner's Sons, 1953, pp. 237–523. An account of the evolution of New Delhi City.

HYDERABAD, Department of Information and Public Relations, *History and Legend in Hyderabad.* Hyderabad: 1953.

HYDERABAD METROPOLITAN RESEARCH PROJECT, *Delimitation of the Hydera-*

bad Metropolitan District. Hyderabad: Osmania University, Institute of Asian Studies, 1966.

INAMDAR, N.R., "Metropolitan Planning in Maharashtra." *Indian Journal of Public Administration*, 14 (July–September, 1968), pp. 670–681.

INDIA, Central Council of Local Self Government, Committee of Ministers, *Report: Augmentation of Financial Resources of Urban Local Bodies*. New Delhi: 1965.

INDIA, Committee on Plan Projects, *Selected Buildings Projects Team on Slum Clearance: Report*. New Delhi: 1953.

INDIA, Committee on Public Undertakings, *Township and Factory Buildings of Public Undertaking: Eighth Report*. New Delhi: 1965.

INDIA, Ministry of Finance, Department of Economic Affairs, *National Survey, No. 6, Survey of Faridabad Township*. Delhi: 1954.

——, *National Sample Survey, No. 8, Report on Preliminary Survey of Urban Unemployment in September 1953*. Delhi: 1956.

——, *National Sample Survey, No. 9, Report on the Sample Survey of Displaced Persons in the Urban Areas of the Bombay State*. Delhi: 1957.

——, *National Sample Survey No. 14, Report on Some Characteristics of the Economically Active Population*. Delhi: 1958–59.

——, *National Sample Survey, No. 16, Report on Employment*. Calcutta: 1955.

——, *National Sample Survey, No. 17, Report on Sample Survey of Employment in Calcutta in 1953*. Calcutta: 1959.

——, *National Sample Survey, No. 34, Employment and Unemployment*. Calcutta: 1956.

——, *National Sample Survey, No. 53, Tables with Notes on Internal Migration*. Calcutta: 1962.

——, *National Sample Survey, No. 52, Tables with Notes on Employment and Unemployment*. Calcutta: 1961.

——, *National Sample Survey, No. 62, Report on Employment and Unemployment, Supplementary to Report No. 16*. New Delhi: 1962.

——, National Sample Survey, No. 63, Tables with Notes on Employment and Unemployment in Urban Areas. New Delhi: 1962.

——, *National Sample Survey, No. 103, Tables with Notes on Urban Labour Force*. New Delhi: 1966.

——, *National Sample Survey, Special Report on the Survey of Persons on the Live Register of the Delhi Employment Exchange*. Delhi: 1954.

——, *National Sample Survey, Working Paper No. 3, Growth of Urban Delhi*. Delhi: 1957.

INDIA, Ministry of Health, *Interim General Plan for Greater Delhi*. Delhi: 1956.

——, *Report of the Committee on Development of Small Towns in Hill and Border Areas*. New Delhi: 1965.

INDIA, Ministry of Health, Central Regional and Urban Planning Organization, "Criteria for Designating Metropolitan Areas." *Journal of the Institute of Town Planners, India* (January–April, 1961), pp. 43–45.

——, "Trends Toward Urbanization in India." *Journal of the Institute of Town Planners, India* (January–April, 1961), pp. 3–11.

INDIA, Ministry of Health, Committee on Model Planning Legislation, *Its Report*. New Delhi: 1965.

INDIA, Ministry of Health, Committee on Urban Land Policy, *Its Report*. New Delhi: 1965.

INDIA, Ministry of Health, Committee on Urban Land Policy, *Report*. New Delhi: 1965.

INDIA, Ministry of Health, Town and Country Planning Organization, *Town and Country Planning in India*. New Delhi: 1962.

INDIA, Ministry of Health, Town Planning Organization, *Slum Clearance and Urban Renewal in Delhi*. New Delhi: 1958.

INDIA, Ministry of Labour, *Employment in the Public Sector*. New Delhi: 1959.

——, *Study of Trends in the Number and Types of Employment Seekers (1953–57)*. New Delhi: 1958.

——, *Study of Trends in the Number and Types of Employment Seekers*. New Delhi: 1959.

——, *Unemployment in Urban Areas*. New Delhi: 1959.

INDIA, Ministry of Works and Housing, *Report of Ford Foundation Team on Urban Housing*. New Delhi: 1965.

——, *Report of the Five Sub-Groups on Housing and Urban and Rural Planning for the Fourth Plan*. New Delhi: 1965.

INDIA, Ministry of Works, Housing, and Supply, *Proceedings of Housing Minister's Conference, Darjeeling, October, 1958*. New Delhi: 1958.

——, *Proceedings of Housing Minister's Conference, Hyderabad, November, 1959*. New Delhi: 1960.

——, *Low Income Group Housing Scheme*, New Delhi: 1957.

——, *Problem of Housing in India*, New Delhi: 1957.

——, *Proposal on Housing for the Second Five Year Plan*. New Delhi: 1957.

——, *A Review of the Housing Situation in India*. New Delhi: 1960.

——, *Subsidized Housing Scheme for Industrial Workers*. New Delhi: 1955.

——, *Report of the Working Group on Housing and Urban Development in the Third Five Year Plan*. New Delhi: 1960.

INDIA, Planning Commission, Metropolitan Transport Team, Committee on Plan Projects, *Traffic and Transport Problems in Metropolitan Cities (Interim Report)*. New Delhi: 1967. Mimeo.

INDIA, Registrar General, *Papers Contributed by Indian Authors to the World Population Conference, Belgrade, 1965*. New Delhi: 1965.

——, *A Selection of Statistics of Small Towns in India: 1961*. New Delhi: 1962.

INDIAN INSTITUTE OF ECONOMICS, *A Socio-Economic Survey of Hyderabad–Secunderabad City Area*. Hyderabad: Government Press, 1957.

INDIAN INSTITUTE OF PUBLIC ADMINISTRATION, *Delhi Municipal Bus Transport: A Study of Some Aspects*. New Delhi: 1959.

——, *Improving City Government: Proceedings of a Seminar, 13–14 September, 1958*. New Delhi: Indian Institute of Public Administration, 1959.

——, "Metropolitan Government System in India," in *Improving City Government: Proceedings of a Seminar 13–14 September, 1958*. New Delhi: Indian Institute of Public Administration, 1959.

——, "Policies and Programmes of City Planning Movement," in *Improving City Government: Proceedings of a Seminar 13–14 September, 1958*. New Delhi: Indian Institute of Public Administration, 1959, pp. 118–157.

——, "Role of the Executive in City Government," in *Improving City Government: Proceedings of a Seminar 13–14 September, 1958*. New Delhi: Indian Institute of Public Administration, 1959, pp. 9–22.

——, "Socio-Economic Surveys of Cities," in *Improving City Government: Proceedings of a Seminar 13–14 September, 1958*. New Delhi: Indian Institute of Public Administration, 1959, pp. 194–202.

——, "Trends in Municipal Finances of Cities," in *Improving City Government Proceedings of a Seminar 13–14 September, 1958*. New Delhi: Indian Institute of Public Administration, 1959, pp. 28–77.

——, "Urban Community Development; Some Administrative Aspects," in *Improving City Government: Proceedings of a Seminar 13–14 September, 1958*. New Delhi: Indian Institute of Public Administration, 1959, pp. 170–178.

——, Bombay Regional Branch, *Problems of Urban Housing*. Bombay: Indian Institute of Public Administration, 1960.

INDIAN INSTITUTE OF PUBLIC OPINION, "The Plight of the Middle Class—The Measurement of Urban Economic Distress." *Monthly Public Opinion Surveys of the Indian Institute* (December, 1967).

——, "Savings in Urban India." *Monthly Statistical Commentary on Indian Economic Conditions* (January, 1962), pp. 19–22.

——, "Structure of Urban and Rural Income Inequalities." *Quarterly Economic Report of the Indian Institute of Public Opinion*, 7 (July, 1960), pp. 48–50.

——, *A Study of Urban Saving Habits*. New Delhi: 1955.

INDIAN INSTITUTE OF TECHNOLOGY, KANPUR, "Urban and Industrial Growth of Kanpur Region." Report of the internation seminar, 29 January to 4 February, 1967.

INDIAN STATISTICAL INSTITUTE, BOMBAY, *Report on the Survey into the Economic Conditions of Middle Class Families in Bombay City*. Bombay: 1952.

INSTITUTE OF TOWN AND COUNTRY PLANNING, INDIA, *Seminar on the National Capital Region*. New Delhi: Institute of Town and Country Planning, India, 1968. Papers are listed under authors' names.

INSTITUTE OF TOWN PLANNERS, *Eleventh Annual Town and Country Planning Seminar, Bhopal, October, 1962* (General theme: planning problems of small and medium size towns and cities). New Delhi: Institute of Town Planners in India, 1962. Papers No. 1–42, Mimeo.

INTERNATIONAL PERSPECTIVE PLANNING TEAM ON SMALL INDUSTRIES, "Report on Development of Small-Scale Industries in India—Prospects, Problems and Policies." Submitted to the Ministry of Commerce and Industries, Government of India. New Delhi: Ford Foundation, 1963.

ISAAC, B.L.U. and P. MARI, "Unauthorized Layouts in the City of Bangalore." *Journal of the Institute of Town Planners, India* (January–April, 1961), pp. 90–92.

IYENGAR, S. KESAVE, *A Socio-Economic Survey of Hyderabad–Secunderabad City Area*. Hyderabad: Government Press, 1957.

JADHAV, M.C., "Pilot Projects for Urban Community Development." *Bharat Sevak* (February, 1961), p. 37.

JAGAT NARAYAN, "City of Dashi." *March of India,* 9 (December, 1957), pp. 30–32.

JAGMOHAN, "Housing and Slum Clearance." *Indian Journal of Public Administration,* 14 (July–September, 1968), pp. 691–708.

JAIN, S.P. "Mortality Trends in India," in U.N. Proceedings of the World Population Conference, Rome, 1954, pp. 439–450.

———, "Note on the Nature and Quality of Indian Demographic Statistics," in U.N. Proceedings of the World Population Conference, Rome, 1954, pp. 121–131.

———, "Religion and Social Differentiation—A Town in Uttar Pradesh." *Economic and Political Weekly,* 3 (January 20, 1968), pp. 201–202.

JAKOBSON, LEO and VED PRAKASH, "Urbanization and Regional Planning in India." *Urban Affairs Quarterly,* 2 (March, 1967), pp. 36–65.

JAMES, R.C., "Labour Mobility, Unemployment and Economic Changes: An Indian Case." *Journal of Political Economy,* 67 (December, 1959), pp. 545–559.

JANAKI, V.A., "Functional Classification of Urban Settlements in Kerala." *Journal of Maharaja Sayajirao University of Baroda,* 3 (October, 1954), pp. 81–114.

JANAKI, V.A. and M.H. AJWANI, "Urban Influence and the Changing Face of a Gujarat Village." *Journal of the Maharaja Sayajirao University of Baroda,* 10 (November, 1961), pp. 59–87.

JANAKI, V.A. and M.C. GHIA, "The Tributary Area of Baroda," *Journal of the Maharaja Sayajirao University of Baroda,* 11 (No. 3, 1962).

JANAKI, V.A. and SAYED, *Geography of Padra Town.* Baroda: Maharaja Sayajirao University of Baroda, Geographical series No. 1, 1962.

JAUHARI, A.S., "Growth of Urban Settlements in the Sutlej-Yamuna Divide." *National Geographical Journal of India,* 8 (March, 1962), pp. 1–24; 8 (June, 1962), pp. 114–135.

———, "Trends of Urbanization in North West India." Paper for 21st International Geographical Congress, held at New Delhi, November–December, 1968, in S.P. Das Gupta and T.R. Lakshmanan (eds.), *Abstract of Papers.* Calcutta: National Committee for Geography, 1968, p. 267.

———, "Trends of Urbanization in the Punjab Plain." *National Geographical Journal of India,* 3 (September–December, 1957), pp. 125–136.

———, "Urban Development in the Sutlej-Yamuna Divide." Paper for 21st International Geographical Congress, held at New Delhi, November–December, 1968, in S.P. Das Gupta and T.R. Lakshmanan (eds.), *Abstract of Papers.* Calcutta: National Committee for Geography, 1968, p. 267.

JAUHARI, A.S. and D.M. BANERJEE, "Trends of Urbanization in Rajasthan State." Paper for 21st International Geographical Congress, held at New Delhi, November–December, 1968, in S.P. Das Gupta and T.R. Lakshmanan (eds.), *Abstract of Papers.* Calcutta: National Committee for Geography, 1968, p. 267.

JAYARAMMAN, R., "Linguistic Groups in Urban Areas." *Economic Weekly,* 16 (24 October, 1964), pp. 1915–1916.

JAYASWAL, S.N.P., "Hierarchy of Service Centers of the Eastern Ganga-

Yamuna Doab, Uttar Pradesh." Paper for 21st International Geographical Congress, held at New Delhi, November–December, 1968, in S.P. Das Gupta and T.R. Lakshmanan (eds.), *Abstract of Papers*. Calcutta: National Committee for Geography, 1968, p. 268.

JAYASWAL, S.N.P. and R.V. VERMA, "Industrial Land Use in Kanpur." Paper for 21st International Geographical Congress, held at New Delhi, November–December, 1968, in S.P. Das Gupta and T.R. Lakshmanan (eds.), *Abstract of Papers*. Calcutta: National Committee for Geography, 1968, p. 268.

JENA, D.D., "Urban Development Trends in Orissa." Paper submitted to the 39th Conference of Indian Economic Association, Cuttack, December, 1956.

JENA, KRISHNACHANDRA, "Modern Impact of Urbanism on Rural Life." *Indian Journal of Social Work,* 21 (September, 1960), pp. 177–179.

JHA, B.N., "Recent Trends of Urbanization in Bihar." *Journal of Ranchi University,* 2 (4), 1964.

JHABVALA, C.S.H., "Middle Class Housing." *Seminar,* 79 (March, 1966), pp. 25–28.

JOSHI, C.B., "Historical Geography of the Islands of Bombay." *Bombay* Geographical Magazine, 4 (December, 1956), pp. 5–13.

JOSHI, D.D., "Stochastic Models Utilized in Demography," in *Papers Contributed by Indian Authors to the World Population Conference, Belgrade, 1965*. New Delhi: Registrar General, 1965, pp. 37–45.

JOSHI, R.V., "Urban Structure in Western India, Poona: a Sample Study." *Geographical Review of India,* 14 (March, 1952), pp. 7–19.

JUNG, FAZAL NAWAZ, "Rural and Urban Cooperatives in India." *A.I.C.C. Economic Review,* 13 (7 March, 1961), pp. 15–18, 32.

KAINTH, MOHAN and M.B. MATHUR, "Urban Renewal: Programme and Policy for the National Capital." Paper for 7th Annual Town and Country Planning Seminar, New Delhi, October, 1968. Mimeo.

KAKADE, R.G., *Socio-Economic Survey of the Lower Stratum of Employees in the City of Bombay*. Bombay: University of Bombay, 1951. Unpublished thesis.

——, *Socio-Economic Survey of Working Communities in Sholapur*. Poona: Gokhale Institute of Politics and Economics, Publication No. 14, 1947.

KALDATE, SUDHA, "Urbanization and Distintegration of Rural Joint Family." *Sociological Bulletin,* 11 (March–September, 1962), pp. 103–111.

KALE, B.D., "Desirability of Presenting Census Data for Effective Urban Sector Separately." *Journal of the Institute of Economic Research,* 3 (July, 1968), pp. 58–68.

——, "Growth of Towns in Mysore State." *Journal of Institute of Economic Research,* 2 (July, 1967), pp. 47–67.

KALRA, B.R., "Occupational Structure of Cities, 1901–61," "in Ashish Bose (ed.), *Patterns of Population Change in India, 1951–61*. Delhi: Allied Publishers, 1967, pp. 287–312.

KAMATH, S.G., "Metropolitan Areas and National Capital Region." Paper for 7th Annual Town and Country Planning Seminar, New Delhi, October, 1968. Mimeo.

KANSAL, J.B., "State of Urbanization in Uttar Pradesh." *A.I.C.C. Economic Review,* 21 (1 August, 1969), pp. 19–23.

KANWAR LAL, *Holy Cities of India.* Delhi: 1961.

KAPADIA, K.M., "The Growth of Townships in South Gujarat: Maroli Bazar." *Sociological Bulletin,* 10 (September, 1961), pp. 69–87.

——, "Rural Family Pattern: A Study in Rural-Urban Relations." *Sociological Bulletin,* 5 (September, 1956), pp. 111–126.

KAR, N.R., "Calcutta als Weltstadt," in J.M. Schultze (ed.), *Zum Probleme der Weltstadt.* Berlin: 1959.

——, "Pattern of Urban Growth in Lower West Bengal." *Geographical Review of India,* 24 (September–December, 1962), pp. 42–59.

——, "Population Dynamics, Internal Migration and Regional Economic Development in India." Paper for 21st International Geographical Congress, held at New Delhi, November–December, 1968, in S.P. Das Gupta and T.R. Lakshmanan (eds.), *Abstract of Papers.* Calcutta: National Committee for Geography, 1968, p. 238.

——, "Social Area Analysis of Calcutta." Paper for 21st International Geographical Congress, held at New Delhi, November–December, 1968, in S.P. Das Gupta and T.R. Lakshmanan (eds.), *Abstract of Papers.* Calcutta: National Committee for Geography, 1968, p. 269.

——, "Urban Characteristics of the City of Calcutta." *Indian Population Bulletin,* 1 (April, 1960), pp. 34–67, 34a–34e.

——, "Urban Hierarchy and Central Functions Around the City of Calcutta and Their Significance in Problems of Urban Geography," in K. Norborg (ed.), *Proceedings of International Geographical Union—Symposium in Urban Geography.* Lund, Sweden: 1960.

——, "Urban Land Use, Functional Activities and Social Morphology of Darjeeling." Paper for 21st International Geographical Congress, held at New Delhi, November–December, 1968, in S.P. Das Gupta and T.R. Lakshmanan (eds.) *Abstract of Papers.* Calcutta: National Committee for Geography, 1968, pp. 269–270.

KARAN, P.P., "Changes in Indian Industrial Location." *Annals of the Association of American Geographers,* 54 (September, 1964), pp. 336–354.

——, "Patna and Jamshedpur." *Geographical Review of India,* 14 (June, 1952), pp. 25–32.

——, "Pattern of Indian Towns—A Study in Urban Morphology." *Journal of the American Institute of Planners,* 23 (March, 1957), pp. 70–75.

KARANJGAOKAR, D.G., "Problems of Planning and Redevelopment of Bhopal: A Case Study." *Journal of the Institute of Town Planners, India* (January–April, 1961), pp. 92–97.

KARIMI, SALAHUDDIN MOHD, "Capitals of Late Medieval Bengal," Paper for 21st International Geographical Congress, held at New Delhi, November–December, 1968, in S.P. Das Gupta and T.R. Lakshmanan (eds.), *Abstract of Papers.* Calcutta: National Committee for Geography, 1968, p. 296.

KARVE, IRAWATI and J.S. RANDIVE, *The Social Dynamics of a Growing Town in Its Surrounding Area.* Poona: 1965.

KATTI, A.P., "A Study of In-Migrants in Dharwar Area." *Journal of Institute of Economic Research,* 2 (July, 1967), pp. 21–40.

KAYASTHA, S.L., "Kandla: A Study in Port Development." *National Geographical Journal of India,* 9 (March, 1963), pp. 12–24.

——, "Kangra, Himalayan Town." *National Geographical Journal of India,* 4 (June, 1958), pp. 89–94.

KESAVA, IYENGAR S., *Socio-Economic Survey of Hut Dwellers in Hyderabad City.* Hyderabad: Indian Institute of Economics, 1959.

——, *Socio-Economic Survey of Hyderabad-Secunderabad City Area.* Hyderabad: Indian Institute of Economics, 1957.

KHAN, ABDUL MAJEED, "Bangalore: Then and Now." *March of India,* 14 (January, 1962).

KHAN, K.A., "Rapid Transit System for Greater Delhi." *Civic Affairs* (June, 1963), pp. 55–61.

KHAN, NASIR AHMAD, *Middle Classes in India.* Lucknow: Uttar Pradesh, Information Department, 1958.

KHANNA, KRISHEN, "Kanpur." *Illustrated Weekly of India,* 82 (December 31, 1961).

KHANNA, R.L., *Municipal Government and Administration in India.* Chandigarh: Mohindra Capital Publishers, 1967.

KHANNA, TEJBIR, "Mass Transport System in Urban Area." *Urban and Rural Planning Thought,* 1 (October, 1958), pp. 264–268.

KHOSLA, G.D., "Chandigarh." *Illustrated Weekly of India,* 82 (October 1, 1961).

KHOSLA, J.N. and ABHIJIT DATTA, "Local Government Administration in Delhi." Paper for 7th Annual Town and Country Planning Seminar, New Delhi, October, 1968. Mimeo.

KIDWAI, IKRAMUDDIN, *Lucknow, Past and Present.* Lucknow: 1961.

KING, A.D., "India's Urban Past: The Case for Urban History in India." Paper for the 6th Annual Conference of the Institute of Historical Studies, Srinagar, 7 to 12 October, 1968. Mimeo.

——, "A Note on the Development of Urban Consciousness." Paper for 7th Annual Town and Country Planning Seminar, New Delhi, October, 1968. Mimeo.

KOENIGSBERGER, OTTO H., "New Towns in India." *Town Planning Review,* 23 (July, 1952), pp. 94–132.

——, "Town Planning in India." *Eastern World,* 7 (August, 1953), pp. 33–35.

KOLHATKAR, V.Y. and C.T. SHAH, "A Survey of Unemployment and Underemployment in the City of Baroda." *Journal of the Maharaja Sayajirao University of Baroda,* 5 (March, 1956), pp. 75–121.

——, "Unemployment in the City of Baroda: A Pilot Survey." *Journal of the Maharaja Sayajirao University of Baroda,* 3 (March, 1954), pp. 29–44.

"Kotla Mubarakpur: An Urban Village." *Urban and Rural Planning Thought,* 1 (January, 1958), pp. 41–54.

KRISHNAN, M., "Madras." *Illustrated Weekly of India,* 82 (September 17, 1961).

KULKARNI, G.S., "Industrial Landscape of Greater Poona." *Bombay Geographical Magazine,* 9 (December, 1961), pp. 85–94.

KULKARNI, M.R., "Small Industry in Two Big Cities: Delhi and Bombay." *Indian Economic Journal,* 11 (April–June, 1964), pp. 452–459.

KULKARNI, P.D., "Social Policy and Strategy for Urban Development." *Social Welfare,* 14 (January, 1967), pp. 6–8.

KULKARNI, S.D., "A Study of In-Migration into Malegaron." *Journal of the Institute of Economic Research,* 3 (July, 1968), pp. 44–57.

KULSHRESTHA, S.K., "Image of Delhi, Capital of India." Paper for 7th Annual Town and Country Planning Seminar, New Delhi, October, 1968. Mimeo.

KUMAR, DHARMA, "Transfer of Surplus Labour from the Rural Sector." *Indian Economic Journal,* 4 (April, 1957), pp. 355–370.

KUMAR, JOGINDER, "The Pattern of Internal Migration in India during 1951–61." Paper presented to the International Union for the Scientific Study of Population, Sydney Conference, 1967.

KUMARAPPA, J.M., "Village Labour Force in the City." *Indian Journal of Social Work* (December, 1947), pp. 213–219.

"Labour Conditions in Municipalities." *Indian Labour Gazette* 12 (February, 1955), pp. 809–824.

LAHIRI, D.B., "Population Data and the Indian National Sample Survey," in *Papers Contributed by Indian Authors to the World Population Conference Belgrade, 1965.* New Delhi: Registrar General, 1965, pp. 47–50.

LAHIRI RANAJIT, "Jabbalpore." *Calcutta Geographical Review,* 7 (June–September, 1945), pp. 67–72.

LAHIRI, T.B., "Urbanization Potential of Village." *Geographical Review of India,* 28 (June, 1966), pp. 29–34.

LAKDAWALA, D.T., "Some Difficulties in the Application of International Trade Theory of Rural-Urban Relationship." *Indian Economic Journal,* 4 (January, 1957), pp. 241–247.

LAKDAWALA, D.T. and B.V. MEHTA, "Small and Medium Scale Engineering Factories in Bombay City." *Journal of the University of Bombay: History, Economics and Sociology Series,* 26 (January, 1958), pp. 56–103.

LAKDAWALA, D.T. and J.C. SANDESARA, "Shops and Establishments in Greater Bombay." *Journal of the University of Bombay: History, Economics, and Sociology Series,* 25 (January, 1957), pp. 56–103.

LAKDAWALA, D.T. et al., *Work, Wages and Well-Being in an Indian Metropolis: Economic Survey of Bombay City.* Bombay: University of Bombay, Economic Series, No. 11, 1963.

LAKDAWALA, D.T. and J.C. SANDESARA, *Small Industry in a Big City: A Survey in Bombay.* Bombay: University of Bombay, Economic Series, No. 10, 1961.

LAL, AMRIT, "Age and Sex Structure of Cities of India." *Geographical Review of India,* 24 (March, 1962), pp. 7–29.

——, "Growth of Industrial Towns in India." *Eastern Economist,* 28 (January 11, 1957), pp. 52–54.

——, "Patterns of In-Migration in India's Cities." *Geographical Review of India,* 23 (September, 1961), pp. 16–23.

——, "Some Aspects of the Functional Classification of Cities and a Proposed Scheme for Classifying Indian Cities." *National Geographical Journal of India,* 5 (March, 1959), pp. 12–24.

——, *Some Characteristics of Indian Cities of Over 100,000 Inhabitants in 1951 with Special Reference to Their Occupational Structure and Func-*

tional Specialization. Bloomington, Ind.: Indiana University, Department of Geography, 1958. Unpublished Ph.D. Thesis.

LAL, P., "Calcutta." *Illustrated Weekly of India,* 82 (August 20, 1961).

LAL, PARMANAND, "Evolution of Port Blair: A Study in Urban Geography." *National Geographical Journal of India,* 8 (June, 1962), pp. 93–113.

LAL, RAMA SHANKAR, "Siwan, a Subregional Service Centre in Bihar." Paper for 21st International Geographical Congress, held at New Delhi, November–December, 1968, in S.P. Das Gupta and T.R. Lakshmanan (eds.), *Abstract of Papers.* Calcutta: National Committee for Geography, 1968, p. 271.

LAMBERT, RICHARD D., "Factory Workers and the Non-Factory Population in Poona." *Journal of Asian Studies,* 18 (November, 1958), pp. 21–42.

——, "The Impact of Urban Society Upon Village Life," in Roy Turner (ed.), *India's Urban Future.* Bombay: Oxford University Press, 1962, pp. 117–140.

——, *Workers, Factories and Social Change in India.* Princeton: Princeton University Press, 1963.

LA TOUCHE, THEODORE W., "Master Plan for Greater Hyderabad." *Civic Affairs* (August, 1963), pp. 43–46.

LEARMONTH, A.T.A., "Regional Planning in India." *Economic Weekly,* 12 (January, 1960), pp. 241–244.

——, et al., *Mysore State, Vol. II, a Regional Synthesis.* Bombay: Asia Publishing House, 1962.

LEARMONTH, A.T.A. and L.S. BHAT, *Mysore State, Vol. I, an Atlas of Resources.* Bombay: Asia Publishing House, 1960.

LEVYVELD, JOSEPH, "Can India Survive Calcutta?" *Imprint* (November, 1968), pp. 7–16.

LEWIS, JOHN P., *Quiet Crisis in India.* New York: Doubleday, 1964.

LIPTON, MICHAEL, "Strategy for Agriculture: Urban Bias and Rural Planning," in Paul Streeten and Michael Lipton (eds.), *The Crisis of Indian Planning.* London: Oxford University Press, 1968.

LOBO, MICHAEL, "Ahmedabad." *Illustrated Weekly of India,* 82 (December 10, 1961).

LOPO, LISBET DE CASTRO, "An Analysis of Internal Migration in Bihar, North India." Paper for 21st International Geographical Congress, held at New Delhi, November–December, 1968, in Niels Kingo Jacobsen and Ruth Helkiaer Jensen (eds.), *Collected Papers–Denmark.* Denmark: Kobenhaavns Universitets Geografiske Institut, 1968, pp. 70–90.

LUCKNOW, Improvement Trust, *Lucknow of Our Dreams.* Lucknow: 1958.

MacDOUGALL, JOHN, *Ancillary Industries in Asansol-Durgapur: A Preliminary Study.* New York: Institute of Public Administration, n.d.

MADHAB, JAYANTA, "Controlling Urban Land Values." *Economic and Political Weekly* 4 (July, 1969), pp. 1197–1202.

——, "Resources for Urban Development." *Indian Journal of Public Administration,* 14 (July–September, 1968), pp. 682–690.

MADHAVA RAU, N., "City Planning." *Public Affairs,* 7 (January, 1963), pp. 11–13.

MADHYA PRADESH, Directorate of Economics and Statistics, *Survey of Educated Unemployed in Nagpur City*. Nagpur: 1956.

MADHYA PRADESH, BHOPAL, Directorate of Economics and Statistics, *Socio-Economic Survey of Bhilai Region*. Bombay: Part I, 1959; Part 2, 1968.

MADRAS, Department of Information and Publicity, *Madras: The Land of Temples*. Madras: 1960.

"Madras City: A Planning Study." *Urban and Rural Planning Thought*, 3 (April, 1960).

MAHALANOBIS, P.C., "Some Concepts of Sample Surveys in Demographic Investigations," in *Papers Contributed by Indian Authors to the World Population Conference, Belgrade, 1965*. New Delhi: Registrar General, 1965, pp. 51–55.

MAHARASTRA, Ministry of Urban Development, Public Health and Housing, *Report of the Committe Appointed for Regional Plans for Bombay-Panvel and Poona Regions*. Bombay: 1966.

MAHESHWARI, B., "Municipal Government in Rajasthan." *Quarterly Journal of the Local Self Government Institute* (July, 1963), pp. 25–38.

MAHESWARI, SHRI RAM, "Inter-Governmental Relations in Urbanization and Urban Development." *Indian Journal of Public Administration,* 14 (July, September, 1968), pp. 568–581.

MAJAMDAR, D.M., *Unemployment Among the University Educated: A Pilot Enquiry in India*. Cambridge, Mass.: M.I.T. Communications Programme India, 1957.

MAJUMDAR, D.N., *Social Contours of an Industrial City: Social Survey of Kanpur, 1954–56*. Bombay: Asia Publishing House, 1960.

MAJUMDAR, M., "Estimation of Vital Rates in the Indian National Sample Survey," in *Papers Contributed by Indian Authors to the World Population Conference, Belgrade, 1965*. New Delhi: Registrar General, 1965, pp. 57–60.

MALENBAUM, WILFRED, "Urban Unemployment in India." *Pacific Affairs,* 30 (June, 1957), pp. 138–150.

MALHOTRA, P.C., *Socio-Economic Survey of Bhopal City and Bairagarh*. Bombay: Asia Publishing House, 1964.

MALIK, HARJI, "The State of Capital." Paper for 7th Annual Town and Country Planning Seminar, New Delhi, October, 1968. Mimeo.

MALKANI, H.C., *Socio-Economic Survey of Baroda City*. Baroda: Maharaja Sayajirao University of Baroda, 1957.

MAMORIA, C.B., "Rural and Urban Composition of Indian Population." *Modern Review,* 99 (February, 1956), pp. 118–124; 99 (March, 1956), pp. 195–202.

MANDAL, G.C., "Rural Versus Urban Sector in India's Economic Planning." *A.I.C.C. Economic Review,* 13 (22 September, 1961), pp. 23–26.

MANICKAM, B.R., "Impact of Industrialization in Bangalore." *Journal of the Institute of Town Planners, India* (January–April, 1961), pp. 80–84.

MANICKAM, T.J., "Indian City Patterns." *Urban and Rural Planning Thought,* 3 (July, 1960), pp. 110–129.

——, "Planning Organization in India." *Urban and Rural Planning Thought,* 3 (July, 1960), pp. 100–109.

———, "Urban Planning in India." Paper for an International Conference on Urbanization in India, held at the University of California, Berkeley, 1960. Mimeo.

MANICKAM, T.J. and B. MISRA, "Urban and Regional Planning." *Indian Journal of Public Administration,* 14 (July–September, 1968), pp. 596–605.

MANICKAM, T.J. and L.R. VAGALE et al., "New Towns in India," in *Public Administration Problems of New and Rapidly Growing Towns in Asia.* New York: United Nations, 1962.

MANN, HAROLD H., "The Housing of the Untouchable Classes in an Indian City (Poona)" in Daniel Thorner (ed.), *The Social Framework of Agriculture.* Bombay: Vora Publishers, 1967, pp. 192–203.

———, "The Untouchable Classes of an Indian City (Poona)," in Daniel Thorner (ed.), *The Social Framework of Agriculture.* Bombay: Vora Publishers, 1967, pp. 175–191.

MANOHAR, SHRI, "Coming to Terms with Reality on Housing in Metropolitan Delhi." Paper for 7th Annual Town and Country Planning Seminar, New Delhi, October, 1968. Mimeo.

———, "On Delhi's Master Plan—A Vision of the National Capital in 1981." Paper for 7th Annual Town and Country Planning Seminar, New Delhi, October, 1968. Mimeo.

———, "Urban Design Process." Paper for 7th Annual Town and Country Planning Seminar, New Delhi, October, 1968. Mimeo.

MARRIOTT, MC KIM, "Some Comments on William L. Kolb's 'The Structure and Function of Cities,' in the light of India's urbanization." *Economic Development and Cultural Change,* 3 (October, 1954), pp. 50–52.

"Master Plan for Hyderabad." *Civic Affairs* (September, 1961), pp. 51–52.

MATHAR, MANSINGH, "Industrial Town Changes a Nearby Village." *Yojana,* 10 (April 3, 1966), pp. 26–27.

MATHUR, J.S., "Urban Community Development," *Indian Journal of Social Work,* 18 (March, 1957), pp. 272–275.

MATHUR, M.V. et al., *Economic Survey of Jaipur City.* Jaipur: University of Rajasthan, 1965.

MATHUR, O.S., "Delhi Housing—Facts and Fictions." Paper for 7th Annual Town and Country Planning Seminar, New Delhi, October, 1968.

MATHUR, PRAKASH, *Internal Migration in India, 1941–1951.* Chicago: University of Chicago, 1961. Unpublished Ph.D. Thesis.

MATHUR, R.R., "Thana (a Study in Urban Landscape)." *Bombay Geographical Magazine,* 14 (December, 1966), pp. 65–95.

MATHUR, S.P., "Growth and Functional Structure of Dehradun." Paper for 21st International Geographical Congress, held at New Delhi, November–December, 1968, in S.P. Das Gupta and T.R. Lakshmanan (eds.), *Abstract of Papers.* Calcutta: National Committee for Geography, 1968, p. 274.

MATTOO, P.K., *A Study of Local Self Government in Urban India.* Jullundur: Jain General House, 1959.

MAYER, ALBERT, "National Implications of Urban-Regional Planning," in Roy Turner (ed.), *India's Urban Future.* Bombay: Oxford University Press, 1962, pp. 335–346.

——, "The New Capital of the Punjab." *American Institute of Architects Journal,* 14 (October, 1950), pp. 166–175.

——, "Some Operational Problems in Urban and Regional Planning and Development," in Roy Turner (ed.), *India's Urban Future.* Bombay: Oxford University Press, 1962, pp. 197–412.

MAYFIELD, ROBERT C., "An Urban Research Study in North India," in Forrest R. Pitts (ed.), *Urban Systems and Economic Development.* Eugene, Ore.: University of Oregon School of Business Administration, 1962, pp. 45–52.

MEIER, RICHARD L., "Relations of Technology to the Design of Very Large Cities," in Roy Turner (ed.), *India's Urban Future.* Bombay: Oxford University Press, 1962, pp. 299–323.

MEHTA, ASOKA, "The Future of Indian Cities: National Issues and Goals," in Roy Turner (ed.), *India's Urban Future: Selected Studies from an International Conference on Urbanization in India, held at the University of California in 1960.* Berkeley: University of California Press, 1962, pp. 413–424.

MEHTA, B.H., "Changing Concepts of Urban Community Development." *Indian Journal of Social Work,* 25 (April, 1964), pp. 35–41.

——, "Social Approach and Basis of Metropolitan Planning," *Journal of the Institute of Town Planners, India* (January–April, 1961), pp. 16–20.

——, "Social Aspects of Urban Redevelopment." *Journal of the Institute of Town Planners, India* (January–April, 1959), pp. 7–10.

——, "Urban Community Organization and Development." *Indian Journal of Social Work,* 18 (March, 1957), pp. 260–272.

——, "Villager in the City: A Study of Rural Urban Relationships." *Indian Journal of Social Work,* 1 (December, 1940), pp. 380–392.

MERTA, M.M., *Location of Indian Industries.* Allahabad: 1952.

MEHROTRA, S.N., "Urbanization in Madhya Pradesh." *Geographical Review of India,* 23 (December, 1961), pp. 29–46.

MELEDINA, M.H. (Comp.), *History of Poona Contonment, 1818–1953.* Poona: 1953.

MENON, V.P., "Bangalore and Its Corporation." *Civic Affairs,* (March, 1963), pp. 18–19.

"Middle Class in Bombay City: Results of Survey of Economic Conditions." *Commerce* (February 25, 1956), p. 362.

MISHRA, B., "Problem of Urban Unemployment." *Indian Journal of Commerce,* 7 (December, 1954), pp. 47–54.

MISHRA, M.L., "Anatomy of Urban Unemployment in India." *Indian Journal of Commerce,* 8 (March, 1955), pp. 1–10.

MISRA, A.B., "Some Trends in Urbanization in India." Paper submitted to the 39th Conference of the Indian Economic Association, Cuttack, December, 1956.

MISRA, BIDYADHAR, "Problems of Urbanization in India." Paper submitted to the 39th Conference of the Indian Economic Association, Cuttack, December, 1956.

MISRA, B.B., *The Indian Middle Classes: Their Growth in Modern Times.* London: Oxford University Press, 1961.

MISRA, B.R., *Report on Socio-Economic Survey of Jamshedpur City*. Patna: Patna University, Department of Applied Economics and Commerce, 1959.

——, "Socio-Economic Survey of Jamshedpur." *Economic Papers*, 2 (November, 1957), pp. 1–26.

MISRA, HAREKRISHANA, "Growth of Urban Population—Its Impact on Urban Development Trends in India." Paper submitted to the 39th Conference of the Indian Economic Association, Cuttack, December, 1956.

MISRA, S.D., "Note on the Socio-Historical Geography of Mathura." *National Geographical Journal of India*, 4 (December, 1958), pp. 189–199.

——, "Social Geography of Mathura." *Indian Geographical Journal*, 34 (January–March and April–June, 1959), pp. 1–24.

MITRA, ASOK, "Bye-Products of the Census of 1961." *Economic Weekly*, 13 (July, 1961), pp. 616–626.

——, *Calcutta: India's City*. Calcutta: New Age Publishers, 1963.

——, "Calcutta, the City with Hundred Gates." *Civic Affairs* (March, 1963), pp. 9–17.

——, "A Functional Classification of India's Towns (Part I)," in Ashish Bose (ed.), *Patterns of Population Change in India, 1951–61*. Delhi: Allied Publishers, 1967, pp. 261–286; "List of Towns with Functional Classification (Part II)." Delhi: Institute of Economic Growth, 1964. Mimeo.

——, "Indian Experience in Recording Economically Active Population: 1961 Population Census," in *Papers Contributed by Indian Authors to the World Population Conference, Belgrade, 1965*. New Delhi: Resgistrar General, 1965, pp. 203–240.

——, "Internal Migration and Urbanization in India." Paper prepared for the ECAFE Expert Working Group on Internal Migration and Urbanization, Bangkok, 1967.

——, "Internal Migration and Urbanization in India, 1961." Paper presented to International Union for the Scientific Study of Population, Sydney Conference, 1967.

——, "1961 Census." *Seminar*, 18 (February, 1961), pp. 15–21.

——, "A Note on Internal Migration and Urbanization in India, 1961," in P.S. Gupta and Galina Sdasyak, *Economic Regionalization of India: Problems and Approaches*. New Delhi: Census of India 1961, Monograph No. 8, 1968, pp. 251–257.

MITRA, S., "The Impact of Patterns of Population Concentration on Urbanization." Paper for International Union for the Scientific Study of Population Conference, London, September, 1969.

MODAK, M.V. and A. MAYER, *Outlines of a Master Plan for Greater Bombay*. Bombay: 1948.

MOHSIN, MOHAMMAD, *Chittaranjan—A Study in Urban Sociology*. Bombay: 1964.

MOOKERJEE, S., "The Growth of Nagpur City." *Geographical Outlook*, 2 (No. 1, 1958).

MOOKHERJEE, D., "West Bengal: Its Urban Pattern." *Geographical Review of India*, 19 (December, 1957), pp. 67–72.

MOOKHERJEE, DEBNATH, "Urban Pattern of Siliguri." *Geographical Review of India*, 19 (September, 1957), pp. 15–20.

MOOKHERJEE, D.N., "Calcutta and Madras Industrial Regions." Paper for 21st International Geographical Congress, held at New Delhi, November–December, 1968, in S.P. Das Gupta and T.R. Lakshmanan (eds.), *Abstract of Papers.* Calcutta: National Committee for Geography, 1968, p. 275.

——, "Urbanization in India." Paper for 21st International Geographical Congress, held at New Delhi, November–December, 1968, in S.P. Das Gupta and T.R. Lakshmanan (eds.), *Abstract of Papers.* Calcutta: National Committee for Geography, 1968, pp. 275–276.

MOORTHY, M. VASUDEVA, "Metropolitan City—A Social Entity." *Journal of the Institute of Town Planners, India* (January–April, 1961), pp. 20–24.

MUHAMMAD, U. JAMAL, "Urban Geography of Coimbatore." *Geographer* (Aligarh), 6 (Summer, 1954), pp. 53–64.

MUKERJEE, RADHA KAMAL (ed.), *City in Transition: A Survey of Social Problems of Lucknow.* Lucknow: Lucknow University, Monograph No. 2, 1952.

MUKERJEE, RADHA KAMAL and BALJIT SINGH, *A District Town in Transition: Social and Economic Survey of Gorakhpur.* Bombay: Asia Publishing House, 1965.

——, *Social Profile of a Metropolis: Social and Economic Structure of Lucknow, Capital of Uttar Pradesh, 1954–56.* Bombay: Asia Publishing House, 1961.

MUKERJI, A.B., "The Bi-Weekly Market at Modinagar." *Indian Geographer,* 2 (December, 1957), pp. 271–294.

——, "The Umland of Modinegar." *National Geographical Journal,* 8 (September–December, 1962), pp. 250–269.

MUKERJI, V., "Application of Some Simple Multi-Regional Growth and Migration Models to District Level Census Data in Maharashtra." *Artha Vijnana,* 6 (September, 1964), pp. 187–205.

MUKHARJI, GIRIPATI, "Urban Land Policy." *Indian Journal of Public Administration,* 14 (July–September, 1968), pp. 582–595.

MUKHERJEE, B., "Ranchi: A Study in Urban Morphology." *National Geographical Journal of India,* 2 (June, 1956), pp. 97–105.

MUKHERJEE, C., "Land Utilization Planning in Metropolitan Calcutta." *Economic Affairs,* 7 (July–August, 1962), pp. 375–386.

MUKHERJEE, CHITTAPRIYA, "Business Trends in Rural Towns of Birbhum." *Khadi Gramodyog,* 11 (October, 1965), pp. 70–92.

——, *Buyers and Sellers of Land in a Rural Town.* Santiniketan: Visvabharati, 1967. Mimeo.

——, "Trend of Land Price in a Rural Town." *Khadi Gramodyog,* 12 (October, 1966), pp. 103–127.

MUKHERJEE, MAHAMAYA, "Functional Association of Towns in Bihar." Paper for 21st International Geographical Congress, held at New Delhi, November–December, 1968, in S.P. Das Gupta and T.R. Lakshmanan (eds.), *Abstract of Papers.* Calcutta: National Committee for Geography, 1968, p. 276.

MUKHERJEE, NILMANI, "Port Labour in Calcutta, 1870–1953: Some Trends of Change." Paper presented in Seminar on Trends of Socio-Economic Change in India, 1871–1961, 11 to 23 September, 1967. Simla: Indian Institute of Advanced Study, 1967.

MUKHERJEE, RAMAKRISHNA, *An Outline for Social Research on Contempory India.* Calcutta: India Statistical Institute, 1968. Mimeo.

——, "On Rural-Urban Differences and Relationships in Social Characteristics." Paper presented to Unesco Seminar on Urban Rural Differences and Relationships with Reference to the Role of Small Towns in Planned Development, Delhi, 27–30 December, 1962. Mimeo.

——, *Sociologist and Social Change in India Today.* New Delhi: 1965.

——, "Urbanization and Social Transformation in India." *International Journal of Comparative Sociology,* 4 (September, 1963), pp. 178–210.

MUKHERJEE, S., "Villages and Towns of West Bengal." *Geographical Review of India,* 19 (June, 1957), pp. 45–48.

MUKHERJEE, SUDHANSU BHUSAN "Urbanization in Burdwan Division," *Calcutta Statistical Association Bulletin* (March, 1965), pp. 1–16.

MUKHERJI, A.B., "Modinagar: Study in Indian Urban Landscapes." *Bombay Geographical Magazine,* 1 (October, 1953), pp. 56–63.

——, "Modinagar: Study in Urban Geography." *Geographical Review of India,* 15 (March, 1953), pp. 8–20.

MUKHOPADHYAYA, CHITTAPRIYA, "Rural-Urban Income Disparity and Taxation Policy," in Nalin Mehta (ed.), *Fiscal Policies and Economic Growth: Building from Below.* Ahmedabad: 1965.

"Municipal Corporations in U.P.: A Historical Note." *Civic Affairs* (February, 1963).

"Municipal Water Supply Services in Cities and Towns of India." *Civic Affairs* (May, 1962), pp. 79–83.

MURPHEY, R., "The City in the Swamp; Aspects of the Site and Early Growth of Calcutta." *Geographical Journal,* 130 (June, 1964), pp. 241–256.

NAG, A., "Estimation and Projection of the Working Force in Greater Bombay." *Asian Economic Review,* 12 (February, 1969), pp. 160–172.

NAGABHUSHAM, K., "Urbanization of Visakhapatnam." Waltair: Andhra University, 1961. Mimeo.

NAIDU, M.V. (ed.), *City of Secunderabad (Deccan).* Secunderabad: Secunderabad Municipal Corporation, 1955.

NADU, V.H.R., "Housing in the Second Plan." *Urban and Rural Planning Thought,* 2 (January, 1959), pp. 2–6.

NAIK, J.P., "Educational Administration in Urban Areas." *Indian Journal of Public Administration,* 14 (July–September, 1968), pp. 736–745.

NAIK, S.S., *Problems of Regulating Buildings in Towns and Cities.* Bombay: Local Self Government Institute, 1953.

NAIR, B.N., "Urbanization and Corruption." *Sociological Bulletin,* 9 (September, 1960), pp. 15–33.

NAIR, K. BALAKUMARAN, "Economic-Demographic Characteristics of Trivandrum City," in R.S. Kurup and K.A. George (eds.), *Population Growth in Kerala.* Trivandrum: 1965.

NAIR, L.R., "Chandigarh, India's City of Tomorrow." *New Commonwealth,* 24 (October 13, 1952), pp. 366–369.

——, *Why Chandigarh.* Simla: Punjab Government, Publicity Department, 1950.

NAIR, P.A., *Employment Market in an Industrial Metropolis—A Survey of*

Educated Unemployment in Bombay. Bombay: Lalvani Publishing House, 1968.

NANDA, A.K., "Pull for Cities? Or Push from Villages?" *Yojana,* 8 (October 11, 1964), p. 27.

NAQVI, H.V., "Industrial Towns of Hindustan in the Eighteenth Century." Paper presented in Seminar on Trends of Socio-Economic Change in India, 1871–1961, 11 to 23 September, 1967. Simla: Indian Institute of Advanced Study, 1967. Mimeo.

———, *Urban Centres and Industries in Upper India, 1553–1803.* Bombay: Asia Publishing House, 1968.

NARAIN, D., "Urbanization and Some Social Problems." *Sociological Bulletin,* 9 (September, 1960), pp. 1–6.

NARAIN, VATSALA, "Migrants in the Metropolitan Areas of India." Paper presented to International Union for the Scientific Study of Population, Sydney Conference, 1967.

NARASIMHAMURTHY, B.S., "Studies in Migration in Mysore State." *Journal of Institute of Economic Research,* 2 (July, 1967), pp. 15–20.

NARAVANE, V.S., "Allahabad." *Illustrated Weekly of India,* 82 (November 12, 1961).

NATARAJAN, D., "1961 Census Schedule." *Economic Weekly,* 12 (16 April, 1960), pp. 616–626.

NATH, KAMLA, "Urban Women Workers: A Preliminary Study." *Economic Weekly,* 17 (11 September, 1965), pp. 1405–1412.

NATH, V., "Planning for Urban Growth." *Indian Journal of Social Work,* 27 (July, 1966), pp. 119–145.

NATH, VISHWAMBHAR, "Urbanization in India with Special Reference to Growth of Cities," in *Proceedings of the World Population Conference, Rome, 31 August–10 September, 1954,* pp. 843–853.

———, "The Village and the Community," in Roy Turner (ed.), *India's Urban Future: Selected Studies from an International Conference on Urbanization in India, held at the University of California in 1960.* Berkeley: University of California Press, 1962, pp. 141–156.

NATIONAL BUILDINGS ORGANIZATION, *Building Materials and Housing in India.* New Delhi: 1957.

———, *Monograph on Housing Situation in India.* New Delhi: 1959.

———, *Some References to Low Cost Housing Mainly in India.* New Delhi: 1958.

NATIONAL COUNCIL OF APPLIED ECONOMIC RESEARCH, *Commodity Disposition Survey in Delhi.* Bombay: Asia Publishing House, 1959.

———, *Development of Beypore Port.* New Delhi: 1959.

———, *Development of Paradeep Port.* New Delhi: 1963.

———, *Market Towns and Spatial Development in India.* New Delhi: 1962.

———, *Saving in India.* New Delhi: 1961.

———, *Some Aspects of Goods Transport by Road in the Delhi Region.* Bombay: Asia Publishing House, 1959.

———, *Tax Incidence on Housing.* New Delhi: 1967.

———, *Traffic Survey of Beypore Port.* New Delhi: 1960.

———, *Traffic Survey of Mangalore and Malpe Ports.* New Delhi: 1961.

——, *Traffic Survey of the Port of Tuticorin*. New Delhi: 1959.

NAYAK, P.R., "The Challenge of Urban Growth to Indian Local Government," in Roy Turner (ed.), *India's Urban Future: Selected Studies from an International Conference on Urbanization in India, held at the University of California, in 1960*. Berkeley: University of California Press, 1962, pp. 361–381.

NEALE, W.C., H. SINGH, and J.P. SINGH, "Kurali Market: A Report on the Economic Geography of Marketing in Northern Punjab." *Economic Development and Cultural Change*, 8 (January, 1965), pp. 129–168.

NIEHOFF, ARTHUR, *Factory Workers in India*. Milwaukee: Milwaukee Public Museum Publications in Anthropology, No. 5, 1957, pp. 102–110.

NIGAM, M.N., "Evolution of Lucknow," *National Geographical Journal of India*, 6 (March, 1960), pp. 30–46.

——, "Functional Regions of Lucknow." *National Geographical Journal of India*, 10 (March, 1964), pp. 38–52.

OOMMEN, T.K., "The Rural-Urban Continuum Re-Examined in the Indian Context." *Sociologia Ruralis*, 7 (No. 1, 1967), pp. 30–48.

OTURKAR, R.V. (ed.), *Poona: Look and Outlook*. Poona: Poona Corporation, 1957.

PACHAURI, A.K., "Delhi-Meerut Urban Corridor." Paper for 7th Annual Town and Country Planning Seminar, New Delhi, October, 1968. Mimeo.

PADKI, M.B., "Out Migration from a Kankan Village to Bombay." *Artha Vijnana*, 6 (March, 1964), pp. 27–37.

PAL, ANIDYA KUMAR et al., "Siliguri—A Case Study for Urban Redevelopment." Geographical Review of India, 27 (December, 1965), pp. 189–199.

PAL, SAROJ KUMAR, "Urban Trends of Metropolitan Cities of India." Paper for 21st International Geographical Congress, held at New Delhi, November–December, 1968, in S.P. Das Gupta and T.R. Lakshmanan (eds.), *Abstract of Papers*. Calcutta: National Committee for Geography, 1968, pp. 277–278.

PANDEY, P., "Urban Hierarchy in Chota Nagpur." Paper for 21st International Geographical Congress, held at New Delhi, November–December, 1968, in S.P. Das Gupta and T.R. Lakshmanan (eds.), *Abstract of Papers*. Calcutta: National Committee for Geography, 1968, p. 278.

PANJABI, R.M., "Chandigarh: India's Newest City." *Geographical Magazine*, 31 (December, 1958), pp. 401–414.

PANKAJ, T., "A Study of the Interland Limits and Traffic Flow Patterns of the Port of Cochin." *Artha Vijnana*, 10 (June, 1968), pp. 229–252.

PANT, PITAMBAR, "Urbanization and the Long-Range Strategy of Economic Development," in Roy Turner (ed.), *India's Urban Future*. Bombay: Oxford University Press, 1962, pp. 182–191.

PARASHAR, JAWAHAR, "Structuring the National Capital Rationale of Strategy." Paper for 7th Annual Town and Country Planning Seminar, New Delhi, October, 1968. Mimeo.

——, "Transportation Planning for the National Capital." Paper for 7th Annual Town and Country Planning Seminar, New Delhi, October, 1968. Mimeo.

PARK, RICHARD L., "The Urban Challenge to Local and State Government:

West Bengal, with Special Attention to Calcutta," in Roy Turner (ed.), *India's Urban Future: Selected Studies from an International Conference on Urbanization in India, held at the University of California, in 1960.* Berkeley, University of California Press, 1962, pp. 382–396.

PARLEKAR, S.H., "Application of Town Planning in the Cities and Rural Areas and Their Effects on Housing Problems." *Quarterly Journal of Local Self Government Institute,* 16 (April, 1955), pp. 761–774.

PARS RAM, *A UNESCO Study of Social Tensions in Aligarh, 1950–51.* Ahmedabad: New Order Book Co., 1955.

PATEL, FRAMJI J., *Poona: A Sociological Study.* Poona: Deccan College, 1957. Unpublished Ph.D. Thesis.

PATEL, KUNJ, *Rural Labour in Industrial Bombay.* Bombay: 1963.

PATEL, SHTIRISH B., "Transport Planning for Indian Cities." *Economic and Political Weekly,* 4 (July, 1969), pp. 1203–1208.

PATHAK, C.R., "Regional Planning in India—A Case Study of the Damodar Valley Region." Paper for 21st International Geographical Congress, held at New Delhi, November–December, 1968. Unpublished.

PATIL, C.S., "A Socio-Economic Survey of the Middle Class in Bombay." *Journal of the University of Bombay; History, Economics and Sociology Series,* 25 (July, 1956), pp. 20–24.

PATIL, P.C., *Regional Survey of Economic Resources, India, Kolhapur: A Typical Study of the Resources and Utility Services of a Region of the Indian Dominion.* Bombay: Bureau of Economics and Statistics, 1950.

PATIL, R.K. and K.M. TALATI, "Trends in Urbanization of Surat City: A Case Study," in *Papers Read at the 39th Annual Conference of the Indian Economic Association, 1958, Bombay.*

PATNA, Improvement Trust, *Outline of the Draft Master Plan of Patna.* Patna: 1961.

PATNAIK, KHETRA MOHAN, "Urban Development in Orissa," in *Papers Read at the 39th Annual Conference of the Indian Economic Association, 1958, Bombay,* pp. 89–95.

PATNAIK, R.C., "Urban Development Trends in India." Paper submitted to the 39th Conference of the Indian Economic Association, Cuttack, December, 1956.

PETHE, V.P., "Changes in Economic Conditions of an Urban Community (Scholapur)," *Artha Vijnana,* 3 (June, 1961), pp. 169–177.

———, "Cities of India." Paper for International Union for the Scientific Study of Population Conferenec, London, September, 1969.

———, *Demographic Profiles of an Urban Population (Sholapur).* Bombay: Popular Prakashan, 1964.

———, "Income Inequalities in Urban Communities." *A.I.C.C. Economic Review,* 13 (April 22, 1962), pp. 26–29.

PEARSON, ROGER, *Eastern Interlude: A Social History of the European Community in Calcutta.* Calcutta: 1954.

PHILIPOS, V.A., "Town and Country Planning, Most Urgent Need of India." *Quarterly Journal of the Local Self-Government Institute* (January, 1961), pp. 197–216.

PIGGOT, STUART, *Some Ancient Cities of India.* Bombay: 1954.

"Planning and Dreaming." *Marg,* 18 (June, 1965). Issue contains a series of articles on Bombay.

"Planning the Metropolis." *Journal of the Institute of Town Planners, India* (January–April, 1961), pp. 1–2.

"Policy for Urban Housing: When?" *Eastern Economist,* 30 (13 November, 1959), p. 759.

POONA, Suburban Municipality, *Poona Suburban Municipality: Sixty Years History (1884–1944).* Poona: 1945.

POTI, S.J., "Study of the Indian Population Growth." *Indian Population Bulletin,* 1 (April, 1960), pp. 82–128.

PRABHU, PANDHARINATH, "Bombay: A Study on the Social Effects of Urbanization on Industrial Workers Migrating from the Rural Areas to the City of Bombay," in *The Social Implications of Industrialization and Urbanization.* Calcutta: UNESCO Research Centre on Social Implications of Industrialization in Southern Asia, 1956, pp. 49–106.

——, "Social Effects of Urbanization on Industrial Workers in Bombay." *Sociological Bulletin,* 5 (March, 1956), pp. 29–78; 5 (September, 1956), pp. 127–143; 6 (March, 1957), pp. 14–33.

PRABHU, V.R., "Dharwar: Study in Indian Urban Landscapes." *Bombay Geographical Magazine,* 1 (October, 1953, pp. 56–63.

PRACH, G.C.K., "Urbanization in India," in R.P. Beckinsale and J.M. Houston (eds.), *Urbanization and Its Problems.* Oxford: Blackwell, 1968, pp. 297–303.

PRAKASA RAO, V.L.S., "Methodology in Urban Research." in M.R. Chaudhuri (ed.), *Essays in Geography (S.P. Chatterjee Volume).* Calcutta: Geographical Society of India, 1965, pp. 38–45.

——, "Problem of Metropolitan Planning: Geographer's Point of View." *Journal of the Institute of Town Planners, India* (January–April, 1961), pp. 12–15.

——, *Regional Planning—Principles and Case Studies.* Calcutta: Indian Statistical Institute, 1963.

——, "Spatial Patterns of Population Distribution and Growth in India." *Applied Economic Papers* (September, 1962), pp. 45–60.

——, "Urban Geography." *Indian Geographical Journal,* 32 (July–December, 1957). Special urban issue.

——, *Towns of Mysore State: A Sample Survey; A Contribution to Macro-Urban Geography.* Bombay: Asia Publishing House, 1965.

——, "Urban Survey for Regional Planning: The Need for Regional Approach." *Journal of the Institute of Town Planners, India* (January–April, 1959), pp. 27–32.

——, "Urban Telangana." Paper presented at the 14th Annual Town and Country Planning Seminar, Hyderabad, 1965.

PRAKASA RAO, V.L.S. and L.S. BHAT, "Delineation of Metropolitan Regions: Method." *Journal of the Institute of Town Planners, India* (January–April, 1961), pp. 61–63.

——, *Planning Regions in the Mysore State: The Need for Readjustment of District Boundaries.* Calcutta: Indian Statistical Institute, 1960.

PRAKASA RAO, V.L.S. and A.T.A. LEARMONTH, "Industrialization and Urbanization in Mysore State." *Applied Economic Papers* (September, 1961), pp. 41–52.

PRAKASH, VED, *Financing New Towns in India*. Ithaca, N.Y.: Cornell University, 1966. Unpublished Ph.D. Thesis.

——, *New Towns in India*. Chapel Hill, N.C.: Duke University Program in Comparative Studies on Southern Asia, 1969.

PRASAD, A., "Double Towns of India: A Problem in Urban Morphology." *Geographical Outlook,* 2 (No. 1, 1958).

PRASAD, BHAGWAN, *Socio-Economic Study of Urban Middle Classes*. Jullundur: Sterling Publishers, 1968.

PRASAD, BRAHAMRAND, "The Problems of Disguised Unemployment in Cities." *Yojana,* 11 (September 17, 1967), pp. 9–10, 27.

——, "Unemployment in Urban Areas." *Indian Journal of Commerce* (December, 1954), pp. 55–62.

PRASAD, L., "Urban Welfare Service." *Social Welfare,* 3 (January, 1956), pp. 24–25.

"Price of Unwise Urbanization—A Proposal for Urban Growth." *Quarterly Economic Report of Indian Institute of Public Opinion,* 13 (January, 1967), pp. 43–50.

PUNJAB, Capital Administration, *Project Estimate of the New Capital of Punjab, Chandigarh*. Simla: 1953.

PUNJAB, Economic and Statistical Organization, *Chandigarh Socio-Economic Survey Conducted in May, 1957*. Simla: 1958.

——, *Report on Employment and Unemployment Survey of Patiala City*. 1960.

PUNJAB, Public Relation Department, *Chandigarh*. Chandigarh: 1960.

RAFIULLAH, S.M., "A New Approach to Functional Classification of Towns." *Geographer* (Aligarh), 12 (1965).

RAGHAVA RAO G., "Livelihood Trends in the Towns of Uttar Pradesh." *Indian Journal of Economics,* 38 (January, 1958), pp. 297–301.

RAI, P.B. and O.P. GUPTA, "Planning for the National Capital Region—Objectives and Strategy of Development." Paper for 7th Annual Town and Country Planning Seminar, New Delhi, October, 1968. Mimeo.

RAI, P.B. and P.G. VALASANGKAR, "Methodology for Preparation of Regional Development Programmes, the National Capital Region—A Case Study." Paper for 7th Annual Town and Country Planning Seminar, New Delhi, October, 1968. Mimeo.

RAJ KRISHNA, "Impact of Industrialism on Indian Culture." *Indian Journal of Adult Education,* 29 (June, 1958), pp. 23–24, 37–38.

RAJA, K.C.K.E., "Problems of Demographic Research in India," in S.N. Agarwala (ed.), *India's Population: Some Problems in Perspective Planning*. Bombay: Asia Publishing House, 1960, pp. 161–172.

RAJAGOPALAN, C., "Bombay, a Study in Urban Demography and Ecology." *Sociological Bulletin,* 9 (March, 1960), pp. 16–38.

——, *The Greater Bombay: A Study in Suburban Ecology*. Bombay: Popular Book Depot, 1962.

——, "Housing Problem in Bombay." *Indian Journal of Social Work,* 25 (April, 1964), pp. 59–71.

RAJAN, B., "Delhi," *Illustrated Weekly of India,* 82 (September 24, 1961).

RAMAKRISHNA, M., "Census Economic Classification." *Economic Weekly,* 11 (3 January, 1959), pp. 2–3; 11 (10 January, 1959), pp. 48–50.

RAMALINGAM, S., "Traffic Problems in Madras." *Civic Affairs* (August, 1963), pp. 99–100.

RAMAN, A.S., "Maduai," *Illustrated Weekly of India,* 83 (February 4, 1962).

RAMAN, K.S., "Bombay City's Water Supply." *Civic Affairs* (June, 1963), pp. 25–29, 62.

RAMAN, K.V., *The Early History of the Madras Region.* Madras: 1959.

RAMANADHAM, V.V. and Y. VANKETESAWARLU, "Economic Aspects of Town Formations in Andhra Pradesh." *Indian Geographical Journal,* 32 (July–December, 1957), pp. 63–65.

RAMASUBRAMANIAM, K.A., "The Case for New Towns in India." *Journal of the Institute of Town Planners, India* (March, 1964), pp. 9–11.

——, "High Land Values Hinder Urban Planning." *Yojana,* 9 (August 29, 1965), pp. 2–3, 28.

——, "Steep Rise in Values of Urban Land." *Yojana,* 10 (January 26, 1966), pp. 55–56.

RAMEGODA, K.S., "Role of Water Supply in the Master Plan for Bangalore." *Civic Affairs* (April, 1952), pp. 75–77.

——, "The State Capital Region of Bangalore." Paper for 7th Annual Town and Country Planning Seminar, New Delhi, October, 1968. Mimeo.

RAMESH, A., "Evolution of Ootacamund." *National Geographical Journal of India,* 10 (March, 1964), pp. 16–28.

RANADE, S.N., "Urban Community Development: Its Nature and Scope." *Economic Weekly,* 11 (July–December, 1959), pp. 1501–1502.

RANDHAWA, M.S., *Beautifying Cities of India.* New Delhi: 1961.

RAO, A.N. MOORTHY, "Bangalore." *Illustrated Weekly of India,* 82 (November 26, 1961).

RAO, B.P., "Evolution of Visakhapatnam." *National Geographical Journal of India,* 6 (December, 1960), pp. 242–259.

RAO, C.B., "Local Elections and Politics." *Indian Journal of Public Administration,* 14 (July–September, 1968), pp. 533–537.

RAO, D.V.R., "Housing in the National Capital." Paper for 7th Annual Town and Country Planning Seminar, New Delhi, October, 1968. Mimeo.

RAO, M.S.A., "Economic Change and Rationality in a Fringe Village." *Economic Weekly,* 14 (September 29, 1962), pp. 1545–1549.

RAO, M.S.V., "Traffic and Transportation in Delhi: Some Aspects of the Problem and Possible Solutions." Paper for 7th Annual Town and Country Planning Seminar, New Delhi, October, 1968. Mimeo.

RAO, N. BASKARA, "Estimate of Migration in the Four Major Cities of India, 1941–51 to 1951–61." *Asian Economic Review,* 9 (November, 1966), pp. 53–63.

RAO, RAGHAVA G., "Livelihood Trends in the Towns of Uttar Pradesh." *Indian Journal of Economics,* 38 (January, 1958), pp. 297–301.

RAO, RAGAHARA, G., "Problems of Urbanization in the Metropolitan Cities of India." *Journal of Social Research,* 5 (March, 1962), pp. 133–44.

RAO, RAJA, "Trivandrum." *Illustrated Weekly of India,* 83 (February 25, 1962).

——, "Varanasi." *Illustrated Weekly of India,* 82 (September 3, 1961).

RAO, V., "Rural Reactions to Urbanization." *Journal of University of Bombay (History, Economics and Sociology Series),* 26 (January, 1958), p. 120.

RAO, V.K.R.V., *An Economic Review of Refugee Rehabilitation in India (A Study of Faridabad Township).* Delhi: Delhi School of Economics, Monograph No. 3, 1955.

——, *An Economic Review of Refugee Rehabilitation in India (A Study of Kingsway Camp).* Delhi: Delhi School of Economics, Monograph No. 5, 1955.

——, *An Economic Review of Refugee Rehabilitation in India (A Study of Nilokheri Township).* Delhi: Delhi School of Economic, Monograph No. 1, 1954.

——, *An Economic Review of Refugee Rehabilitation in India (a Study of Rajpura Township).* Delhi: Delhi School of Economics, Monograph No. 6, 1955.

——, *An Economic Review of Refugee Rehabilitation in India (a Study of Tripura Township).* Delhi: Delhi School of Economics, Monograph No. 7, 1955.

RAO, V.K.R.V. et al., *Papers on National Income and Allied Topics.* Bombay: Asia Publishing House, 1960.

RAO, V.K.R.V. and P.B. DESAI, *Greater Delhi—A Study of Urbanization, 1940–1957.* Delhi: 1965.

RAU, CHALAPATHI, "Lucknow." *Illustrated Weekly of India,* 82 (August 27, 1961).

RAY, SIB NARAYAN, "India: Urban Intellectuals and Rural Problems," in Leopold Labedz (ed.), *Revisionism: Essays on the History of Marxist Ideas.* London: 1962, pp. 374–386.

RAYAPROL, SRINIVAS, "Hyderabad." *Illustrated Weekly of India,* 82 (August 27, 1961).

RAZA, MOONIS, "Urbanization in Pre-Historic India." *Geographer* (Aligarh), 4 (May, 1951), pp. 15–29.

REDDY, B. GOPALA, "The Problem of Slums." *Civic Affairs* (February, 1962), pp. 62–65.

REED, WALLACE, E., A Real Interaction in India: Commodity Flows of the Bengal-Bihar Industrial Area. Chicago: University of Chicago, Department of Geography, Research Paper No. 110, 1967.

"Re-Planning of Metropolitan Calcutta: C.M.P.O.'s First Report." *Civic Affairs* (July, 1963), pp. 9–14; (August, 1963), pp. 13–24; (September, 1963), pp. 9–17.

Report on the Development Plan for Greater Bombay. Bombay: 1964.

"Review of Urban Geography," in S.P. Chatterjee, *Fifty Years of Science in India; Progress of Geography.* Calcutta: Indian Science Congress Association, 1964, pp. 193–208.

RIZVI, S.Q.A., "The Role of Rivers in the Evolution of Urban Centres." *Geographer* (Aligarh), 8, 9 (1956, 1957), pp. 45–52.

Ross, Aileen D., *Hindu Family in Its Urban Setting*. Bombay: Oxford University Press, 1961.

Roy, D.M., *Small Scale Industries in Meerut*. Meerut: Meerut College, 1961. Mimeo.

Roy, Kamelendu, "Central Area of Calcutta—Its Growth and Development." *Journal of the Institute of Town Planners, India* (January–April, 1961), pp. 74–77.

Roy, S.K., "Urban Transportation in India." *Indian Journal of Public Administration*, 14 (July–September, 1968), pp. 716–735.

Rudolph, Lloyd I., "Urban Life and Populist Radicalism—Dravidian Politics in Madras." *Journal of Asian Studies*, 20 (May, 1961), pp. 283–297.

Sabavala, Sharokh, "Bombay." *Illustrated Weekly of India*, 82 (August 13, 1961).

Sachdev, H.R., "Housing Problem in Urban Areas: Private Initiative Requires to Be Stimulated." *Commerce* (December, 1959), pp. 130, 132.

Sah, J.P., "Five Year Plans for Our Cities and Towns." *Journal of the Institute of Town Planners, India* (March, 1964), pp. 12–15.

——, *Municipal Revenues in India: A Perspective*. New Delhi: Town and Country Planning Organization, 1965. Mimeo.

Sah, J.P. and S.S. Dutta, "Economic Development and Spatial Planning in India." *Ekistics*, 24 (January, 1967), pp. 33–39.

——, *Economic Development and Spatial Planning in India*. New Delhi: Town and Country Planning Organization, 1966. Mimeo.

Sah, J.P. and L.H. Marathe, *Some Aspects of Financial Administration of Local Bodies in India*. New Delhi: Town and Country Planning Organization, 1965. Mimeo.

Saksena, R.N., *Refugees—A Study in Changing Pattern (A Survey Conducted in Dehradun)*. Bombay: Asia Publishing House, 1961.

Samena, N.P., "Non-Residential Establishments in Towns of Upper Ganga-Yamuna Doab." Paper for 21st International Geographical Congress, held at New Delhi, November–December, 1968, in S.P. Das Gupta and T.R. Lakshmanan (eds.), *Abstract of Papers*. Calcutta: National Committee for Geography, 1968, p. 280.

Samena, N.P., "Occupational Structure, Population, Size and Central Place Considerations Regarding Urban Centres in India." *Geographical Observer* (Meerut), 3 (March, 1967), pp. 1–14.

Sandesara, J.C., "Migration and Metropolitan Living: A Study of Indian Cities." *Economic Weekly*, 16 (9 May, 1964), pp. 807–810.

Sangave, Vilas A., "Changing Pattern of Caste Organization in Kolhapur City." *Sociological Bulletin*, 11 (March–September, 1962), pp. 36–61.

Saran, T.K. "Some Salient Features of the Master Plan for Patna." *Journal of the Institute of Town Planners, India* (January–April, 1961), pp. 116–119.

Sarma, Jyotirmoyee, "Role of the Census in Regional Studies of Cities." *Calcutta Review*, 151 (May, 1959), pp. 124–132.

Sastry, D.U., "Some Aspects of Economic and Social Life in Devangere: A Study in Urbanization." *Asian Economic Review*, 2 (August, 1960), pp. 467–489; 3 (November, 1960), pp. 12–24.

SASTRY, S.M.Y., "Visakhapatnam." *Illustrated Weekly of India* 83 (January 21, 1962).

SAWHNEY, J., "The New Metropolis of Punjab-Chandigarh: A Case Study in Town Planning and Administration." *Quarterly Journal of the Local Self Government Institute,* (April, 1963), pp. 313–333; (July, 1963), pp. 1–13.

SAYASTHA, S.L., "Trend of Urbanization in the Beas River Basin in Himalaya." Paper for 21st International Geographical Congress, held at New Delhi, November–December, 1968, in S.P. Das Gupta and T.R. Lakshmanan (eds.), *Abstract of Papers.* Calcutta: National Committee for Geography, 1968, pp. 270.

SAXENA, S., "Demographic Aspects of Urbanization." Paper for International Union for the Scientific Study of Population Conference, London, September, 1969.

SCHWARTZBERG, JOSEPH E., "Four Geographical Cross Sections Through Indian Economic History." Paper for 21st International Geographical Congress, held at New Delhi, November–December, 1968, in S.P. Das Gupta and T.R. Lakshmanan (eds.), *Abstract of Papers.* Calcutta: National Committee for Geography, 1968, p. 240.

SEHGAL, J.M., "The Population Distribution in Greater Bombay." *Asian Economic Review,* 8 (February, 1966), pp. 185–195.

SEHGAL, KULDIP CHANDER, "Migration into Bombay—A Factual Study." *Economic Times,* 8 (September 7, 1969), pp. 5, 6.

SEN, AMAL KUMAR, "Bankura: A Study of the Cultural Landscape of an Urban Area." *Geographical Review of India,* 18 (March, 1956), pp. 9–14.

——, "City Regionalism in India." *Geographical Review of India,* 19 (March, 1957), pp. 12–19.

——, "Techniques of Classifying the Functional Zone of a City." *Geographical Review of India,* 21 (March, 1959), pp. 37–42.

SENGUPTA, P. and GALIN SDASYUK, *Economic Regionalization of India: Problems and Approaches.* New Delhi: 1961 Indian Census, Monograph No. 8, 1968.

SENGUPTA, P., "Effects of Internal Emigration and Immigration in India During 1951 to 1961." Paper for 21st International Geographical Congress, held at New Delhi, November–December, 1968, in S.P. Das Gupta and T.R. Lakshmanan (eds.), *Abstract of Papers,* Calcutta: National Committee for Geography, 1968, p. 233.

SEN, S.N., "Calcutta's Lonely Crowd." *Economic Weekly,* 11 (28 February, 1959), pp. 282–284.

——, *City of Calcutta: A Socio-Economic Survey, 1954–55 to 1957–58.* Calcutta: Bookland, 1960.

——, *Delhi and Its Monuments.* Calcutta: 1948.

SEN, S.N. et al. (eds.), *Calcutta.* Calcutta: Indian Science Congress Association, 1952.

SHAFI, SAYED S., "Census Data and Urban Planning Requirements." *Journal of the Institute of Town Planners, India* (January–April, 1959), pp. 33–39.

——, "The Concept of Group Housing in the Delhi Plan." *Journal of the Institute of Town Planners, India* (March, 1964), pp. 6–8.

——, "Problems and Criteria in the Delineation of Metropolitan Regions for Comprehensive Planning." *Journal of the Institute of Town Planners, India,* (January–April, 1961), pp. 50–60.

——, "Urban Growth and Metropolitan Regional Planning." *Journal of the Institute of Town Planners, India* (January–April, 1961), pp. 50–60.

SHAH, B.B., "Planning Proposals for Abmedabad." *Journal of the Institute of Town Planners, India* (January, 1961), pp. 98–103.

SHAH, M.C., "Policy Making and Municipal Administration." in *Improving City Government: Proceedings of a Seminar 13–14 September, 1958.* New Delhi: Indian Institute of Public Administration, 1959, pp. 3–6.

SHAHDEO, N.K.N., "Urban Growth and Agricultural Changes in Ranchi." Paper for 21st International Geographical Congress, held at New Delhi, November–December, 1968, in S.P. Das Gupta and T.R. Lakshmanan (eds.), *Abstract of Papers.* Calcutta: National Committee for Geography, 1968, pp. 281–282.

SHARMA, B.D., "Urbanization and Economic Development." *Indian Journal of Public Administration,* 14 (July–September, 1968), pp. 374–490.

SHARMA, O.P., "Changing Pattern of Rural Society; Rural Urban Interaction Pattern in India," a paper presented in Seminar on Trends of Socio-Economic Change in India 1971–1961, 11–23 September, 1967 Simla: Indian Institute of Advanced Study, 1967.

SHARMA, SURENDER K. "Image of the Capital." Paper for 7th Annual Town and Country Planning Seminar, New Delhi, October, 1968. Mimeo.

SHARMA, T.R., *Location of Industrries in India.* Bombay: Hind Kitabs, 1954.

SHETH, N.R., "Census and Social Reality." *Economic Weekly,* 11 (21 March, 1959), pp. 417–418.

SHILS, EDWARD, *The Intellectual Between Tradition and Modernity—The Indian Situation.* Netherlands: 1961.

SHREEVASTAVA, M.P. and K.K. TANDON, "Economy of Small Towns in Orissa." *A.I.C.C. Economic Review,* 20 (15 November, 1968), pp. 18–24, 33.

SHRINIVASAN, K.N. and B.R. ARORA, "Recent Urbanization in India in Retrospect." *A.I.C.C. Economic Review,* 18 (15 July, 1966), pp. 31–35.

SHRINIVASAN, N.S., "Traffic Planning: An Urgent Need of Urban Areas." *Yojana,* 10 (February, 1966), pp. 20–21.

SHUKLA, R.C., "Urban Unemployment in India—Its Causes and Cure." *Indian Journal of Commerce,* 7 (December, 1954), pp. 1–13.

SIDDIQUI, N.A., "The Composition of Population of Moradabad City." *Geographer* (Aligarh), 10 (1958), pp. 43–57.

——, "Daily Commuting: In the Context of Urban Geographical Studies." *Geographical Observer,* 2 (March, 1966), pp. 3–7.

SILIGURI, Planning Organization, *Siliguri—Interim Development Plan.* Calcutta: 1965.

SINGH, AYODHYA, "Industrial Housing in India: A Review." *A.I.C.C. Economic Review,* 14 (22 March, 1962), pp. 12–15.

SINGH, A.N., "Functional Zones of Itarsi Town." *National Geographical Journal of India,* 13 (June, 1967).

SINGH, BALJIT, *Investigations Regarding Small Scale Industries in Moradabad.* Lucknow: Lucknow University, 1961.

———, *A Report on Unemployment in the City of Lucknow.* Lucknow: Lucknow University Press, 1955.

———, *Urban Middle Class Climbers: A Study in Social Mobility.* Lucknow: Lucknow University, I.K. Institute of Sociology and Human Relations, Monograph No. 7, 1959.

SINGH, B.N., "Some Aspects of Urbanization in Damodar Valley Region." *Ekistics,* 21 (No. 2, 1966), p. 342.

———, "Urban Land Policy in India." *National Geographical Journal of India,* 11 (September–December, 1965), pp. 131–143.

SINGH, B. SATWANT, "Housing of Squatters in Delhi." Paper for 7th Annual Town and Country Planning Seminar, New Delhi, October, 1968. Mimeo.

SINGH, HARI HAR, "Impact of Industrialization on Cities of India." Paper for 21st International Geographical Congress, held at New Delhi, November–December, 1968, in S.P. Das Gupta and T.R. Lakshmanan (eds.) *Abstract of Papers.* Calcutta: National Committee for Geography, 1968, p. 282.

———, "Residential Structure of Kanpur in Uttar Pradesh." Paper for 21st International Geographical Congress, held at New Delhi, November–December, 1968, in S.P. Gupta and T.R. Lakshmanan (eds.), *Abstract of Papers.* Calcutta: National Committee for Geography, 1968, p. 282.

———, "Urban Land Use of Jaunpur." *National Geographical Journal of India,* 6 (June, 1960), pp. 115–123.

SINGH, KASHI PRASAD, "The Pedestrian and the Motor Age City." Paper for 7th Annual Town and Country Planning Seminar, New Delhi, October, 1968. Mimeo.

SINGH, KHUSHWANT, "Agra." *Illustrated Weekly of India,* 83 (March 11, 1962).

———, "Amritsar." *Illustrated Weekly of India,* 82 (October 29, 1961).

SINGH, KUMOD KUMAR, "Historical Study of Patna, Bihar." Paper for 21st International Geographical Congress, held at New Delhi, November–December, 1968, in S.P. Das Gupta and T.R. Lakshmanan (eds.), *Abstract of Papers.* Calcutta: National Committee for Geography, 1968, p. 285.

SINGH, K.N., "Barhaj: A Study in the Changing Pattern of a Market Town." *National Geographical Journal of India,* 7 (March, 1961), pp. 21–36.

———, "Changes in Functional Sturcture of Small Towns in Eastern U.P." *Indian Geographer,* 6 (August, 1961), pp. 21–40.

———, "Function and Functional Classification of Towns of Uttar Pradesh." *National Geographical Journal of India,* 5 (September, 1959), pp. 121–148.

———, "Morphology of the Twin-Township of Dehri-Dalmianagar." *National Geographical Journal of India,* 3 (September–December, 1957), pp. 169–179.

———, "Toward Developing a Rational Central Place System in Varanasi Region." Paper for 21st International Geographical Congress, held at New Delhi, November–December, 1968, in S.P. Das Dupta and T.R. Laksmanan (eds.), *Abstract of Papers.* Calcutta: National Committee for Geography, 1968, p. 283.

SINGH, LAL, "Umland of Agra." *National Geographical Journal of India,* 2 (September, 1956), pp. 149–152.

SINGH, M., "Evolution of Meerut." *National Geographical Journal of India,* 11 (September–December, 1965), pp. 144–158.

SINGH, M.M., *Municipal Government in the Calcutta Metropolitan District: A Preliminary Survey.* New York: Institute of Public Administration, 1966.

SINGH, M.M. and ABHIJIT DATTA, *Metropolitan Calcutta: Special Agencies for Housing, Planning and Development.* New York: Institute of Public Administration.

SINGH, NAND K., "Some Trends in Urbanization in India." *A.I.C.C. Economic Review,* 18 (1 December, 1966), pp. 35–40.

SINGH, OM PRAKASH, "Functional Hierarchy of Central Places in Uttar Pradesh." Paper for 21st International Geographical Congress, held at New Delhi, November–December, 1968, in S.P. Das Gupta and T.R. Lakshmanan (eds.), *Abstract of Papers.* Calcutta: National Committee for Geography, 1968, p. 283.

SINGH, ONKAR, "Hierarchy of Towns in Uttar Pradesh." Paper for 21st International Geographical Congress, held at New Delhi, November–December, 1968, in S.P. Das Gupta and T.R. Lakshmanan (eds.), *Abstract of Papers.* Calcutta: National Committee for Geography, 1968, pp. 283–284.

——, "Urban Morphology of Babraich, Uttar Pradesh." Paper for 21st International Geographical Congress, held at New Delhi, November–December, 1968, in S.P. Das Gupta and T.R. Lakshmanan (eds.), *Abstract of Papers.* Calcutta: National Committee for Geography, 1968, p. 284.

SINGH, R.L., "Ballia: A Study in Urban Settlement." *National Geographical Journal of India,* 2 (March, 1956), pp. 1–6.

——, *Banaras: A Study in Urban Geography.* Banaras: Nand Kishore, 1955.

——, *Bangalore: A Study in Urban Geography.* Banaras: 1964.

——, *Bangalore: An Urban Survey.* Varanasi: 1964.

——, "Development of Twin-Township of Dehri Dalmianagar." *National Geographical Journal of India,* 2 (September, 1956), pp. 121–127.

——, "Evolution of Bangalore City." *National Geographical Journal of India,* 7 (December, 1961), pp. 232–244.

——, "Faizabad-cum-Ayodhya." *National Geographical Journal of India,* 4 (March, 1958), pp. 1–6.

——, "Gorakhpur: A Study in Urban Morphology." *National Geographical Journal of India,* 1 (September, 1955), pp. 1–10.

——, "Mirzapur: A Study in Urban Geography." *Geographical Outlook,* 1 (January, 1956), pp. 16–27.

——, "Trend of Urbanization in the Umland of Banaras." *National Geographical Journal of India,* 2 (June, 1956), pp. 75–83.

——, "Two Small Towns of Eastern U.P.: Sultanpur and Chunar." *National Geographical Journal of India,* 3 (March, 1957), pp. 1–10.

SINGH, R.L. and S.M. SINGH, "Evolution of the Medieval Towns in Sarjupar Plain of the Middle Ganga Valley: A Case Study." *National Geographical Journal of India,* 9 (March, 1963), pp. 1–11.

——, "Mungra—Radshapur: A Urban Settlement." *National Geographical Journal of India,* 6 (December, 1960), pp. 199–206.

SINGH, R.L. and B. MUKHERJEE, "Functional Zones of Ranchi." *National Geographical Journal of India,* 13 (September–December, 1957), pp. 117–124.

SINGH, R.P., "Bokaro, the Future Steel Town." *National Geographical Journal of India,* 7 (December, 1961), pp. 244–256.

SINGH, RAJA RAM, "Functional Structure of Jaunpur City." Paper for 21st International Geographical Congress, held at New Delhi, November–December, 1968, in S.P. Das Gupta and T.R. Lakshmanan (eds.), *Abstract of Papers.* Calcutta: National Committee for Geography, 1968, p. 284.

SINGH, SUPRIYA, "What Future for Delhi's Transport?" *Shakti,* 5 (February, 1968), pp. 35–36.

SINGH, S.C., "Evolution of Azamgarh." *National Geographical Journal of India,* 9 (September, 1958), 9 (September–December, 1963), pp. 175–186.

SINGH, S.P., "Demographic Features of Lucknow." *National Geographical Journal of India,* 5 (September, 1959), pp. 157–175.

SINGH, TARLOK, "Problems of Integrating Rural, Industrial, and Urban Development," in Roy Turner (ed.), *India's Urban Future.* Bombay: Oxford University Press, 1962, pp. 327–334.

SINGH, UJAGIR, "Allahabad—A Study in Its Planning and Development." *National Geographical Journal of India,* 7 (June, 1961), pp. 99–115.

——, *Allahabad—A Study in Urban Geography.* Varanasi: 1962.

——, "Banaras: A Note on Its Urban Geography." *Indian Geographical Journal,* 27 (January–June, 1952), pp. 26–31.

——, "Bombay: A Study in Historical Geography, 1667–1900 A.D." *National Geographical Journal of India,* 6 (March, 1960), pp. 19–29.

——, "Calcutta Conurbation." *National Geographical Journal of India,* 4 (June, 1958), pp. 95–102.

——, "Cultural Zones of Allahabad." *National Geographical Journal of India,* 6 (June, 1960), pp. 95–104.

——, "Demographic Structure of Allahabad." *National Geographical Journal of India,* 4 (December, 1958), pp. 165–187.

——, "Evolution of Allahabad." *National Geographical Journal of India,* 4 (September, 1958), pp. 109–129.

——, "Functional Regions of Allahabad." *National Geographical Journal of India,* 5 (June, 1959), pp. 67–90.

——, "Geographical Analysis of the Essential Services of Allahabad." *National Geographical Journal of India,* 6 (September, 1960), pp. 176–193.

——, "Geographical Zones of Allahabad." *National Geographical Journal,* 2 (March, 1956), pp. 36–47.

——, "Growth of Transport and Communication in Allahabad." *National Geographical Journal of India,* 5 (December, 1959), pp. 186–204.

——, "Industrial Landscape of Allahabad." *National Geographer,* 11 (May, 1960), pp. 29–37.

——, " 'Kaval' Towns: A Comparative Study in Functional Aspects of Urban Centres in Uttar Pradesh." *National Geographical Journal of India,* 8 (September–December, 1962), pp. 238–249.

——, "New Delhi—Its Site and Situation." *National Geographical Journal of India,* 5 (September, 1959), pp. 113–120.

——, "The Origin and Growth of Kanpur." *National Geographical Journal of India,* 5 (March, 1959), pp. 1–11.

——, "Umland of Allahabad." *National Geographic Journal of India,* 7 (March, 1961), pp. 37–51.

SINGH, U., "Changes in the Builtup Area of Kaval Towns in the Ganga Plain." *National Geographical Journal of India,* 12 (December, 1966), pp. 203–217.

——, "Moghal Sarai—A Study of Land Use." *National Geographical Journal of India,* 10 (September–December, 1964), pp. 136–145.

SINGH, U. MOGHAL SARAI, "Geographical Analysis of Slum Areas in India Cities with Special Reference to Kanpur." *National Geographical Journal of India,* 12 (December, 1966).

SINGH, VISHWA NATH PRASAD, "Historical Development of Deoghar Town." Paper for 21st International Geographical Congress, held at New Delhi, November–December, 1968, in S.P. Das Gupta and T.R. Lakshmanan (eds.), *Abstract of Papers.* Calcutta: National Committee for Geography, 1968, p. 301.

——, "Urbanization in Chota Nagpur Plateau, Bihar." Paper for 21st International Geographical Congress, held at New Delhi, November–December, 1968, in S.P. Das Gupta and T.R. Lakshmanan (eds.), *Abstract of Papers.* Calcutta: National Committee for Geography, 1968, p. 285.

SINGH, V. MOGHAL SARAI, "Distribution and Character of Cities of Ganga Plain." *National Geographical Journal of India,* 11 (March, 1965), pp. 1–12.

SINGH, V.R., "Changes in the Functional Landscape of Chunar, Uttar Pradesh." Paper for 21st International Geographical Congress, held at New Delhi, November–December, 1968, in S.P. Das Gupta and T.R. Lakshmanan (eds.), *Abstract of Papers.* Calcutta: National Committee for Geography, 1968, pp. 284–285.

SINHA, B., "Urban Geography of Orissa." *Indian Geographical Journal,* 32 (July–December, 1957), pp. 86–94.

SINHA, J.N., "Comparability of 1961 and 1951 Census Economic Data." *Artha Vijnana,* 6 (December, 1964), pp. 273–288.

SINHA, PRADIP, "Urbanization and the Bengali Middle Class." Paper presented in the Seminar on Trends of Socio-Economic Change in India 1871–1961, 11 to 23 September, 1967. Simla: Indian Institute of Advanced Study.

SINHA, R.L.P., "Optimum City Size." *Indian Geographer,* 9 (December, 1964), pp. 83–116.

SIRKIN, GERALD, "The Strange Case of City Versus Planner." *Economic Weekly,* 14 (February 1962), pp. 147–148.

SIVARAMAKRISHNAN, K.S., "Planning for a Resource Region—A Case Study (Asansol and Durgapur Region)." Paper for 21st International Geographical Congress, held at New Delhi, November–December, 1968. Unpublished.

SIVARAMA KRISHNAN, K.C., "Urban Housing: Challenge and Response." *Economic and Political Weekly,* 4 (September 4, 1969), pp. 1443–1445.

SIVAYYA, K.V., *Trade Union Movement in Vidakhapatnam.* Waltair: Andhra University Press, 1966.

"Slow Development of Chandigarh." *Civic Affairs,* (September, 1961), pp. 52–53.

SMAILES, A.E., "Indian City Structure." Paper for 21st International Geographical Congress, held at New Delhi, November–December, 1968, in

S.P. Das Gupta and T.R. Lakshmanan (eds.), *Abstract of Papers*. Calcutta: National Committee for Geography, 1968, p. 286.

"The Small Town." *Yojana*, 5 (June 11, 1961), pp. 13, 26.

"Social Factor in the Morphogenesis of Varanasi." Paper for 21st International Geographical Congress, held at New Delhi, November–December, 1968, in S.P. Das Gupta and T.R. Lakshmanan (eds.), *Abstract of Papers*. Calcutta: National Committee for Geography, 1968, p. 284.

SOKJEE, A.H. (ed.), *Politics of a Peri-Urban Community in India*. Bombay: Asia Publishing House, 1964.

SOM, R.K., "Population Trends and Problems in India," in S.N. Agarwala (ed.), *India's Population: Some Problems in Perspective Planning*. Bombay: Asia Publishing House, 1960, pp. 59–68.

——, "Response Biases in Demographic Enquiries." *Papers Contributed by Indian Authors to the World Population Conference, Belgrade, 1965*. New Delhi: Registrar General, 1965, pp. 73–84.

"Some Economic Issues in Urban Planning." *Civic Affairs*, (November, 1962), pp. 69–73.

SONDHI, M.L., "Revitalizing Delhi." *Shakti*, 5 (February, 1968), pp. 5–8.

SOVANI, N.V., "The Analysis of Over-Urbanization." *Economic Development and Cultural Change*, 12 (January, 1964), pp. 113–122.

——, "Internal Migration and the Future Trend of Population in India," in *Papers Contributed by Indian Authors to the World Population Conference, Belgrade, 1965*. New Delhi: Registrar General, 1965, pp. 163–166.

——, "Potential Out-Migrants and Removable Surplus Population in Three Districts of Orissa (India)." Paper for International Population Conference, Vienna, 1959.

——, *Social Survey of Kolhapur City*. Poona: Gokhale Institute of Politics and Economics, Vol. I, 1948; Vol. II, 1952.

——, "The Structure of Urban Income in India." *Artha Vijnana*, 6 (September, 1964), pp. 145–179.

——, "Trend of Urbanization in India" Paper read at the 39th Annual Conference of the Indian Economic Association, 1958, Bombay.

——, "The Urban Social Situation in India." *Artha Vijnana*, 3 (June, 1961), pp. 85–106; 3 (September, 1961), pp. 195–224.

——, *Urbanization and Urban India*. Bombay: Asia Publishing House, 1966.

SOVANI, N.V. et al., *Poona: A Resurvey, the Changing Pattern of Employment and Earnings*. Poona: Gokhale Institute of Politics and Economics, 1956.

SOVANI, N.V. and KUSUM PARDHAN, "Occupational Mobility in Poona City Between Three Generations." *Indian Economic Review* (August, 1955), pp. 23–36.

SOVANI, N.V. and NILAKANTH RATH, *Economics of a Multi-Purpose River Dam: Report of an Enquiry into the Economic Benefits of the Hirakud Dam*. Poona: Gokhale Institute of Politics and Economics, 1960.

SPATE, O.H.K., "Two Federal Capitals—New Delhi and Canberra." *Geographical Outlook*, 1 (1956), pp. 1–8.

SPATE, O.H.K. and E. AHMAD, "Five Cities of the Gangetic Plain: A Cross Section of Indian Culture History." *Geographical Review*, 40 (April, 1950), pp. 260–278.

SPEAR, PERCIVAL, "Mughal Delhi and Agra," in Arnold Toynbee ,(ed.), *Cities of Destiny*. London: Thames and Hudson, 1967.

SREEVASTAVA, M.P. and RAI, P.B., "Ghaziabad—Functions of Ring Town Analysed." Paper for the 7th Annual Town and Country Planning Seminar, New Delhi, October, 1968. Mimeo.

SRIDHARAN, K.V., "Urban Community Development." *Social Welfare*, 9 (July, 1962), pp. 2–4.

SRINIVAS, M.N., "Industrialization and Urbanization of Rural Areas." *Sociological Bulletin*, 2 (September, 1956), pp. 79–88.

——, "Social Anthropology and the Study of Rural and Urban Societies." *Economic Weekly*, 11 (January, 1959), pp. 133–140.

——, *Social Change in Modern India*. New Delhi: Allied Publishers, 1966.

SRINIVASACHARI, C.S., "Stages in the Growth of the City of Madras." *Journal of Madras Geographical Association*, 2 (October, 1927), pp. 79–105.

SRINIVASAN, N.S., ISHWAR CHANDRA, and B.L. SURI, "Some Aspects of Traffic and Transportation Problems of Delhi." Paper for 7th Annual Town and Country Planning Seminar, New Delhi, October, 1968. Mimeo.

SRIVASTAVA, V.K., "Sidhi-District Headquarters in Transition." *Geographical Thought* (Gorakhpur), 3 (No. 1, 1967).

STEIN, J.A., "Neighborhood Planning for Modern Industrial Towns." *Urban and Rural Planning Thought*, 1 (January, 1958), pp. 29–40.

STEPANEK, J.E. et al., *Industrialization Beyond the Metropolis: Current Development in India*. Hyderabad: Small Industries Extention Training Institute, 1963. Mimeo.

STOLNITZ, GEORGE J., *An Analysis of the Population of India*. Bloomington, Ind.: Indiana University, 1967. Mimeo.

"Structure of Urban and Rural Income Inequalities." *Quarterly Economic Report of the Indian Institute of Public Opinion*, 7 (July, 1960), pp. 48–50.

SUBRAMANIAN, R., "Electric Trains for Calcutta Suburbs: A.E.I.'s Notable Contribution." *Commerce* (December, 1958), p. 112.

SUBRAMANYAM, N., "Regional Distribution and Relative Growth of Cities of Tamilnad." *Indian Geographical Journal*, 16 (January–March, 1941), pp. 71–83.

SUCHIT, KUMAR, "Urban Landscape of Chandannagar, West Bengal." Paper for 21st International Geographical Congress, held at New Delhi, November–December, 1968, in S.P. Das Gupta and T.R. Lakshmanan (eds.), *Abstract of Papers*. Calcutta: National Committee for Geography, 1968, pp. 286–7.

SUKLABAIDYA, L.M., "A Survey on Urbanization in North Bengal." *Economic Studies*, 2 (June, 1967), pp. 680–684.

SUNDARAM, K.V., "Planning and Plan Implementation in the National Capital Region." Paper for 7th Annual Town and Country Planning Seminar, New Delhi, October, 1968. Mimeo.

SUNDARAM, K.V. and V.L.S. PRAKASA RAO, "Regional Planning: A Review." Paper for 21st International Geographical Congress, held at New Delhi, 27 to 30 November, 1968. Unpublished.

SUNDHARAM, K.P., "Urban Unemployment in India." Indian Journal of *Commerce*, 8 (March, 1955), p. 17–21.

Suri, K.B., "Towns: Size, Economic Structure and Growth." *Economic and Political Weekly*, 3 (August 10, 1968), pp. 1247–1251.

Suryakantham, K.V. et al., *Intensive Enquiries into Working Class Cost of Living in Hyderabad City*. Hyderabad: Hyderabad Economic Association, 1951.

Swamy, V.S., "A Note on the Application of Pareto Law to the Cumulative Distribution of Towns by Size Class of Population." *Indian Population Bulletin*, 2 (August, 1961), pp. 333–345.

———, "Some Aspects of Migration in Kerala," in R.S. Kurup and K.A. George (eds.), *Population Growth in Kerala*. Trivandrum: 1965.

———, "Some Aspects of Urban Population." *Indian Population Bulletin*, 2 (August, 1961), pp. 321–331.

Swarnakar, G.P., "Agro-Urban Development in Tamluk Area in West Bengal." Paper for 21st International Geographical Congress, held at New Delhi, November–December, 1968, in S.P. Das Gupta and T.R. Lakshmanan (eds.), *Abstract of Papers*. Calcutta: National Committee for Geography, 1968, p. 287.

Tahir Razvi, S.M., *Regional Development of Aligarh City*. Aligarh: Aligarh Muslim University, 1961. Unpublished report.

Tangri, Shanti S., *Urban Growth, Housing and Economic Development* Ann Arbor, Mich.: University of Michigan, Center for South and Southeast Asian Studies, Working Paper No. 6, 1966.

———, "Urban Growth, Housing and Economic Development: The Case of India." *Asian Survey*, 8 (July, 1968), pp. 519–538.

———, "Urbanization, Political Stability," in Roy Turner (ed.), *India's Urban Future*. Bombay: Oxford University Press, 1962, pp. 192–211.

Tewari, A.R., "Mathura—A Study in Site and Development." *National Geographical Journal of India*, 9 (March, 1963), pp. 48–59.

Tewari, Abinash Chander, *Municipalities and City Takers in the Punjab, 1963*. Delhi: Adarsh Publishing House, 1963.

Tewari, T.R., "Publishing Health Administration in Urban Areas." *Indian Journal of Public Administration*, 14 (July–September, 1968), pp. 709–715.

Thacker, M.S., *India's Urban Problem*. Mysore: University of Mysore, 1966.

———, "Planning for Urban Development." *Yojana*, 9 (31 January, 1965), pp. 2–4.

Thadani, Jaya, "City of Chandigarh." *March of India*, 11 (January, 1959), pp. 19–24.

Thakore, M.P., "Sixteen Sites in Delhi." *Indian Geographer*, 8 (December, 1963), pp. 84–119.

Thawani, V.D., *Survey of Urban Employment and Unemployment in Assam*. Gauhati: University of Assam, 1961.

Thirunaranan, B.N. "Tiruttani: A Study of a Temple Town." *Indian Geographical Journal*, 32 (July–December, 1957).

Thomas, P.T., "Urban Community Needs." *Indian Journal of Social Work*, 14 (March, 1953), pp. 265–269.

Thorner, Daniel and Alice Thorner, "Economic Concepts in the Census of India, 1951." *Indian Population Bulletin*, 1 (April, 1960), pp. 68–81.

——, "Economic Recommendations for the Census of 1961." *Economic Weekly,* 11 (5 September, 1959), pp. 1230–1242.

THYAGARAJAN, V., "Tuticorin: A Town Study." *Indian Geographical Journal,* 16 (April–June, 1941), pp. 179–192.

TIKEKAR, S.R., *Bhilai: A New Steel City.* Bombay: 1952.

TIWARI, A.R., "Urban Regions of Agra." *Agra University Journal of Research,* 6 (January, 1958), pp. 101–114.

TOWN PLANNING ORGANIZATION, *Redevelopment Plan for Kotla Mubarakpur.* New Delhi: 1958.

TRYWHITT, JACQUELINE, "Chandigarh," *Royal Architecture Institute of Canada Journal,* 31 (January, 1954), pp. 22–46.

TURNER, ROY (ed.), *India's Urban Future: Selected Studies from an International Conference on Urbanization in India, held at the University of California in 1960.* Berkeley: University of California Press, 1962.

TYSEN, FRANK J., *District Administration in Metropolitan Calcutta.* New York: Institute of Public Administration, 1966.

UNESCO, Research Centre on Social and Economic Development in Southern Asia, *Small Industries and Social Change: Four Studies in India.* New Delhi: Allied Publishers, 1966.

——, *Social Aspects of Small Industries in India: Studies in Howrah and Bombay.* Delhi: UNESCO, 1962.

——, *Status Image in Changing India:* Bombay: Manaktalas: 1967.

UNESCO, Research Centre on the Social Implications of Industrialization in Southern Asia, *Public Administration Problems of New and Rapidly Growing Towns in Asia.* New York: UNESCO, 1962.

——, *Report on a Preliminary Enquiry on the Growth of Steel Towns in India: Study on Problems of Urbanization.* Calcutta: UNESCO, 1959.

UNITED NATIONS, *Urbanization: Development Policies and Planning.* New York: United Nations, 1968.

UNITED NATIONS, Department of Economic and Social Affairs, *Mysore Population Study: Report of a Field Survey Carried out in Selected Areas of Mysore.* New York: United Nations, 1961.

UNNITHAN, T.K.N., "Social Aspects of Planning of Metropolitan Cities." *Journal of the Institute of Town Planners, India* (January–April, 1961), pp. 25–29.

UPADHYAYA, P.K., "Some Aspects of the Problem of Urban Sanitation." *Agra University Journal of Research,* 2 (January, 1954), pp. 17–21.

Urban Development Trends in India: Papers Submitted to the 39th Conference of the Indian Economic Association, Cuttack.

"Urban Housing." *Commerce* (April 8, 1961), p. 619.

"Urban Housing and the Plan." *Economic Weekly,* 9 (9 November, 1957), pp. 1453–1455.

"Urban Housing Gap." *Quarterly Economic Report of the Indian Institute of Public Opinion,* 8 (January, 1962), pp. 47–50.

"Urban Land Policy." *Economic Weekly,* 14 (11 August, 1962), pp. 1290–1291.

"Urban Unemployment." *Eastern Economist,* 30 (27 March, 1959), p. 644.

"Urban Unemployment in India (Symposium)." *Indian Journal of Commerce,* 7 (December, 1954), pp. 1–62; 8 (March, 1955), pp. 1–10.

USEEM, JOHN and RUTH HILL USEEM, *The Western Educated Man in India: A Study of His Social Roles and Influence.* New York: Dryden Press, 1955.

UTTAR PRADESH, Economics and Statistics Department, *Rural and Urban Wages in Uttar Pradesh.* Lucknow: 1950.

VAGALE, L.R., "Bangalore: A Study of Its Planning Problems." *Journal of the Institute of Town Planners, India* (January–April, 1961), pp. 84–90.

———, "Basic Issues in Planning of Small Urban Communities and Case Studies of a Few Towns in India." Paper for 11th Annual Town and Country Planning Seminar, Bhopal, October, 1962. New Delhi: Institute of Town Planners, India.

———, "A Case Study of Chandigarh and Environs in the Regional Setting." Paper for Seminar on Planning for Urban and Regional Development Including Metropolitan Areas, New Towns and Land Policies, 10 to 20 October, 1966, Nagoya, Japan.

———, "Metropolitan Cities in India: Urban Development Problems and Local Administration." Paper for Seminar on Municipal Government. Chandigarh: Punjab University, Public Administration Division, 1965. Mimeo.

———, "Neighbourhood Planning and Its Relation to Housing." *Urban and Rural Planning Thought,* 9 (January–April, 1966), pp. 27–32.

———, "Structure of Industrial Cities in India." *Indian Builder,* 1965 Annual.

———, "Structure of Metropolitan Cities in India." *Urban and Rural Planning Thought,* 9 (January–April, 1966), pp. 2–26.

———, "Trends in Housing Needs of India." *Urban and Rural Planning Thought,* 2 (April, 1959), pp. 50–58.

———, "Thoughts on Urban and Regional Planning Education in India." *Urban and Rural Planning Thought,* 9 (January–April, 1966), pp. 33–39.

VAGALE, L.R. et al., "Faridabad—A Critical Study of the New Town." *Urban and Rural Planning Thought,* 2 (July, 1959), pp. 84–108.

VAIDYA, G.D. and M. OSMAN, *Survey of Housing Conditions of Working Class Families in Hyderabad City.* Hyderabad: Hyderabad Economic Association, 1951.

VAIDYANATHAN, K.E., "Components of Urban Growth in India, 1951–61." Paper for International Union for the Scientific Study of Population Conference, London, September, 1969.

———, *Population Redistribution and Economic Change, India, 1951–61.* Philadelphia: University of Pennsylvania, 1967. Unpublished Ph.D. Thesis, 2 Vols.

VAKIL, C.H. and PERIN H. CABINETMAKER, *Government and the Displaced Persons: A Study in Social Tensions.* Bombay: 1956.

VAKIL, C.H. and USHA MEHTA, *Government and the Governed: A Study in Social Tensions.* Bombay: 1956.

VANKATRAMALAH, P., "Input Structure of Residential Buildings—A Case Study." *Artha Vijnana,* 6 (June, 1964), pp. 77–90.

VARMA, K.N., "Jabalpur: A Study in Urban Geography and Town Planning." *Geographer* (Aligarh), 1 (December, 1962).

VARMA, P.L., "Chandigarh—The City of Tomorrow." *Nirman,* (April–June, 1954), pp. 10–11.

VASANTA, A., "A Method to Delimit Areas of Urban Concentration," *Indian Geographical Journal,* 32 (July–December, 1957), pp. 95–100.

VAZIFDAR, J.P., "Economy of Housing." *Indian Builder,* (January, 1966), pp. 47–49.

VENKATARANGAIYA, M., "Bombay and Calcutta," in William A. Robson (ed.), *Great Cities of the World.* London: Allen & Unwin, 1957, pp. 137–164.

VENKATARAYAPPA, K.N., *Bangalore: A Socio-Ecological Study.* Bombay: University of Bombay Press, 1957.

——, "Urban Land Value and Land Utilization Trends in India." *Sociological Bulletin,* 9 (September, 1960), pp. 34–47.

VENUGOBAL, E., "Commercial Structure of Warangal City Andhra Pradesh." Paper for 21st International Geographical Congress, held at New Delhi, November–December, 1968, in S.P. Das Gupta and T.R. Lakshmanan (eds.), *Abstract of Papers.* Calcutta: National Committee for Geography, 1968, p. 287.

VEPA, RAM K., "Urbanization and Industrial Development." *Indian Journal of Public Administration,* 14 (July–September, 1968), pp. 491–497.

VIDYARTHI, L.P., *Cultural Configuration of Ranchi.* Calcutta: Bookland, 1968.

——, *The Sacred Complex in Hindu Gaya.* Bombay: Asia Publishing House, 1961.

VIPRA, "Indian Census of 1961: Need for Checks and Follow-Up." *Economic Weekly,* 13 (10 June, 1961), pp. 879–884.

VISARIA, PRAVIN, "Growth of Greater Bombay, 1951–1961." *Economic and Political Weekly,* 4 (July, 1969), pp. 1185–1190.

VISARIA, PRAVIN M., *The Pattern of Out Migration from Coastal Maharashtra in India.* Princeton: Princeton University, 1968. Mimeo.

——, *The Sex Ratio of the Population of India.* Bombay: University of Bombay, Department of Economics, 1968. Mimeo.

VISHWANATH, M.S., "Functional Classification of Small Urban Centres in Mysore State." Paper for 21st International Geographical Congress, held at New Delhi, November–December, 1968, in S.P. Das Gupta and T.R. Lakshmanan (eds.), *Abstract of Papers.* Calcutta: National Committee for Geography, 1968, p. 287.

WEINER, MYRON, "Urbanization and Political Protest." *Civilizations,* 17 (No. 1/2, 1967), pp. 44–50.

——, "Violence and Politics in Calcutta." *Journal of Asian Studies,* 20 (May, 1961), pp. 275–281.

WEINER, MYRON and RAJNI KOTHARI, *Indian Voting Behavior.* Calcutta: Firna K.L. Mukhopadhyay, 1965.

WEST BENGAL, *Report on a Sample Enquiry into the Living Conditions in the Bustees of Calcutta and Howrah, 1948–49.* Calcutta: 1949.

——, *Scheme for Development of Calcutta and Its Hinterland.* Alipore: 1960.

WEST BENGAL, Statistical Bureau, *Calcutta Bustees: A Statistical Survey.* Calcutta: 1956.

——, *Economic Survey of Small Industries, 1954, Calcutta Industrial Area.* Calcutta: 1956.

——, *Economic Survey of Small Industries of Calcutta, 1952–53.* Calcutta: 1954.

——, *Report on a Sample Enquiry into the Living Conditions in the Bustees of Calcutta and Howrah, 1948–49.* Calcutta: 1949.

WOOD, J., "Development of Urban and Regional Planning in India." *Urban and Rural Planning Thought,* 2 (April, 1959), pp. 72–77.

——, "Development of Urban and Regional Planning in India." *Land Economics,* 34 (November, 1958), pp. 310–315.

——, "Town Planning in India: Status and Education." *Urban and Rural Planning Thought,* 1 (October, 1958), pp. 223–243.

WOODRUFF, GERTRUDE M., An Adidrarida Settlement in Bangalore India: A Case Study of Urbanization. Boston: Radcliffe College, 1959. Unpublished Ph.D. Dissertation.

——, "Family Migration in Bangalore." *Economic Weekly,* 12 (January–June, 1960), pp. 163–172.

WURSTER, CATHERINE BAUER, "Urban Living Conditions, Overhead Costs, and the Development Pattern," in Roy Turner (ed.), *India's Urban Future.* Bombay: Oxford University Press, 1962, pp. 277–298.

YADEVA, D.N., "Progress of Housing Schemes in Bihar." *Indian Builder* (January, 1961), p. 10.

ZACHARIAH, K.C., "Bombay Migration Study: A Pilot Analysis of Migration to an Asian Metropolis." *Demography,* 3 (No. 2, 1966), pp. 378–392.

——, "Historical Study of Internal Migration in the Indian Sub-Continent 1901–1931." Philadelphia: University of Pennsylvania, 1962. Unpublished Ph.D. Thesis.

——, *Historical Study of Internal Migration in the Indian Sub-Continent, 1901–1931.* Bombay: Demographic Training and Research Centre, 1965.

——, "Internal Migration in India from the Historical Stand-Point," in *Proceedings of the International Statistical Institute.* Ottawa: 1963.

——, Internal Migration in India, 1941–51. Bombay: Demographic Training and Research Centre, 1960. Mimeo.

——, "Migration and Population Growth in Kerala," in R.S. Kurup and K.A. George (eds.), *Population Growth in Kerala.* Trivandrum: 1965.

——, *Migration in Greater Bombay.* Bombay: Demographic Training and Research Centre, 1964.

ZACHARIAH, K.C. and J.P. AMBANNAVAR, "Population Redistribution in India: Inter-State and Rural-Urban," in Ashish Bose (ed.), *Patterns of Population Change in India, 1951–61.* New Delhi: Allied Publishers, 1967, pp. 93–106.

ZACHARIA, K.C. and A. SEBASTIAN, "Juvenile Working Migrants in Greater Bombay." *Indian Journal of Social Work,* 27 (October, 1966), pp. 255–262.

ZINKIN, TAYA, "India's Most Modern City: Chandigarh." *American Institute of Architects Journal,* 22 (November, 1954), pp. 87–94.

INDONESIA

BRUNER, EDWARD M., "Urbanization and Ethnic Identity in North Sumatra."
    American Anthropologist, 63 (June, 1961), pp. 508–521.
CASTLES, LANCE, "The Ethnic Profile of Djakarta." *Indonesia*, 2 (April, 1967)
    pp. 153–204.
"Cities of Indonesia: Djakarta." *Indonesian Review*, 1 (1951), pp. 157–164.
"Djakarta's Satellite Town." *Asian Review*, 51 (January, 1955), pp. 82–84.
GEERTZ, CLIFFORD, *Peddlars and Princes: Social Development and Economic
    Change in Two Indonesian Towns.* Chicago: University of Chicago Press,
    1963.
——, "The Social History of an Indonesian Town." *American Sociological
    Review,* 31 (June, 1966), pp. 425–426.
GOANTIANG, T., "Growth of Cities in Indonesia, 1930–1961." *Tijdschrift Voor
    Economische en Sociale Geografie* (Rotterdam), 56 (May–June, 1965),
    pp. 103–108.
HEEREN, H.J., "Internal Migration in Indonesia." Paper for the International
    Union for the Scientific Study of Population, Sydney Conference, Austra-
    lia, 21 to 25 August 1967, pp. 591–596.
—— (ed.), "Urbanization of Djakarta." *Urbanisasi Djakarta Economi dan
    Keuangan Indonesia,* 8 (1955).
MILONE, P.D., "Contemporary Urbanization in Indonesia." *Asian Survey,* 4
    (August, 1964), pp. 1000–1012.
——, *Urban Areas in Indonesia: Administrative and Census Concepts.* Berke-
    ley: University of California, Institute of International Studies, 1966.
TANGOANTIAN, "Some Notes on Internal Migration in Indonesia." Paper pre-
    pared for the ECAFE Expert Working Group on Problems of Internal
    Migration and Urbanization, 24 May to 5 June, 1967, Bangkok. Mimeo.
WERTHEIM, W.F., *Indonesian Society in Transition.* The Hague: W. Van
    Hoeve, 1959.
WERTHEIM, W.F. et al. (eds.), *The Indonesian Town: Studies in Urban So-
    ciology.* The Hague: Selected Studies on Indonesia, Number 4, 1958.
WITHINGTON, W.A., "The Kotapradja or 'King Cities' of Indonesia." *Pacific
    Viewpoint,* 4 (March, 1963), pp. 75–86.
——, "Medan: Primary Regional Metropolis of Sumatra." *Journal of Geog-
    raphy,* 61 (February, 1962), pp. 59–67.

LAOS

HALPERN, JOEL M., *Economy and Society of Laos—A Brief Survey.* New
    Haven: Yale University, South East Asia Studies, 1964.

MALAYSIA

ALLEN, D.F., *The Major Ports of Malaya.* Kuala Lumpur: 1951.
——, *The Minor Ports of Malaya.* Singapore: 1953.
CAIDWELL, J.C., "Urban Growth in Malaya: Trends and Implications." *Popu-
    lation Review,* 7 (January, 1963), pp. 39–50.
COOPER, E., "Urbanization in Malaya." *Population Studies,* 5 (November,
    1951), pp. 117–131.

DOBBY, E.H.G., "Settlement Patterns in Malaya." *Geographical Review,* 32 (April, 1942), pp. 211–232.

McGEE, T.G., "The Cultural Role of Cities: A Case Study of Kuala Lumpur." *Journal of Tropical Geography,* 17 (May, 1963), pp. 178–196.

———. *Malays in Kuala Lumpur City; A Geographical Study of the Process of Urbanization.* Victoria: University of Wellington, New Zealand, 1969. Unpublished Ph.D. Thesis.

McGEE, T.G. and W.D. McTAGGART, *Petaling Jaya: A Socio-Economic Survey of a New Town in Selangor, Malaysia.* Wellington, New Zealand: 1967.

NEVILLE, R.J.W., "An Urban Study of Pontian Kechil, Southwest Malaya." *Journal of Tropical Geography,* 16 (October, 1962), pp. 32–56.

OOI JIN-BEE, *Land, People and Economy in Malaya.* London: Longmans, Green, 1964.

SILCOCK, T.H. (ed.), *Readings in Malayan Economics.* Singapore: Eastern Universities Press, 1961.

SENDUT, H., "Patterns of Urbanization in Malaya." *Journal of Tropical Geography,* 16 (October, 1962), pp. 114–130.

———, "The Structure of Kuala Lumpur, Malaysia's Capital City," in Gerald Breese (ed.), *The City in Newly Developing Countries: Readings on Urbanism and Urbanization.* London: Prentice-Hall, 1969, pp. 461–473.

SWEE-HOCK, SAW and CHENG-SIOK HWA, "Migration Policies in Malaya and Singapore." Paper for International Union for the Scientific Study of Population Conference, London, September, 1969.

NEPAL

"Evaluation and Utilization of Population Census Data: Nepal." Seminar on Evaluation and Utilization of Census Data in Asia and the Far East, Bombay, June–July, 1960. Mimeo.

PAKISTAN

AFZAL, M., "Migration to Urban Areas in Pakistan." Paper for the International Union for the Scientific Study of Population, Sydney Conference, Australia, 21 to 25 August, 1967, pp. 684–695.

AHMAD, N., "The Urban Pattern in East Pakistan." *Oriental Geographer* 1 (January, 1957), pp. 37–39.

AHMED, KAZI S., "Urban Population in Pakistan." *Pakistan Geographical Review,* 10 (August, 1955), pp. 1–16.

AHMID, NAFIS and A.K.M. HAFIZUR RAHMAN, "Development of Industry in Chittagong." *Oriental Geographer* (July, 1962), pp. 139–158.

AKHTAR, A.U., "Impact of Industrialization on Socio-Economic Life in the Urban Areas of Pakistan." *Pakistan Labour Gazette,* 7 (January–March, 1959), pp. 18–25.

ALSPACH C., "Urban Community Development." *Social Welfare in Pakistan,* 5 (September, 1958), pp. 4–7.

ANWAR, ABDUL AZIZ, *A Socio-Economic Survey of Industrial Labour in Selected Centres.* Lahore: Board of Economic Enquiry, Punjab (Pakistan).

BOSE, ASHISH, "Internal Migration in India, Pakistan and Ceylon." Paper for the World Population Conference, Belgrade, 30 August to 10 September, 1965.

——, "The Role of Small Towns in the Urbanization Process of India and Pakistan." Paper presented at the International Union for the Scientific Study of Population Conference, London, September, 1969.

DAVIS, KINGSLEY, *Population of India and Pakistan*. Princeton: Princeton University Press, 1951.

HAZEER, M.M., "Urban Growth in Pakistan." *Asian Survey*, 6 (June, 1966), pp. 310–318.

HUSAIN, HYDER, "Some Aspects of the Rural Urban Composition of Population in East Pakistan." *Pakistan Geographical Review*, 13 (1950), pp. 24–27.

KHAN, F.K. and M.H. KHAN, "Delimitation of Greater Dacca." *Oriental Geographer*, 5 (July, 1961), pp. 95–120.

KHAN, FAZLE KARIM and MOHAMMAD MASOOD, "Urban Structure of Comilla Town." *Oriental Geographer*, 6 (July, 1962), pp. 109–138.

KHAN, MOHAMMAD IRSHAD, "Industrial Labour in Karachi." *The Pakistan Development Review*, 3 (Autumn, 1963), pp. 598–599.

KHAN, N.M., "The Rise of Satellite Towns." *Pakistan Quarterly*, 6 (August, 1956), pp. 8–11.

KROTKI, KAROL J., "First Release from the Second Population Census of Pakistan, 1961. *"Pakistan Development Review*, 1 (Autumn, 1961), pp. 67–77.

KUREISHY, K.U., "Economic Basis of Urban Development in West Pakistan." *Proceedings of the 9th Pakistan Science Conference*, 3 (1957).

KUX, D., "Growth and Characteristics of Pakistan's Population." *Population Review*, 6 (January, 1962), pp. 42–58.

MOHAMMAD, HUSSAIN, "Impact of Industrialization on Socio-Economic Life in the Urban Areas of Pakistan." *Pakistan Labour Gazette*, 7 (January–March, 1959), pp. 10–17.

PHILLIPS, W., JR., "Urbanization and Social Change in Pakistan." *Phylon* (Atlanta, Ga.), 25 (Spring, 1964), pp. 33–43.

RAHIM, A.M.A., "Impact of Industrialization on Socio-Economic Life in the Urban Areas of Pakistan." *Pakistan Labour Gazette*, 7 (January–March, 1959), pp. 2–9.

RANIS, G., *Urban Consumer Expenditure and the Consumption Function*. Karachi: Institute of Development Economics, Monographs in the Economics of Development, Number 6, 1961.

RUDDOCK, G., "Capital of East Pakistan (Dacca)." *Pakistan Quarterly*, 7 (Spring, 1957), pp. 49–58, 64–68.

SAIDUDDIN AHMAD, K., "Some Aspects of the Rural-Urban Composition of Population in East Pakistan." *Pakistan Geographical Review*, 13 (1958), pp. 1–17.

SALIMA, O., "Urban Community Development: The Experience in Pakistan." *International Social Service Review*, 6 (March, 1960), pp. 25–33.

PHILLIPPINES

AMYOT, JACQUES, *The Chinese Community on Manila*. Chicago: University of Chicago Press, 1960.

McIntyre, W.E., "The Retail Pattern of Manila." *Geographical Review,* 45 (January, 1955), pp. 66–80.

Ramos, Carlos P., "Manila and Surburban Towns." Paper presented to Regional Seminar on Public Administration Problems of New and Rapidly Growing Towns in Southern Asia, New Delhi, December, 1960.

Spencer, J.E., "The Cities of the Philippines." *Journal of Geography,* 57 (September, 1958), pp. 288–294.

Ullman, E., "Trade Centres and Tributary Areas of the Philippines." *Geographical Review,* 50 (April, 1960), pp. 203–218.

United Nations, *Population Growth and Manpower in the Philippines.* New York: 1960. A joint study by the United Nations and the Government of the Philippines.

Wernstedt, F.L., "Cebu-Focus of Philippine Inter-Island Trade." *Economic Geography,* 33 (October, 1957), pp. 336–346.

SINGAPORE

Dobby, E.H.G., "Singapore, Town and Country." *Geographical Review,* 30 (January, 1940), pp. 84–109.

Fraser, J.M., "The Character of Cities, Singapore: A Problem in Population." *Town and Country Planning,* 23 (November, 1955), pp. 505–509.

——, "Town Planning and Housing in Singapor." *Town Planning Review,* 23 (April, 1952), pp. 5–25.

Gamba, Charles, "Housing and Town Planning in Singapore." *Eatsern World,* 7 (August, 1953), pp. 35–38.

Hooi, Tan Jake and Alan F.C. Choe, "Expert Report on Planning Metropolitan Singapore." Seminar on Metropolitan Planning in Asia, Tokyo, June 7 to 14, 1964.

Kay, Barrington, *Upper Nankin Street, Singapore: A Sociological Study of Chinese Households Living in a Densely Populated Area.* Singapore: University of Malaya Press, 1960.

Singapore, Department of Social Welfare, *Social Survey of Singapore: A Preliminary Study of Some Aspects of Social Conditions in the Municipal Area of Singapore.* Singapore: Singapore Department of Social Welfare, 1947.

Swee-Hock and Cheng-Siok Hwa, "Migration Policies in Malaya and Singapore." Paper for International Union for the Scientific Study of Population Conference, London, September, 1969.

Yeh, S.H.K., "Urbanization and Public Housing in Singapore." Paper for International Union for the Scientific Study of Population. Sydney Conference, Australia, 21 to 25 August, 1967, pp. 696–707.

You, P. Seng, "The Population Growth of Singapore." *Malayan Economic Review,* 4 (October, 1959), pp. 56–69.

THAILAND

Evers, H.D., "The Formation of a Social Class Structure: Urbanization, Bureaucratization and Social Mobility in Thailand." *American Sociological Review,* 31 (August, 1966), pp. 480–488; and *Journal of South East Asian History,* 7 (September, 1966), pp. 100–115.

GOLDSTEIN, SIDNEY, "Urban Growth in Thailand, 1947–67." *Journal of Social Sciences,* 12 (April, 1969), pp. 100–118.

——, "Urbanization in Thailand, 1947–67." Paper for International Union for the Scientific Study of Population Conference, London, September, 1969.

NAWARAT, PRASERT, "Expert Report on Planning Metropolitan Bangkok." Seminar on Metropolitan Planning in Asia, Tokyo, June 7 to 14, 1964.

STERNSTEIN, LARRY, "Contemplating a Hierarchy of Centres in Thailand." *Pacific Viewpoint,* 7 (September, 1966), pp. 229–235.

VIETNAM

DINH, TRAN VAN, "Administration of the Saigon Municipality." Paper presented to Regional Seminar on Public Administration Problems of New and Rapidly Growing Towns in Southern Asia, New Delhi, December, 1960.

VIETNAM, Secretariat d'Etat a l'Economic Nationale, Institut National de la Statistique, *Enquetes Demographiques au Vietnam en 1958.* Saigon: 1960.

ABOUT THE CONTRIBUTORS

BRIAN J.L. BERRY is Professor of Geography and Director of the Training Program of the Center of Urban Studies at the University of Chicago. In addition, he is a member of the university's Committee on Southern Asian Studies. He received his early education in economics at the University of London and his graduate degrees in geography at the University of Washington. He is the author of many books and articles in the fields of location theory, regional science, spatial analysis and urban studies.

ASHISH BOSE is Senior Fellow at the Institute of Economic Growth, Delhi University, where he received his graduate degrees. As a specialist in urban demography he has contributed papers on different aspects of urbanization at many international conferences and has lectured at a number of universities in the United States, the United Kingdom and Japan. Dr. Bose is author of *Urbanization in India: An Inventory of Source Materials* (1970) and several other books and articles on urbanization and urban development problems in India.

GERALD M. DESMOND is Economic Advisor to the United Nations Center for Housing, Building and Planning, which is responsible for the formulation and implementation of the U.N.'s program in Housing and Urban Development. Mr. Desmond received his B.A. from Columbia University and his M.A. from the University of California at Berkeley in 1960. Prior to joining the Center he was with the U.S. Bureau of the Budget and also served on missions with the World Bank. He has published several reports and articles, mainly on the financial aspects of housing and urban development.

LEO JAKOBSON is Professor of Urban and Regional Planning at the University of Wisconsin. He received his M. Arch. from the Technical University at Helsinki. He also studied at the Royal Academy of Arts in Stockholm and at the University of Pennsylvania. Professor Jakobson has practiced planning in Finland, Sweden, Israel, Puerto Rico and the U.S. In 1964–65 he served as a consultant on regional planning with the Ford Foundation Advisory Planning Group in Calcutta. He has taught at the University of Wisconsin since 1958 and has lectured at several universities in the United States, India and Finland. In addition to professional reports his publications include monographs and articles on urban development, urban design and planning education.

APRODICO A. LAQUIAN is Director of Research of the International Association for Metropolitan Research and Development (INTERMET) with headquarters in Toronto, Canada. He is currently on leave from the University of the Philippines, where he is an associate professor of public administration and deputy director of the Local Government Center. Dr. Laquian has a B.A. in Public Administration from the University of the Philippines and a Ph.D. in Political Science from Massachusetts Institute of Technology. He is the author of the *City in Nation Building* (1966) and *Slums are for People* (1969).

TERRENCE GARY MCGEE is Senior Lecturer in Geography at the University of Hong Kong. He was educated at Auckland University and Victoria University of Wellington, New Zealand. Prior to joining the University of Hong Kong, Dr. McGee was Lecturer in Geography at the University of Malaya and Victoria University of Wellington. He has traveled extensively in South East Asia, carrying out detailed research on Malay Migrants in Kuala Lumpur City, Federation of Malaysia. At present he is working on a research project on the problems and characteristics of hawkers (street vendors) in Hong Kong and other Asian cities. He is the author of *The Southeast Asian City,* and numerous articles on urbanization in Asia.

VED PRAKASH is Associate Professor of Urban and Regional Planning at the University of Wisconsin. He obtained his B.A. and M.A. degrees from the University of Lucknow, India, and his Ph.D. from Cornell University. Prior to joining the University of Wisconsin he was a Lecturer at the University of Lucknow and at the Indian Institute of Management, Calcutta. He has also been with the Town Planning Organization and Tax Research Unit, Government of India. Most recently Dr. Prakash has been a consultant to the United Nations and the U.S. AID Mission to Colombia. His publications include *New Towns in India* and articles on municipal investment planning and urban development.

TARLOK SINGH is Deputy Executive Director for Planning, UNICEF. From 1950–67 he was closely associated with the preparation of India's Five-Year Plans. Since relinquishing office as Member of India's Planning Commission in 1967, he has been Visiting Senior Research Economist at the Woodrow Wilson School of Public and International Affairs, Princeton University, and Fellow and Visiting Professor at the University of Stockholm, Institute for International Economic Studies. His writings include *Poverty and Social Change* (1945 and 1969), *Towards An Integrated Society* (1969), *The Planning Process* (1963), *Land Resettlement Man-*

ual for Displaced Persons (1952) and a forthcoming study, *India's Development Experience: An Evaluation.*

STANISLAW WELLISZ is Professor of Economics at Columbia University where he teaches economic development and planning. He received his early education in Poland and England and obtained his Ph.D. degree in economics from Harvard. He has taught at Williams College, the University of Chicago and Warsaw University, where he spent a year as exchange professor. Dr. Wellisz spent the years 1961–63 as economist with the Calcutta Metropolitan Planning Organization, and in 1964 served several months as consultant to the Pakistani Planning Commission on problems of urbanization and physical planning. He has participated in various economic missions to Iran, Liberia, Yugoslavia, and other countries. Publications include *Economies of the Soviet Bloc* (1964) and numerous articles on economic development and related issues.

INDEX